Embodied Knowledge

Perspectives on Belief and Technology

Edited by
Marie Louise Stig Sørensen
and Katharina Rebay-Salisbury

OXBOW BOOKS
Oxford and Oakville

Published by
Oxbow Books, Oxford, UK

ISBN 978-1-84217-490-6

This book is available direct from:

Oxbow Books, Oxford, UK
(Phone: 01865-241249; Fax: 01865-794449)

and

The David Brown Book Company
PO Box 511, Oakville, CT 06779, USA
(Phone: 860-945-9329; Fax: 860-945-9468)

or from our website

www.oxbowbooks.com

Cover image
'The Philadelphia Twins' by Christie Brown 1991
Photograph: Christie Brown

A CIP record for this book is available from the British Library

Library of Congress Cataloging-in-Publication Data

Embodied knowledge : perspectives on belief and technology / edited by Marie Louise Stig Sørensen and Katharina Rebay-Salisbury. -- 1st edition.
pages cm
Includes bibliographical references.
ISBN 978-1-84217-490-6
1. Social archaeology--Case studies. 2. Archaeology and history--Case studies. 3. Human body--Social aspects--History--To 1500. 4. Senses and sensation--Social aspects--History--To 1500. 5. Learning--Social aspects--History--To 1500. 6. Belief and doubt--Social aspects--History--To 1500. 7. Technology--Social aspects--History--To 1500. 8. Material culture--History--To 1500. I. Sørensen, Marie Louise Stig, II. Rebay-Salisbury, Katharina.
CC72.4.E45 2013
930.1--dc23

2012039224

Printed and bound in Great Britain by
Short Run Press, Exeter

Contents

1. Embodied knowledge. Reflections on belief and technology

Marie Louise Stig Sørensen and Katharina Rebay-Salisbury

This book explores the relationship between knowledge and the body through a series of historical and archaeological case studies. More specifically, it considers the concept of embodied knowledge by exploring some of the apparent diverse and yet shared forms of what may be called embodied knowledge. The papers share a focus on knowledge as it is implicit and expressed through the human body and bodily action, and as it formed through intentional practices. But what is this kind of knowledge? Using specific case studies of knowledgeable actions, the book explores *embodied knowledge* through a focus on practice. It does so through two different, yet interconnected aspects of how such knowledge expresses itself: belief and technology.

Challenging dichotomies

The deconstruction of the common juxtaposition of these two forms of knowledge is an essential aspect of our approach to embodied knowledge. Rather than seeing them as distinct categories, and thus reproducing an unhelpful polarisation, the chapters each in their own ways help to underline essential connections between these two forms of knowledge. In some ways, this is part of the social science and humanities challenge to the mind/body dualism, which has been so prevalent in our disciplines and so subjected to theoretical criticism over recent decades (e.g. Turner 2008). At the same time, irrespective of whether the disconnection between beliefs and technologies is part of this fundamental dualism in our epistemologies, it has certainly been furthered by the ways in which academia produces specialists with non-overlapping knowledge. This means that we usually become specialists in one of these major areas, but not in both; we develop and work within different specialist languages and networks. We are rarely pushed to acknowledge how fields such as technology and beliefs overlap and may affect each other. Therefore, rather than pursuing abstract arguments about embodied knowledge, we have selected to use an essentially empirical, one may say an ethnographic, approach to demonstrate the overlapping meanings and at times even dependencies between these two forms of knowledge.

This is why the body must be brought into focus, providing the reason for the title of the volume: 'Embodied knowledge'. We call attention to the shared epistemological and ontological aspects of the knowledge required to make pots and perform rituals, to similarities and convergences that are acknowledged in our routine interpretative languages and expectations. In daily life, we recognise, depend on, and use the connection between these different areas of practice. The body is the main forum for learning about how to do, think and believe, and practices as apparently diverse as belief and technologies are accordingly enacted and performed through the body in similar manners. Moreover, as these practices become embodied knowledge, they come to inhabit and affect the body as motor skills and practiced ways of doing things (Sofaer 2006). The body is thus the forum for the learning and performance of belief just as it is for the learning of metallurgy and how to produce metal objects. This does not mean that the two are the same; rather they are both rooted in the body and may be involved in the constitution and performance of each other. It is these interrelations that are of interest and in need of further scrutiny through case-study based reflections.

Together, the papers collected here emphasise and explore the characteristics that are common to belief and technology: for instance, their organisation around repeated 'ritualised' practices, their transmission through observation and imitation, their involvement in the construction of socio-cultural norms and the differentiation of individuals in terms of knowledge based roles. In the accounts provided in the chapters, belief and technologies are often intertwined and they show how through this relationship epistemological and ontological reasoning is constructed. Moreover, the chapters demonstrate how notions of right and wrong, or good and bad, are extended into practical actions and performed in terms that emphasise how to act in the right manner – this applies to practices as diverse as baptism and spinning. The case studies show how belief needs technologies in order to become expressed and gain forms appropriate for them being practiced and appreciated, and, likewise, how technologies are enmeshed with ideas about right and wrong procedures, and good and bad associations.

Knowledge/embodied knowledge

The question of what knowledge is has a long intellectual history. Knowledge and belief have often been linked or, quite the opposite, seen to be in opposition. Technology, on the other hand, has commonly been associated with practice in a manner that emphasises it as non-discursive and tacit, as 'understood by the body', or as the application of technical knowledge.

Over recent decades, the concept of embodiment and embodied knowledge has challenged this dichotomy arguing that: 'Embodiment is the ensemble of corporeal practices which produce and give "a body" its place in everyday life. Embodiment locates or places particular bodies within a social *habitus* […] embodiment is the lived experience of the sensual or subjected body' (Turner 2000: 496). Apart from providing embodiment in the sense of this quote, the body itself affects discourse and the performance and understanding of what knowledge is, and through this it also affects how beliefs and technologies are perceived. The body as a point of reference, of housing and sheltering experiences, is a very common and strong metaphor; it is a starting point for the granting and forming of many forms of meaning. In the world we create around us, our bodies are the measure of all things. Architecture, tools and objects are adjusted to match our body proportions and sizes. Through such body-sized paraphernalia we overcome our bodily limitations and enlarge the potential spheres of action. Belief similarly extensively explores references to the body.

Another strong tendency is that practices and their outcomes become imagined and perceived in terms of body references, adding a different meaning to the concept of embodiment, as things are literately made into bodies. This is often seen in archaeological and ethnographic studies, but it can also be observed in modern society, as many of our contemporary technological developments are ascribed human traits. In colloquial language these technological products are not merely humanised, they are made understandable by linguistically morphing into bodies: computers go to sleep and catch viruses. This ability to anthropomorphise technology affects many aspects of how we comprehend and respond to our material environment, showing how removed the often-argued distinction between abstract and practical knowledge, between beliefs and technologies, is from observations of the world around us. We routinely use abstractions and metaphors to comprehend the inventions we have made.

To make the exploration of the two dimensions of embodied knowledge explicit, the volume is divided into two parts. The first focuses on beliefs about the body as they are embedded in various societies and preserved in or expressed through material culture and particular practices. The case studies range widely in time, from prehistory to today, and spatially from Western Europe to China. There are beliefs about many things, such as the universe, animals, and consequences. The interrelations between such diverse forms of beliefs would in themselves be fascinating to study, but due to our interest in embodied knowledge, we elected to focus on beliefs which in various ways are about the body. The chapters therefore analyse case studies about birth, death, transformation, sickness and religious beliefs. The second part focuses on technology, on what might be classified as practical or tacit knowledge. Here our concern was to bring into focus the technologies themselves, and the chapters therefore aim to introduce reflections on different material practices, such as carpentry and weaving. This part mainly uses archaeological case studies to benefit from the expertise in technology and material culture traditionally offered by the discipline.

Belief

People understand belief in many ways, and the practising of beliefs use a variety of devices for establishing orthodoxies, or, on the contrary, to absorb change. Belief is often an interesting mix between abstract thinking and mundane matters. It may be perceived as a deeply personal phenomenon or in terms of official doctrines, and it is commonly framed within social norms and systems of sanctions. The chapters show that beliefs are rooted in inherited traditions and practices and at the same time subjected to reinventions and transformations. Beliefs are therefore deeply dependant on processes of learning and the ability to emulate and communicate. A major part of belief is about learning how to perform one's belief, including doing prescribed rituals in the correct manner. Belief is usually ritualistic, these papers show, and this means an emphasis on things being done in the right ways, the appropriate objects being used, and the appointed persons being involved, repeatedly. Beliefs themselves are indeed often not only abstract, but also ill-defined, referring to a feeling more than knowledge. However, through performance, beliefs become physical: the body is used as an instrument of ritualised behaviour and things are used both practically and as symbols. The chapters focus on two areas where the interconnection between the material and beliefs are particularly pronounced: the dead body and its status, and the ideological concerns about the immaterial aspects of a human, such as the soul and the mind.

Katharina Rebay-Salisbury's 'How burial practices are linked to beliefs' (Chapter 3), for instance, looks at the burial practices of cremation and inhumation diachronically to understand how beliefs are associated with the dead body and how the body was understood. In a survey of three case studies ranging from early 20th century Vienna, through the Greek and Roman world to Bronze Age Europe, she asks whether the mode of disposal and treatment of a dead body is a crucial clue to beliefs or if the divergent practices are merely embedded in every-day routines and traditions. While contemporary sources give some explanations for why inhumation or cremation was favoured at times, these are by no means consistent. There is no one and simple explanation of the meaning of cremation or inhumation as a distinct practice. It is clear, however, that associated practices and customs draw on the experience of everyday life, using its metaphors and images, and this in turn may inform us about basic tenets of the underlying beliefs. Furthermore, superficial explanations gloss over much more deeply seated social concerns, in which the discourse over inhumation and cremations is merely used as a stand-in battleground to communicate and deal with other topics, such as politics and power struggles. This gives us an insight into how layered beliefs can be, that they are not straightforward simple, but may be contradictory, vague, and hard to summarise. Burial practices perhaps do not follow beliefs as a result, but customary ways of disposing the body are accompanied by beliefs, explained and justified through them.

Tim Flohr Sørensen's 'Delusion and disclosure: human disposal and the aesthetics of vagueness' (Chapter 4) continues the theme of beliefs about death. He observes, with changing contemporary burial practices in Denmark as a case study, how knowledge of death is worked through on various levels, including the psychological and communal. Dealing with death is so problematic for the living because knowledge of it is fundamentally un-knowable. Before our own death, we can only experience death through the loss of others. Sørensen's case study contrasts the 'traditional' way of dying in Denmark, deeply rooted in Christian belief and with a focus on the integrity of the body as central to the belief in bodily resurrection, with the now much more common practice of cremation in an increasingly secular society. Although beliefs about death and religious beliefs are not the same, they intersect; religious belief provided not only specific ideas and universally accepted explanations of death, but also physical places for farewell rituals and resting places. As cremation became more popular, the time between death and burial became longer and cremation, as well as the interment of the ashes, became further and further removed from even the closest kin, who are often not present at these stages of burial. Furthermore, unmarked resting places instead of family plots are becoming increasingly popular. The ever decreasing material expression of death may be linked to a less fixed idea about death, and an embracing of deliberate vagueness. In terms of knowledge, Sørensen differentiates between 'finding meaning' and 'making sense': the former taken to signify the projection of cognised knowledge, the latter denoting the ways in which sensory stimuli and emotions correlate and create balance in the psyche. For death to make sense, one must come to terms with the loss as well as preserve the memory of loved one, and rituals and aesthetics provide the necessary scaffolding for this task. And yet, *contra* Freudian psychology, traumatic experiences may be better veiled than disclosed – in coping with death, this means a protecting layer of vagueness and as little material contact with the dead body as possible.

Mads Dengsø Jessen takes a closer look at 'Material culture and the construction of religious knowledge' (Chapter 5), situating religious knowledge as it is embodied both in human beings and the material world. By looking at how theological concepts were transmitted in early Christian (late Iron Age and early Medieval) Scandinavia, he zooms in on two particular interesting case studies: the portable altar as a miniature, condensed version of a church, and the role of the Eucharist. Missionaries used portable altars in locations without a church infrastructure – the altars combined the functions of shrine for relics and table and had the power to transform any place into a sacred space. Through this, religious concepts were externalised into the material world, but the altars were also integral to the religious message. The Eucharist, also a material object, did not only embody Christ in the eyes of the believer after having been transformed through sacred power evoked by the priest, but could also be taken into one's own body and work through it, creating a community of believers through partaking in the ritual. In both of these examples, the materiality of human life and its familiarity is cleverly integrated with new religious knowledge. We can think about them in terms of embodied knowledge, extended into the material world and distributed amongst a number of people. Dengsø Jessen argues for a better recognition of the extra-somatic dimensions of religious knowledge, as they are situated in and influenced by context and environment; in turn, they play an important part to be manipulated and integrated in religious concepts.

In Chapter 6, 'Sealed by the cross: protecting the body in Anglo-Saxon England', Helen Foxhall Forbes takes us into the world of the seemingly familiar, yet strange medieval way of theological thinking. A world which is populated by invisible, but omnipresent angels and demons that may all affect us in good and bad ways seems difficult to imagine today. Body and soul had to be protected from demonic powers and may have had angelic powers as helpers. In medieval times, body, soul and mind were linked in such a way that practices and rituals done through the body as a medium may affect the internal and invisible soul – the fate of which was central in medieval Christian belief. Sacraments such as baptism or the anointment of the sick involved treatment of the body, but this was not the aim *per se*; in fact, it was the soul that was addressed by these rituals. Specific gestures, body movements and positions were an integral and essential part of prayer and blessings. The body may have been the medium through which the spirit of a person could be reached, but the body also exhibited the state of the person to the external world, for instance in the case of obsession or illness. The plurality of ways in which sickness was conceptualised and explained in theological terms is again a good example of the complex and fragmented way beliefs work. Beliefs about the body overarched life and death – the dead body, although of much less concern than the soul of the deceased, could, similarly to the living body, be a medium of manipulation and an external sign of inner workings.

Mary Laven's chapter on 'The role of healing in the Jesuit mission to China, 1582–1610' (Chapter 7) continues the theme of transmission of religious knowledge and belief. The Jesuits' missionary efforts to convert the Chinese to Christianity are well documented and give interesting insights into cultural clashes as well as similar understandings between the worlds of Catholicism and Chinese religion (which includes elements of Buddhism, Taoism, and Confucianism). Although highly educated and prepared for intellectual exchange, the Jesuit missionaries did not celebrate the best success in persuading the Chinese of the truth of Christian knowledge through discourse, and they had far more positive effects through their advice on and involvement with much more immediate, bodily concerns. Miracles of healing, exorcism and bringing about fertility, all actions that work directly on or through the body, could be integrated into existing worldviews and were much better attractors to Christianity than teaching and preaching. These emotional, 'embodied' experiences of Christianity responded to profound human needs; the bodies of Christians were seen to be protected by God. The only downside to the successes was that appealing to the popular side of religion did not win over the educated segment of society the Jesuits would have wanted to attract. Another aspect of the deeply embodied nature of Christianity that focussed on Christ's body, the blood, suffering and crucifixion of Christ as well as the meaning of the Eucharist, did not prove easy to explain and transmit. In Mary Laven's chapter, we learn important lessons on the multi-layered facets of embodied belief and how they appeal in different ways to different strata of society.

Jacob Copeman's 'Protest re-embodied: shifting technologies of moral suasion in India' (Chapter 8) takes us into a different world of body manipulations. He explores how bodily techniques are used to articulate protest in India. Classically, fasting is a means of expressing discontent, but more recently, blood and organ donation have become more common. Copeman investigates the complex interrelationships between these new technologies, concepts of the body and religious or spiritual beliefs. He argues that donating blood is a way for the donor to reach multiple recipients as well as their dependents and descendents; as blood donation is both anonymous and done in a way that separates blood into several components used in different patients, donors believe that this technology might benefit a large group of people. In turn, the larger the number of recipients, the more spiritual merit the donor receives. Blood giving is not only motivated medically, but also entangled in a wide range of beliefs. It replaces, for instance, other practices such as offering animal blood, whilst at the same time, gives the donor the social venue to express opposition against it. The process of blood transfusion becomes entirely a political method of self-identification in terms of the religious community and of virtue. Furthermore, the substance of blood is also believed to contain a number of qualities, such as strength, that may be distributed. While fasting withdraws the body from the world, blood donation extends the body into it – the knowledge about the way in which the body is to engage in protest has shifted.

Technology

The second part of the book, dedicated to technology, focuses on material knowledge and in particular the common association between technology and tacit and non-discursive knowledge (Ryle 1949, Polanyi 1966). Issues discussed and theorised include the performance of technology and the interaction of tradition and innovation. The chapters also touch upon the ways in which technology is embodied and such knowledge learnt and transmitted. Technology is usually thought about as 'hard stuff' and is explored in terms of physical properties, inventions and impacts. Its study is often influenced by assumptions about rational reasons, goal-orientated experimentations and 'natural' tendencies towards maximisation. Investigations of society, whether historical or archaeological, commonly take for granted that technological activities are product orientated, and their resulting objects are approached as direct evidence of technical prowess and intensions. Technology studies are thus commonly based on inferences made from the objects themselves – their material properties, their technical and resource demands and the organisation of labour needed for their execution. It is the work of the hand guided by the rational brain. The chapters collected here challenge this by showing technologies as social, discursive and knowledgeable. The chapters also deal with the question of how practical knowledge is represented in non-literate ways and explore alternative forms of knowledge repositories.

In 'The language of craftsmenship' (Chapter 10) Harald Bentz Høgseth takes tool marks left in woodwork as the starting point of his study of embodied knowledge of craftsmanship. The knowledge of a craftsperson is difficult to articulate, as it combines a high level of abstraction. For instance, when picturing the finished object and the way to get there, knowledge about properties and affordances of tools and materials, and the manual skill necessary to put these ideas into practice is needed. This includes drawing on all senses, feeling with fingers, seeing, hearing and smelling; in other words, situating oneself and one's body into the materiality of the world around. In an experimental approach, Høgseth attempts to reconstruct the processes behind artefacts, tool marks, working patterns and actions. Here, the toolmarks play a crucial role: if a tool has damaged edges, it leaves specific traces on the surface of the wood, which can be recorded and used to reconstruct the actions. This includes observations on how the craftsperson stood, when he cut the timber with an axe, which weight the axe had, which angle the timber was cut by, etc; the whole *chaîne opératoire* that leads to the finished object is re-enacted. A replication of this process, however interesting, is not the goal: Høgseth develops a system of notation that translates movements and techniques into an academic format and that simultaneously helps analyse them. The notation system draws on a notation system that has been developed for dance – here again, it is difficult to articulate movement, sounds, choreography in all expressional details, but at least aspects can be recovered through a specific notation system. For carpentry, this includes the movement pattern, action and rhythm of the blows. What Høgseth developed is a way to make tacit knowledge literate, or non-discursive knowledge discursive – in a world in which traditional craftsman's knowledge is about to die, this might be the only way to preserve it.

Sheila Kohring's 'Conceptual knowledge as technologically materialised: a case study of pottery production, consumption and community practice' (Chapter 11) theorises the role of engagement of the body with its social and material environment in construction and transmission of knowledge. She emphasises that technologies are part of our understanding of the world, and are used to communicate social meanings. In this, the 'community of practice' becomes important: it is the social arena in which techniques are learnt and loaded with meanings, and through continuous practice and experiencing routine, both are reinforced. Kohring bases her argument on pottery production in Copper Age Spain. She investigates variability and conformity in the material, focussing on both visual differences and technical choices. Importantly, the visual and the haptic are experienced directly by producers and consumers of a vessel, whilst certain technological aspects of the vessel might not be equally apparent. Looking at all aspects of the *chaîne opératoire*, she observes how social and technical knowledge co-inform each other, and how each step in the production is shaped by knowledge of both production and use within the whole of the society, not just the potters. In this shared social and technological knowledge, a repertoire of acceptable difference is developed which is constructive to the communal identity.

In Chapter 12, 'Many hands make light work: embodied knowledge at the Bronze Age tell at Százhalombatta, Hungary', Joanna Sofaer and Sandy Budden give further insights into pottery production, this time with a case study set in Bronze Age Hungary. Pottery production is scrutinised for evidence of the relationship between individual and society. The authors challenge the universal assumption that each pot is made individually by one person from start to finish. As a very physical action, where the body of the potter is engaged with clay and tools, potting is a kind of embodied knowledge that is acquired through practice, but cannot be un-learnt. The skill level of a potter is apparent in every single piece of work, for instance in the frequency of production errors and the range of variability within a certain type of pottery. It is suggested that simple, small forms like cups require less skill than large, composite vessels, which have to be made by an experienced person. However, when analysing the pottery assemblage it becomes clear that not all steps of the *chaîne opératoire* exhibit similar levels of skill for each individual vessel, for instance, decoration or firing technique may be different from consistency in the wall thickness. It seems that the responsibility for individual vessels is shared, and includes apprentices and experienced potters. Learning to be a potter involves acquiring non-discursive knowledge through the repeated enactment of bodily performance, and this embodied knowledge may be critical to the construction of social categories, such as masters and apprentices.

Lise Bender Jørgensen's 'Spinning faith' (Chapter 13) investigates spindle spinning as a basic craft that was practised more or less continuously in the past. Considerable lengths of yarn were needed for making textiles; one square metre of average quality cloth contains at least 1,000 km yarn. Spinning takes practice to do evenly and efficiently, but once learnt it becomes second nature – a simple, subconscious movement, almost like a sleight-of-hand. Yarn can be spun in two directions: clockwise or anti-clockwise. Once the spinning process has begun it is essential to keep twisting in the same direction, or else fibres unwind and cause the yarn to disintegrate. Jørgensen shows how in most societies, one direction of twist is the norm, and suggests that the other has often been considered the 'devil's direction'. In Pharaonic Egypt, yarns were always twisted anti-clockwise; this continued throughout the Hellenistic, Roman and Byzantine Periods, and after the Arab conquest of Egypt in 641 AD. By the 13th century AD, however, this had been reversed: textiles from Mamluk Egypt are almost exclusively twisted clockwise. In most of Europe, clockwise twisted yarns have been the rule ever since the Bronze Age, except for special effects such as spin patterns and sometimes for weft yarns. Scandinavians deviated from this norm during the 1st millennium BC, preferring anti-clockwise twisted yarns, but around 200 AD they started to conform to continental habits. This chapter explores changes in the practice of textile technology in Europe and around the Mediterranean during the 1st Millenniums BC and AD, investigating if and how they might be related to changing beliefs and perceptions.

Sensory aspects of metalworking technology are at the heart of Maikel Kuijpers' Chapter 14, 'The sound of fire, taste of copper, feel of bronze, and colours of the cast'. Taking the sensual human body as a focal point, through which the world is experienced and understood, he argues that metalworking technology in the Bronze Age was experienced more than rationally understood. He advocates an interpretative turn in archaeological scholarship that takes this change of perspective into account and pays more attention to craft, craftsmanship and skill. Rather than investigating technologies in the framework of modern material science, an embodied perspective exploring the skills involved in metalworking furthers a holistic understanding of the relationships between the body, materials and knowledge in the past. This approach also challenges the deep-seated dichotomy between discursive and non-discursive knowledge and returns to the original meaning of *tekne,* which included the idea of skill and the engagement with material with both the mind and the senses. Kuijpers argues that there was no such thing as metalworking technology in the Bronze Age, but rather metal craft and craftsmanship.

Connections and reflections

The bridges between the two parts are provided by many points that reference themes covered in the other part of the volume – these are often major parts of the observation, but just as often apparently minor points, which traditionally would be considered marginal. One example is Bender Jørgensen's observation of the 'handiness' in Bronze Age yarn spinning. This has no practical explanations, and thus it is of little interest to technological investigations. From another point of view, however, it opens up questions about why people do things in particular ways: was this about superstition and believing that it was better or right to spin in this fashion? These pointers reveal the seepage between the categories of technologies and beliefs. Throughout the chapters there are in fact again and again observations and details that apparently belong in the other part of the volume, and many, if not all, the papers could have been placed in either of the two parts. Sometime the interconnection between technology and beliefs, as in the case of blood transfusion in India discussed by Jacob Copeman, is obviously so strong that the distinction between the two is clearly dissolved even if the rituals discussed are manifestly about religious beliefs and performance. In other cases, such as in Harald Bentz Høgseth's description of the movements of the carpenter, the connection is less apparent, but nonetheless present – the ability to choreograph the carpenter's movement is due to the fact that s/he aims to do it 'right' and works within a 'tradition of practice'. Of particular importance, such practices would include 'non-discursive' but nonetheless transferable knowledge about the properties of wood, the abilities of tools and the connections between these and the body itself. Similarly, both the cross and the portable altar are material things infused by beliefs, and through this they are granted both status and abilities which moves them outside normal material categories. Such crossovers between the thingness (Olsen 2010: 81) and the religious meaning of objects, the ability for the objects to become the essence of abstract beliefs, are discussed with reference to different contexts by Helen Foxhall Forbes, Dengsø Jessen, as well as Mary Laven. The strong link between beliefs and attitudes is especially well illustrated by studies of burial practices. Katharina Rebay-Salisbury's exploration of ways of treating the body through time and Tim Flohr Sørensen's detailed discussion of changes in modern attitudes to death each demonstrates and questions how the two aspects – beliefs and material practice – may mutually reinforce each other during changes of tradition.

Such observations and their implications for how beliefs, technologies, bodies and things are interconnected are an under-explored field. Contemplating the papers collected here in reference to one of the two parts helps to both illustrate a particular dimension of embodied knowledge and, simultaneously, raises the question of how each of the chapters also contributes to the opposite theme. One of the themes emerging from this volume is the question of the moral significance of embodied knowledge. If knowledge enables choice, there is room for good and bad decisions, and for guilt. Does this extend to embodied knowledge, which may be tacit and non-discursive? Several chapters show the linking between the physical and the spiritual, how abstract concepts (such as membership of a group) through a simple gesture may be permanently marked on the body. They show how there is capacity for both outer and inner performance and transformation. Baptism and the absolution of sins in the Christian church are very informative of how such linkages are drawn through a set of prescribed acts including the use of specified gestures and instruments. Such links between the physical or material and the spiritual can be very strong and unquestioned, as illustrated by the cross in the Christian church or by how sickness was conceptualised and understood in Europe during the Early Historic Period.

We may ask what the common distinction between discursive and non-discursive knowledge actually implies? In particular, how can non-discursive knowledge be transferred, and have we tended to mistake non-verbal for non-discursive? Several of the papers investigating technologies bring to the fore the question of handling and learning. Learning has traditionally been associated with either observation or instruction. But if learning is also about recognising and replicating details, judging subtle qualities, how it feels when it is right, then we might have to ask what kind of knowledge is involved? What does it mean that it 'feels right', how does the body know in these instances and what processes of learning have been involved? The papers do not provide final answers to these compelling questions, but we are given sufficient insights into elements of these processes that they do not appear as merely embodied anymore. Such elements are, for instance, learning processes through which knowledge about procedures are acquired, including recipes, methods, and beliefs which together ensure standardisations and compatibility. Sheila

Kohring refers to this as 'community of practice' and Harald Bentz Høgsted as 'the language of the craft'. Whatever label we give this, the studies show a discursive understanding of technologies even if the language used may follow its own format.

The cover image for this volume, *The Philadelphia Twins* by Christie Brown, was chosen to express the complex nature of knowledge, and the interdependency between abstract and technological knowledge (Fig. 1.1). Christie Brown's artwork draws from archaeology, mythology and psychoanalysis, using fragments of the past as 'a way of learning'. Exploring the transformative qualities of clay and the metaphorical associations of the casting process, she views her work as '… embodiments of transition from one state to another, of

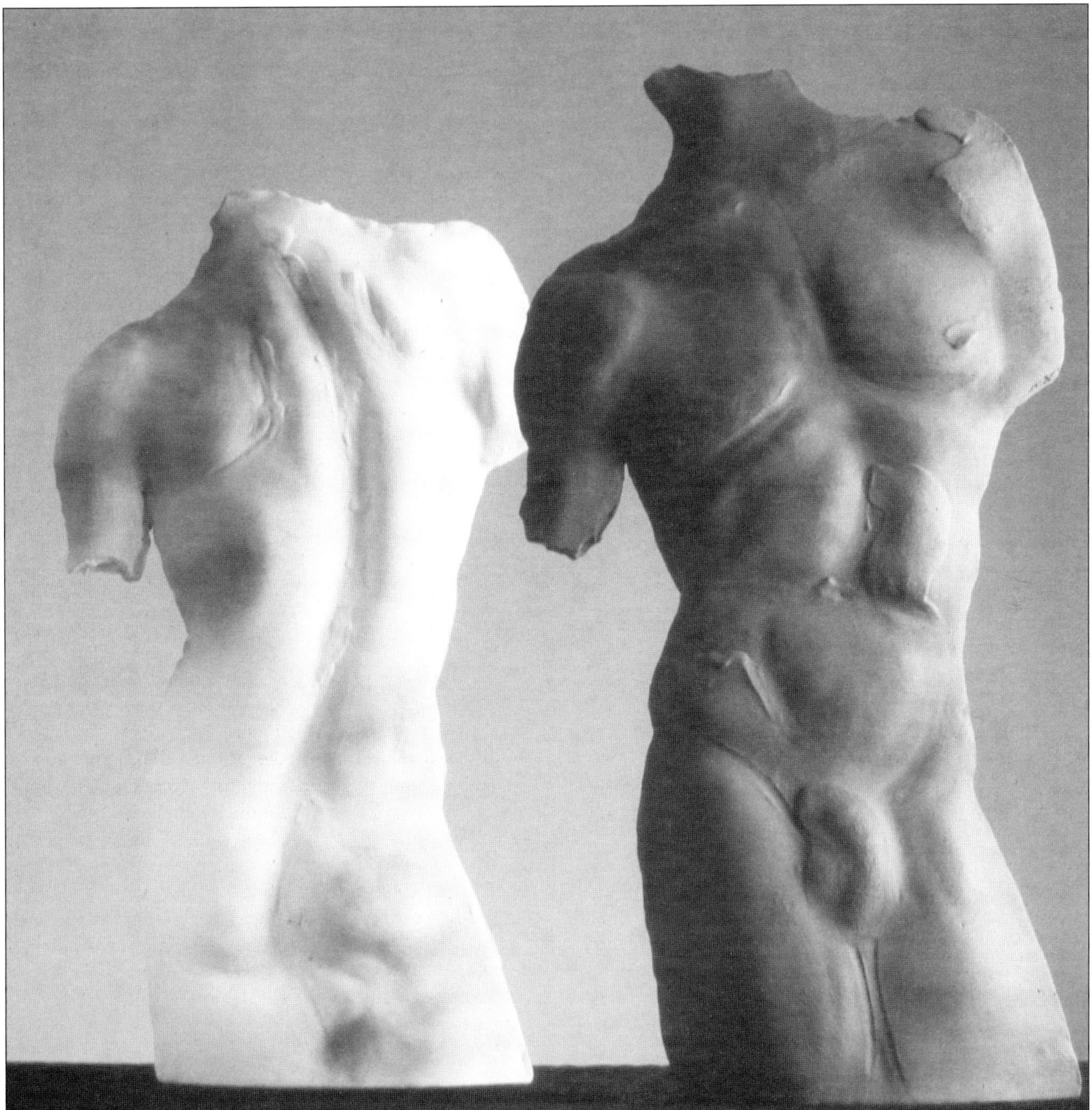

Fig. 1.1: The Philadelphia Twins *by Christie Brown (photo: Christie Brown 1991)*

transformation and individuation'. In *The Philadelphia Twins* 'two figures were conceived as one work and given a title that related to narrative and memory' (Brown 2007). It is the perfect artwork to illustrate belief and technology as aspects of embodied knowledge for this volume.

References

Brown, C. 2007. *Embodying Transformation*. Interpreting Ceramics 8. Retrieved January 17, 2011, from http://www.uwic.ac.uk/icrc/issue008/articles/02.htm.

Olsen, B. 2010. *In Defense of Things: Archaeology and the Ontology of Objects*. Maryland: Altamira.

Polanyi, M. 1966. *The Tacit Dimension*. London: Routledge and Kegan Paul.

Ryle, G. 1949. *The Concept of Mind*. London: Hutchinson.

Sofaer, J. 2006. *The Body as Material Culture. A Theoretical Osteoarchaeology.* Cambridge: Cambridge University Press.

Turner, B. S. 2000. *The Blackwell Companion to Social Theory*. Oxford: Blackwell.

Turner, B. S. 2008. *The Body and Society: Explorations in Social Theory*. Oxford: Blackwell.

Part I

2. Introduction to Part I: belief as practice

Marie Louise Stig Sørensen

The concept of belief has played an important role in many disciplines and informed many kinds of analysis and reflections. It corresponds in various ways to social observations and to the manner in which certain kinds of propositions are made; the concept may be seen as a reflection on human societies. Belief refers to faith, as well as to attitudes we hold and what we regard to be true. Belief is therefore extensively found within how we talk about things in everyday life. It is used as a way of reasoning about actions in one's own familiar environment or to explain observations in unfamiliar contexts and to account for properties that cannot be known. Belief may be used to hide or compensate for a lack of more detailed comprehension in terms of rational insights and reasoning, both by the believer and when observing the action of others. Statements of belief are commonly simultaneously concrete yet abstract. Such statements involve subject and object, and belief is routinely performed throughout our lives, taking on formal as well as informal forms, sometimes acknowledged, other times taken for granted.

There are thus many different aspects to the concept of belief, and a range of understandings are reflected in the ways in which the word is routinely used. Amongst these, two common uses are interesting here insofar as they show the wide-ranging practices and ways of reasoning, which the term refers to. One is the usage of 'belief' in a rather off-handed manner to refer to a way of thinking and reasoning. As such, it is commonly used in everyday conversation; for example, to present a rationale or expectation, as in 'I believe you will get into trouble if you do that' or 'I believe the trains will be late again', where 'belief' is used interchangeably with 'think' or 'predict'. In such usage, belief is partner to the reflections on and statements about relations and expectations. The other main understanding of the concept uses 'belief' to indicate faith in the sense of agreeing to a codified system of belief, which may be religious, philosophical or ideological. Belief in this sense is usually presented through a complex ontology. When used in this manner, belief is a substitute for reasoning, or rather, the belief presents and provides the reasoning, often represented through simple creeds, such as 'I believe in equal rights' or 'I believe in God'. Such beliefs are commonly formed around arguments about the supernatural, sacred, or divine. What these various understandings of belief share is that they involve mental speculations about relationships and causalities and attempts at understanding the world. Belief also commonly results in various forms of representation of the particular ontology.

Although belief and knowledge are different epistemologically, it is interesting to question their inter-relationships and how they are experienced in practice. The connection (and distinction) between knowledge and belief has been a core concern within the philosophy of knowledge since Plato defined knowledge as 'justified true belief' (Chisholm 1982). This definition of knowledge focuses on the requirements for truth by stating that in order to know that a given proposition is true one must not only believe it to be true but also have justification for doing so. Later philosophers of knowledge have questioned this proposition, with particular attention given to situations where the justification is true but the knowledge would be false (Gettier 1963).

Plato's classic statement may therefore also be used to ponder upon what is involved in the transformation from belief to different kinds of knowledge – is belief a step towards, or another form of, knowledge? The interesting question here is how justification is reached, a topic intensely discussed by philosophers (e.g. Ayer 1956, Pollock and Cruz 1999), and in particular what role, if any, 'bodily truth' has – what kind of truth is established, or knowledge experienced and exercised, when a movement feels right or a ritual has been performed appropriately and faithfully? This leads us to the question of different forms of knowledge. Plato and later philosophers of knowledge were mainly concerned with what is now usually referred to as propositional knowledge, whereas the bodily truth I just introduced would be considered procedural knowledge, which is knowledge about how to do something.

To connect procedural knowledge to arguments about truth and belief the next step is to ask how socially, rather than though abstract contemplations, we express and understand the relationship between a sense of being right and that of knowing. In particular, does propositional knowledge play any role in this experience, and, if so, in what manner? The handiness that craftspeople refer to, the sense of things feeling right, what is that? It certainly involves knowledge – however articulated. This is where embodied knowledge comes into our considerations. Embodied knowledge is about practice; it refers to how individuals interact in and make sense of their environment and act meaningfully within it. Theories of embodiment have been important in the social sciences and humanities since the 1980s (e.g. Csordas 1994, Shilling 2004, Turner 2000) and have given rise to much more punctilious attention to various forms and expressions of knowledge within cultural practices. There is often, however, still a tendency of assuming that such knowledge is non-discursive, and when working with embodiment theories we are still inclined to think about humans as embedded in and as an embodiment of their *habitus* (Bourdieu 1990) in such a manner that discursive aspects of embodiment disappear. Moreover, since belief traditionally has been linked to mental speculations, and thus considered to be a discursive practice, belief and embodied knowledge are rarely linked.

This is surprising if we think about how belief is acquired and performed, as this involves several forms of practices in which 'learning to do' is paramount. When considering belief, the distinction between propositional and procedural knowledge (the 'knowing that' and the 'knowing how' coined by Ryle in 1949) therefore seems to break down. This has also been argued within the philosophy of mind (e.g. Dretske 1988), and at a general level Merleau-Ponty has been particularly influential in challenging this kind of distinction. Arguing from a 'being-in-the-world' position, he rejects the idea of the mind separate from the body. To him, thinking is part of an active relationship between embodied humans and the world, and all knowledge is therefore embodied (1962). Embodiment is, he argues '... knowledge in the hands, which is forthcoming only when bodily effort is made, and cannot be formulated in detachment from that effort' (1962: 144). Modern neuroscientists have furthered such arguments by arguing that the mind itself is embodied. A particularly interesting debate has taken place within cognitive linguistics, where, for instance, the works of George Lakoff and Mark Johnson on the language of everyday life have proposed an embodied understanding of language and meaning. They argue that the conceptual metaphors we use are formative for how we think and act (1980, 1999). 'What our bodies are like and how they function in the world thus structures the very concepts we can use to think. We cannot think just anything – only what our embodied brains permit' (Lakoff 1999). Such arguments would suggest that even belief uses concepts based on our embodied experiences, and as such, experience of the world provides the structure for how we reason about it.

Apart from the challenge of understanding the epistemological status of belief and its relationship to embodied knowledge, there has also been a critique of the very concept of belief, questioning its usefulness for cultural analysis. This has especially been voiced within social anthropology. The background to this is found in earlier approaches to religion (and beliefs) which saw them as a mirror of society (Barnard and Spencer 1996). It is easy to see such approaches as limited, although recent arguments about embodiment would resurrect the appreciation of the close link between the ideas that inform systems of beliefs and other aspects of society. This does not mean, however, that belief needs to be reduced to being mere reflections of society. Rather, what seems to be emerging is a way of understanding that embraces how various cultural forms are all interwoven and influenced by embodied understandings of the world. The critique from social anthropology is, however, also based on a concern with the problem that analyses will tend to normalise complex abstract thinking and behaviours by reducing them to 'belief'. It has been argued that transforming belief into a discursive form, which will happen through its analysis, will always alter it and represent it as something different to what it was; it will distort and transform what beliefs are to people (ibid.). It seems, however, that banning the concept of belief from our social studies places at risk the relevance of such studies, as the concept, both as a common-sense reference and as codified systems, matches a range of behaviours which in various ways may be identifiable and which are important to study. Moreover, singling 'belief' out as suffering from the impact of our studies appears to be isolating some parts of human and societal behaviour from other aspects in a manner that is entirely academic. This volume suggests that by embracing the seepage between different concepts and by exploring the connections between different forms of embodied knowledge, such as beliefs and technologies, we can partly overcome the tendency of isolating one kind of knowledge system from other kinds. It also demonstrates that there are ways of analytically engaging with belief that are not dependent on assuming that access to a true rendering of the believer's thoughts and emotions is possible. Rather the case studies demonstrate that belief, in the sense of faith, can be explored through studies of how it is performed, instrumentalised and institutionalised.

For a more explorative study of belief as embodied knowledge, it is helpful to focus on how belief is linked to action. This, however, does not merely refer to how it is practiced; rather the concern must be to understand how it is through its performance that belief acquires recognisable and shared forms and meanings. It is by doing things right in terms framed around variables such as order, content and rhythm that belief becomes substantiated. Belief is therefore not only abstract speculation; it is also generated by knowledge embedded in action – just as technology is. Moreover, the gestures and practices that are involved are commonly experienced and perceived as fully integral to the belief. Such practices are not, therefore, representations; they are an integral part of the belief. It is, therefore, not a matter of belief enacted but rather of a belief becoming and being constituted through the action. In the most literal sense, the belief is experienced and fulfilled in the action (Gilchrist 2000). This is also why ritualisation, with its insistence on and attention to form and repetition is of fundamental importance to the study of belief.

Another important aspect is the instrumentalisation and materialisation of belief, how it comes to reside in objects and uses these as its instruments of ritualisation. Dengsø Jessen's call (Chapter 5) for better recognition of the extra-somatic dimensions of religious knowledge relates directly to this dimension, as beliefs are situated in and influenced by context and environment. In turn, material objects and designated places, such as crosses and temples, are important as the material form of religion. Moreover, it is through the shared understanding of their meanings that such places and objects may be integrated in religious concepts, as well as being manipulated and used within wider socio-political discourses. The Christian church provides numerous examples of such sanctification of material things, where things become different kinds of matter through particular rituals. Most obviously, the wine and bread become the blood and flesh of Christ through the miracle of Eucharist, but other icons within the Church, such as the cross and the holy water, are similarly part of a transfiguration where entities change through performance.

Studies of belief have commonly focused on processes of making meaning, and in particular on the human ability for and tendency of speculative engagement with their environment, as well as their need to understand it. This volume expands on this to explore the relationship between belief and knowledge as expressed within practice, and attempts to draw attention to the interplay between discursive and non-discursive knowledge. Belief is normally considered to be abstract thoughts and discursive, but it is also often deeply emotional, as it is experiential and expressed through performances and the associative dimensions of rituals and the instruments used. Therefore, although at one level belief may be exceedingly abstract, even ill-defined, a feeling more than knowledge, at another level belief also has much more mundane characteristics that are expressed through their dependencies on recognisable and shared forms, and knowledge about these forms, so beliefs employ recipes just like a baker or a carpenter would.

As a way of introducing the first part of this volume, the aims of this preamble have been to point to the practical, performative and material aspects of belief and through that to demonstrate their close links to diverse forms of knowledge. The arguments lead us to embodied knowledge and the interplay between the body (mind and muscles) and the practice of belief, to the realisation of the strong connection between speculations and their representations through staged rituals, conventions and even technology. Belief is not, however, merely reflected or represented through such somatic means; their dynamic interaction is more complex. Furthermore, beliefs permeate our practices widely. On one hand, they take the form of faith, and on the other hand, they constitute ways of reflecting that inform our everyday actions and routines in terms such as 'right' and 'wrong'. The assumption that belief is all abstraction is accordingly challenged. Studies of belief repeatedly return us to how they are performed, what are the right ways of performing procedures, which are the objects to be used, and in what order – and what are the perceived consequences of not following these conventions. Just as with technologies, notational systems (whether written in a book or mnemonic) assist and direct the performance of beliefs. It follows, I suggest, that to engage in depth with the studies of belief we should investigate the practices through which people perform what they call belief, because it is through these same practices that people confirm their understandings of belief and relate them to and compare them against those of others. Insights into how belief involves not just the mind but the entire body might help us to find more nuanced ways of investigating belief as embodied knowledge and practice.

Beliefs may present themselves as abstract, but as a social phenomenon they are dependent on being shared, recognised and appreciated. They need to take specific forms. It is also inbuilt in belief that it follows prescriptive routines both in terms of practice and thinking, even if these change and are transformed – how otherwise would they be recognised as beliefs? Ritualisation of practices, use of specialised tools, the formulation of body expressions – whether verbal or somatic – are some of the tools that are used to make beliefs distinct and part of social life, and in such extensions from thoughts to practices belief invades the fields of technologies. They become dependent on particular objects, objects that not only represent, but in themselves may become the essence of belief by overcoming their material meanings to become something very different. Despite the common equation between belief and abstract thoughts, belief as performed becomes in many ways very concrete. This is why this introduction has singled out performance and instrumentalisation as core elements of belief. The papers following demonstrate some of these aspects through case studies from different points in time and different cultural contexts.

References

Ayer, A. J. 1956. *The Problem of Knowledge*. London: Macmillan.

Barnard, A., and J. Spencer (eds) 1996. *Encyclopedia of Social and Cultural Anthropology*. London: Routledge.

Bourdieu, P. 1990. *The Logic of Practice*. Cambridge: Polity Press.

Chisholm, R. 1982. *Knowledge as Justified True Belief. The Foundations of Knowing*. Minneapolis: University of Minnesota Press.

Csordas, T. L. 1994. *Embodiment and Experience: The Existential Ground of Culture and Self*. Cambridge: Cambridge University Press.

Dretske, F. 1988. *Explaining Behavior: Reasons in a World of Causes*. Cambridge, MA: MIT Press.

Gettier, E. L. 1963. Is justified true belief knowledge? *Analysis* 23: 121–123.

Gilchrist, R. 2000. "Unsexing the Body: The Interior Sexuality of medieval Religious Women," in R.S. Schmidt and B. L. Voss (eds) *Archaeologies of Sexuality*. 89–103. London: Routledge.

Lakoff, G. 1999. *"Philosophy In The Flesh" A Talk With George Lakoff*. Retrieved 18 October 2011, from http://www.edge.org/3rd_culture/lakoff/lakoff_p1.html.

Lakoff, G., and M. Johnson. 1980. *Metaphors We Live By*. Chicago: Chicago University Press.

Lakoff, G., and M. Johnson. 1999. *Philosophy in the Flesh: The Embodied Mind and its Challenge to Western Thought*. New York: Harper Collins.

Merleau-Ponty, M. 1962. *Phenomenology of Perception*. London: Routledge.

Pollock, J. L., and J. Cruz. 1999. *Contemporary Theories of Knowledge*. Lanham: Rowman and Littlefield.

Ryle, G. 1949. *The Concept of Mind*. Chicago: University of Chicago Press.

Shilling, C. 2004. *The Body in Culture, Technology and Society*. London: Sage.

Turner, B. S. (ed.) 2000. *The Blackwell Companion to Social Theory*. Oxford: Blackwell.

3. Inhumation and cremation: how burial practices are linked to beliefs

Katharina Rebay-Salisbury

Choosing whether to inhume or cremate a body after death is a situation people are rarely confronted with; burial practices are normally tied into long standing traditions and non-discursive practices, whose original meanings might not be even apparent to the participants of the burial rituals themselves. Things are done in a particular way, because they are remembered to have always been done so, and this is what makes them the right thing to do. This is the moral of tradition – actions are justified by past practices. Burial practices do change, however, and this is a point where deviant practices have to be justified, explained, discussed, and negotiated. This chapter explores changes from inhumation to cremation and how they might be linked to beliefs. With beliefs, I do not necessarily mean religion in a formalised sense; I mean beliefs about the body, death, and the afterlife. Beliefs are constructed in communication and enacted by society, relating to how people think about the world and how they make it understandable in their own terms, striving for order and reason about aspects of life which are beyond their control. Beliefs do not have to be coherent and consistent, and it is possible to hold conflicting beliefs – with the circumstances determining which beliefs are given priority at any given time.

Understanding the nature of bodies and their status after death is one of the challenges for societies, and it is encountered every time one of their members dies. Both cremation and inhumation are practical responses to the reality a dead body represents, but they are radically different approaches; the dichotomy between inhumation and cremations represents an intrinsic conflict between the urge to preserve the body for as long as possible and the desire for transformation. Sørensen and Bille, for example see a clue to why many societies use cremation as a means to dispose of the dead in the very tactile as well as metaphorical way fire is understood as a transformer (Sørensen and Bille 2008: 256).

Burial rituals are amongst the few archaeological visible forms of practice that may hint at beliefs; in fact, they are often interpreted as synonymous for beliefs about an afterlife (Ucko 1969: 264), although this relationship may not be quite so simple. Many communities provide for the needs of the dead (Ucko 1969), while others visibly express a part of the person's social personality, or put objects in the ground that had particular emotional connotations. The ways people are buried depend on a variety of factors and a number of people, including the mourning community and the deceased, and also on his or her will. It has also recently been argued that the burning corpse on the pyre may have been perceived as having some sort of agency (Williams 2004). Burial practices reflect, as Hertz put it, the relationships between three kinds of personae: the corpse of the deceased, the soul of the deceased, and the remaining society of mourners (Hertz 1907: 176).

Amongst archaeologists, Childe, Piggott and Clark have opposed the view that the change from inhumation to cremation can be equated with a change in religious beliefs (Childe 1944: 85–88, Clark 1960: 232, Piggott 1965), but a cross cultural study of 31 non-state societies using the Human Relations Area Files (Carr 1995) found that mortuary practices clearly take philosophical-religious factors – beliefs – into account (ibid. 188). The practices are determined by a complex mix of factors, which includes social, political, physical and circumstantial ones. In the examples that were collated, a wide range of variables, such as body position and treatment, grave location and characteristics, cemetery organisation, circumstances of death and the social position of the deceased was related to philosophical-religious beliefs (defined as beliefs about sickness and death, physical health and safety, the afterlife, the journey to the afterlife, the nature of the soul, the soul's existence, maturation, waning during life, beliefs about reincarnation, beliefs and myths about universal

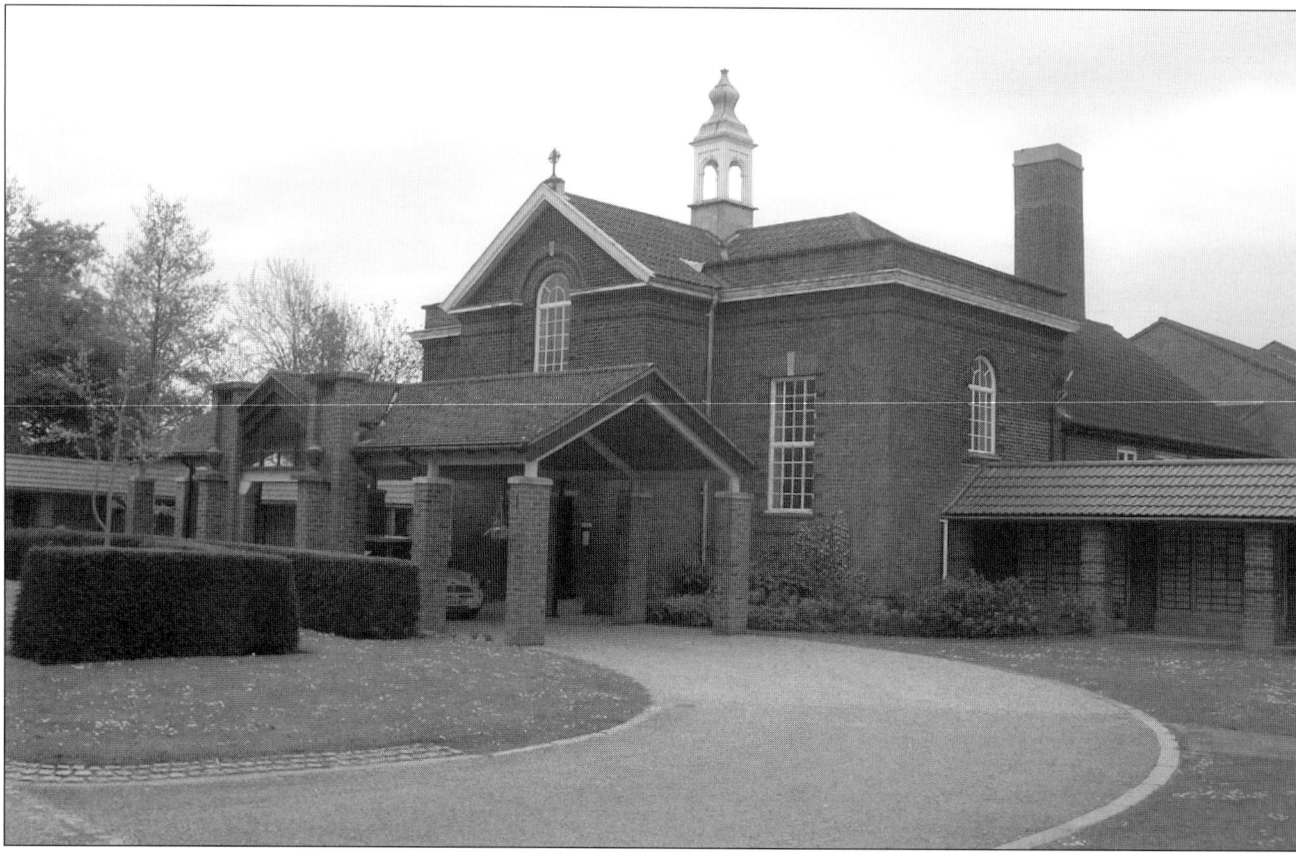

Fig. 3.1: The Cambridge City Crematorium (photo: K. Rebay-Salisbury).

orders and their symbols, the symbolic classification of a person upon death, origin myths, beliefs about third-party souls and spirits, beliefs about responsibilities to or punishments of the soul of the deceased, beliefs about the status and change of status of the person at/after their death and their effect on the living).

The study revealed a number of very interesting associations between burial practices and beliefs (encouraging for archaeologists, who are very often confronted with sceptical views about the feasibility of interpreting burial data). It was found that beliefs were documented to determine variation in mortuary practices more often than any other factors in all types of societies (ibid. 168, 189). Most commonly, these were beliefs about universal orders, beliefs in a soul, afterlife, and journey to afterlife, causes of sickness and death and responsibilities and punishment of souls. The preparation and treatment of the body, its orientation and position as well as the spatial arrangement of grave goods were found most useful for reconstructing philosophical-religious beliefs (ibid. 157), whereas aspects of personal identity like age, gender, and horizontal or vertical social position were reflected less frequently and primarily through grave goods and grave location (ibid. 165). Cremation as a form of body treatment is not separately addressed in this article.

The treatment of a dead body, the building of a grave structure, and the performance of funerary rituals are actions that build on beliefs about the body as well as embodied knowledge. They draw on a range of familiar social practices and techniques used in everyday lives of communities and are loaded with connotations, meanings and metaphors. It is not difficult to see how the transforming power of fire is transferred from the pyrotechnical realm to the social realm and applied to transforming the body through cremation. The way cremation is carried out is, of course, embedded in the technological circumstances of the time. The most common use of pyrotechnological methods include lighting an open-air fire or maintaining a fire in pits, hearth and ovens; but fire has also been used in more complex settings. During prehistory and the Classical period technological knowledge of pyrotechnics was utilised in firing ceramics, preparing food, smelting metals and a range of other production techniques. Pyre building and open-air or pit cremation could draw on this knowledge. In the Industrial Age the utilisation of the transforming power of fire took new shapes and triggered the building of crematories inspired by and borrowing from the aesthetics of the industrial architecture of the time. This often included brick architecture and high chimneys, as seen, for instance, in the Cambridge City Crematorium (Fig. 3.1).

In this article, I will utilise and discuss three case studies as examples of how a variety of different beliefs feed into the elaboration of burial practices and how practical and embodied knowledge plays a role in shaping these beliefs. The three case studies all have one thing in common: they characterise a period of change from inhumation to cremation or the parallel existence of both rites. The re-introduction of cremation after the Enlightenment in Vienna, Classical Greek and Roman burials, and Bronze Age Central Europe represent a diverse range of settings in which body politics are played out through a very simple variable – whether to inhume or cremate the dead.

Case study 1: the re-introduction of cremation after the Enlightenment in Vienna

The re-introduction of cremation in Central Europe during the end of the 19th and beginning of the 20th century is a well documented example of how burials are linked to beliefs. It is also a good example of how small scale, regional dynamics can be part of wider trends, at the same time as they maintain their own trajectories and characteristics. The change in burial rites was based on rational reasoning, but the association of cremation with the liberal movement caused the church to oppose cremation in Catholic countries. A person's private decision about the form of their burial soon became a political statement.

Ever since Charlemagne prohibited cremation as a pagan practice in 789 AD (Mims 1998: 178), cremation was seen as a disbelief in the resurrection, a crucial dogma of Christian belief (ironically, the punishment for disobeying this law was being burnt alive – cremation was apparently only illegal when applied to dead bodies). Cremation had not been practised as a burial rite for about a millennium when it was slowly re-introduced in the course of the Enlightenment. Beliefs about God and the afterlife had slowly changed and given way to scientific reasoning; hygiene and the lack of space in cemeteries were put forward as rational reasons to favour cremation, but a range of other motifs, such as horror of putrification, equally played a role. In Western Europe, cremation was spearheaded by individuals like Percy Shelley, who was cremated in Italy in 1822, or Sir Henry Thompson, who founded the Cremation Society in 1874 (Mims 1998: 179), resulting in the legalisation of cremation in 1884 in Britain (Tarlow 1992: 130). Discussions on the 'right and wrong' of cremation as a burial practice in the 19th century (Jupp 1990, Parsons 2005, Prothero 2001, Sørensen and Bille 2008, Thalmann 1978) were intrinsically linked to visions of the afterlife and quite specific theological discourses of the different strands of Christianity; at the same time, not believing in any form of afterlife became an increasingly accepted option. Protestant churches accepted cremation earlier than the Roman Catholic Church, which only lifted the ban on cremation in 1963. The Eastern Orthodox Church still does not allow cremation, and only in 2008 did the Nation's highest court in Greece approve a government petition to legalise cremation (The Associated Press 2008). Within Europe alone, the percentage of cremation varies widely (Davies and Mates 2005). The highest rates of cremation are found in Great Britain and the Scandinavian countries (almost 70%), while countries with a strong Catholic heritage have very low cremation rates and former communist country rates fall somewhere in between the two extremes. These huge differences cannot simply be explained by one simple cause – for instance if one believes in physical resurrection or not – but have to be understood as one part of the wider history of body politics in any given region.

In the early modern period before the re-introduction of cremation, attitudes towards corporal remains in Central Europe were different than those in Western Europe. In Central Europe it was, for instance, customary that bodies remained in the ground for only a short time; graves would be leased for a few decades before the bodies were exhumed and kept in charnel houses underneath or next to local churches, as space in cemeteries was often limited. The bones unearthed while digging new graves were often dismembered from their anatomical context, and consequently not treated as belonging to an individual, but as part of 'dead matter'. Individual bones of skeletal bodies were often rearranged in patterns, with collections of skulls, femurs, or other body parts arranged together. Examples of this practice can be seen all over central Europe (Westerhoff 1989, Zilkens 1983), notably in the catacombs underneath St. Stephen's Cathedral in Vienna (Gruber and Bouchal 2005). This practice created a new 'community of the dead', dissolving every notion of the individual. Furthermore, these ossuaries functioned as a kind of *memento mori*, constantly reminding the living of their own mortality. In the Sedlec Ossuary in the Czech Republic, for example, about 70,000 human bones have been artistically arranged by František Rint in 1870 to form decorations and furnishings for the chapel, namely bells in the four corners of the chapel, a chandelier in the centre of the nave, two monstrances and the coat-of arms of the Schwarzenberg family (Koubsky 2010). In the famous ossuary of Hallstatt in Austria (Lehr 1979), in contrast, about 2000 skulls were exhumed and painted with ornaments by the families of the deceased. Flowers, leaves and serpents were popular motifs, and the name of the individual and the year of death were often mentioned as well, thus preserving some notion of individuality on the person's bones. From these examples we can infer relaxed attitudes towards the handling of body parts, in which fragmentation of skeletons did not cause major moral concerns.

The emphasis of funerary rituals in Central Europe was not about the body and its integrity; it was about the elaboration of the funeral itself. Particularly in Vienna, funerals were certain to attract a large number of spectators, and for Beethoven's funeral (Davies 2001), for example, a reported

20,000 mourners attended. Post-mortems activities were much more wide spread and accepted in Central Europe than in contemporary Western Europe (Buklijas 2007). Obtaining bodies for medical training did not cause the same public uproar as, for instance, in England (Cherryson 2010); as long as the person was given a 'proper funeral' afterwards, it did not matter if the body was dissected or not. Beethoven even insisted on a post-mortem to establish the cause of his deafness, and interest in the cause of his illness and death lead to two exhumations, during which body parts were removed and replaced by body parts of strangers (Meredith 2005). In the period before the re-introduction of cremation in Central Europe, bodily integrity was much less of a concern, and not linked to the idea of resurrection in the same way as in Protestant countries, even though the official doctrine of the Roman Catholic Church was still favouring inhumation.

In Austria, the re-introduction of cremation after the Enlightenment was pioneered by doctors, scientists and philosophers, who started to campaign for cremation in opposition to the horrendous costs for funerals and the maintenance of graves, but also grounded in concerns about hygiene. Whereas the first crematoria were built in Woking, England, in 1878 and Gotha, Germany, in 1878, several attempts to get permission to build a crematory were not successful in Austria. Cremation societies were founded in Austria, which organised cremation abroad and campaigned for acceptance of this burial form (Hauf 1996). At first these societies started out as politically neutral, but since the Catholic conservative leadership of the monarchy did not permit cremation, the emerging socialist movement adopted cremation as part of their platform and cremation became a means to express anticlerical attitudes. The slogan of the Austrian Worker's Funeral Association was 'A proletarian life, a proletarian death, and cremated in accordance with the progress of culture' (Morris 1992: 34). Only after the First World War and the end of the monarchy was the first crematorium built. The Crematorium of the Central Cemetery in Vienna was planned by Clemens Holzmeister and built in 1922 in the fashionable and modern 'expressionist' style (Bauer 2004), designed to communicate the notion of progress and social change (Fig. 3.2). The building was opened by the socialist mayor of Vienna the following year. To be inhumed

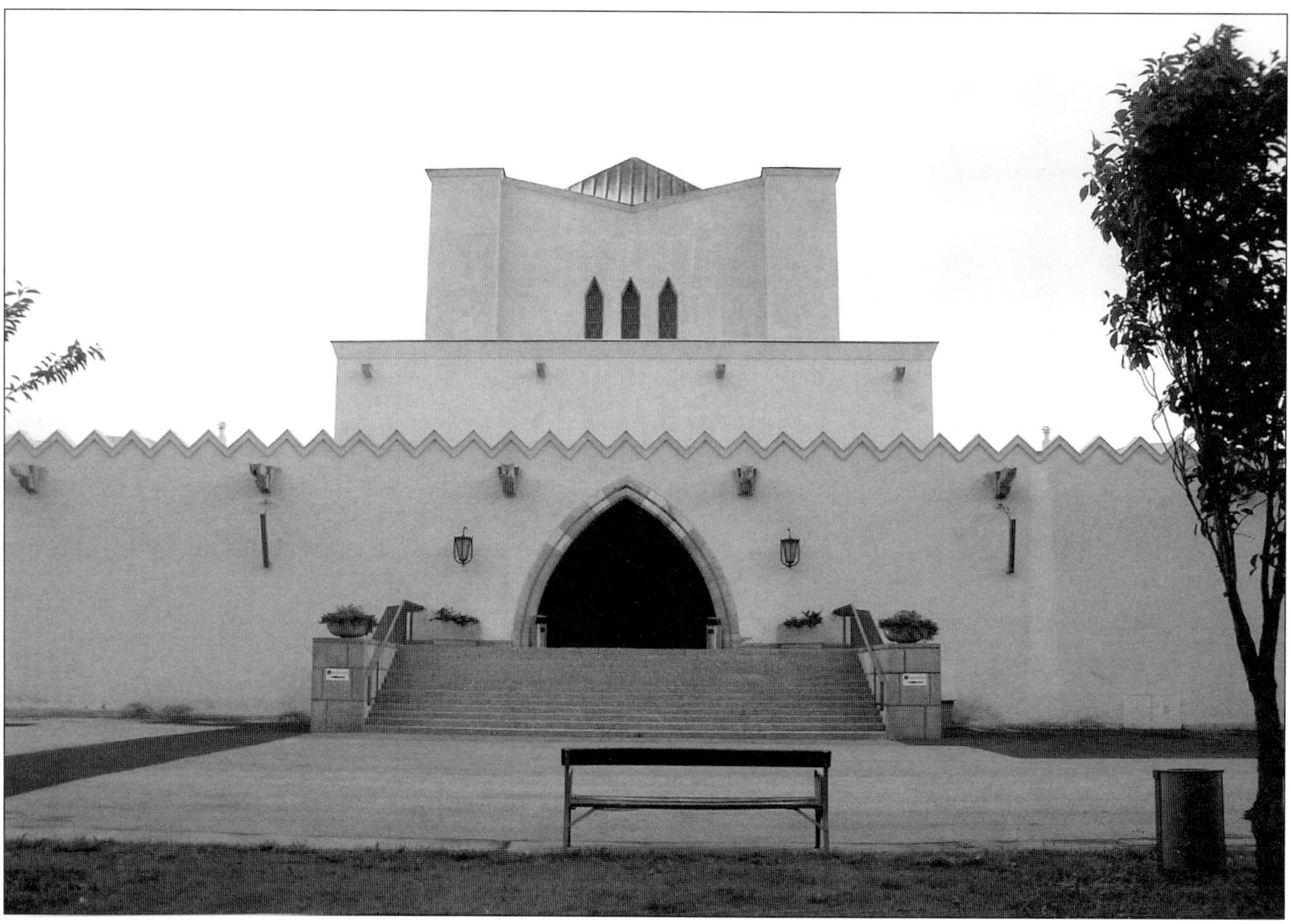

Fig. 3.2: Crematorium Simmering, Vienna (photo: K. Rebay-Salisbury 2011).

or cremated became a matter of political opinion and a public statement in a politically divided country. In the 1920s and 1930s, the battle between conservatives and socialists was fought in all areas of life and included the choice of how one wanted to have one's body treated after death (Ebner 1989).

The choice of body treatment and burial form was therefore clearly linked to beliefs, but the dispute was not actually about religious beliefs or visions of the afterlife: it was entangled within a wider political debate about the significance of and emancipation from religion in everyday life. New scientific understandings of the body played only a minor role in this debate. The choice between inhumation and cremation became utilised to express political opinions and standpoints. Through time and with increasing practice the general acceptance of cremation grew, and particularly since the Roman Catholic Church gave its consent, cremation has become a matter of personal taste and preference. In retrospect, the subtlety of this dispute would be hard to understand if only its archaeological record was available, although non-verbal ways of expressing opinions such as the use of architecture and aesthetic styles would still be a clue to how cremation was understood at the time. But can we compare this situation to Classical Greek, Roman or prehistoric funerary rites?

Case study 2: Classical Greek and Roman funerary rites

The Greek and Roman worlds went through cycles of centuries when cremation or inhumation was the dominant funerary rite, and at times inhumation and cremation were practiced simultaneously. So far, no convincing, simple and clear-cut explanation for this phenomenon has been found. Homer mentions cremation exclusively as a burial rite in the *Iliad* and *Odyssey*, and the funerals of Patroclos (Il. 23, 161), Hector (Il. 24, 778.), Elpenor (Od. 12, 11–15) and Achilles (Od. 24, 65) are described in detail. Illustrations of cremation can be viewed on Greek vases from the Geometric Period onwards (Boardman 1998). These classic texts and images played a major role in shaping the way cremation was understood in later periods, for instance when it was re-discovered in the course of the Enlightenment (Fig. 3.3). The Homeric poems suggest that when the body is cremated, the soul is free to enter Hades, and Richardson holds it possible that the ancient Greeks believed the soul was released from this world more rapidly and effectively through cremation (Richardson 1985: 50). Nevertheless, the soul was also believed to leave the body at the time of death, and to give it peace, employing proper burial rites was crucial – a body left unburied aroused the anger of the dead and brought divine punishment. In addition to this, dead bodies were seen as polluting (Hope and Marshall 2000). The proper and correct burials rites mattered, but whether these involved cremation or inhumation was less important (Richardson 1985: 51).

The afterlife – Hades – was not entirely envisioned as a pleasant place, rather as a gloomy empire where the dead suffered a sad existence, although this view changed in later Greek philosophy to include a system in which mortals were either rewarded or condemned. Alternative beliefs to Hades existed: some individuals could become immortals, heroes, and ancestors – all of which were believed to have the power to influence the fate of the living. Another alternative view was the belief in re-incarnation, that a single person lives out multiple lives, which was already mentioned by Herodotus and probably circulated in Greece by the sixth century BC (Richardson 1985: 61). The nature and character of the soul

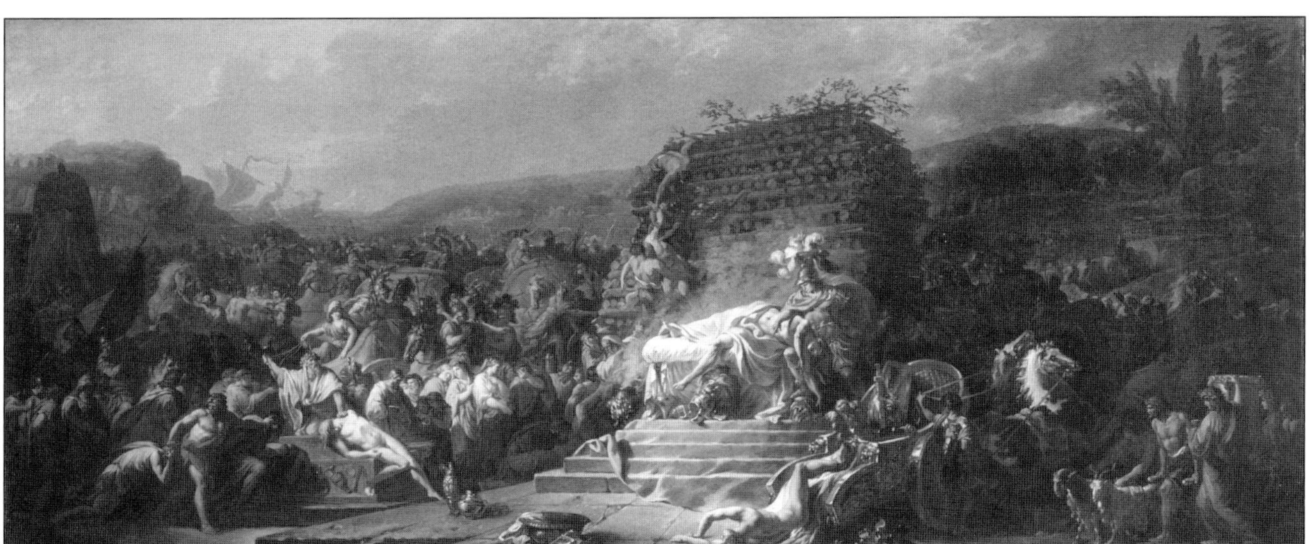

Fig. 3.3: Jacques Louis David's painting of 'The Funeral of Patroclus', 1778 (reproduced with permission from the National Gallery of Ireland, Dublin).

was a topic of much discussion in all aspects of Greek art and philosophy, which had, by Plato, developed into a view that the soul was divine, immortal, and contained the real and enduring personality (ibid. 65). Again, there is no absolutely coherent belief system to be found: ideas about what happens after death were many and diverse. A patchwork of myths and stories created the framework in which a range of practices were played out.

It is interesting that in the Classical Greek world the emphasis is on the right, respectful funerary practice rather than a 'correct' ideological justification or logical underpinning. Ancient Greeks practiced both inhumation and cremation with the dominant practice depending on place and time and both practices could often be found side by side. In Athens, cremation declined in popularity around 400 BC, and in the absence of convincing connections to changes in religion and beliefs, scholars have turned to social and economic explanations. It has been debated whether cremation or inhumation presented the 'cheaper option', and how much grave goods played a role in display, conspicuous consumption, and politics (Fagerstöm 1993, McKinley 2006, Morris 1987). Analyses have shown that it is difficult to estimate whether one rite is associated with 'richer' grave goods than the other. Athenian fifth and fourth century cremations tend to have more pottery, but less metal than inhumations (Morris 1992: 116). Overall, cycles of elite fashion, imitation by commoners, and reactions to these by further differentiation (including noble understatement) seem to dictate inclusion of grave goods and the preferred burial rite.

On the Italian peninsula the Villanovans practiced cremation like other contemporary late Bronze and early Iron Age societies. By the end of the eighth century BC, however, cremation and inhumation were practised simultaneously, despite chronological or spatial preferences for one or the other form of burial. The choice of burial rite seemed to depend on family tradition and individual preference (Toynbee 1971: 15). Cremated bones were placed in an urn, which was then placed in a tomb. Etruscan tombs were often decorated with funerary scenes or lively scenes of feasting, athletic exercise, dancing and music, which can be interpreted as scenes from an afterlife imagined as fairly paradisiacal, and, or alternatively, as the reality of banquets and games held by the living for the honour of the deceased. By the fourth century BC gruesome scenes of death and horror as well as demons appeared and might hint at a change in how the afterlife was imagined – less enjoyable and more threatening, dangerous, and full of grief (ibid. 17).

Like the Greek, Roman ideas about what happens after death were many and diverse, although a belief in an immortal soul – material or immaterial – seems to dominate (ibid. 34). Funerary rights and rituals were closely linked to the belief that spirits of the dead lived on, dwelled in or close to the graves, and might occasionally interfere with the living. Practices of remembering, soothing and satisfying the dead included the lighting of lamps and the offering of small gifts such as flowers and meals at the cemetery site on several occasions throughout the year. Part of the food was then left at the tomb as an offering. Food and drink could also be poured directly onto the bones in the grave through holes and pipes (ibid. 51). The boundary between life and death was envisioned to be permeable to a certain extent, and communication between the living and dead was perceived to be possible. The spirits could step over this permeable boundary to help or harm their descendants, and the living could act to ensure the loyalty of the deceased ancestors.

Inhumation was believed to be the older burial rite in Rome (after Cicero and Pliny, Toynbee 1971: 39), although archaeological evidence exists from the eight century BC for both rites (Momigliano 1963: 101). In the Republic the main practice was cremation. It seemed that some families clung on to the inhumation rite for longer, such as the statesman Sulla's family, of which he was the first to be cremated. The choice seemed to depend on family tradition and individual preference; certainly the status and wealth of individual families played a role in building and maintaining these traditions. Some slaves and the very poor without family or friends were buried in mass graves. Death was thought of as polluting and demanded acts of purification in the Roman world; this is probably one of the reasons why all bodies, cremated or not, had to be buried outside the cities. To leave the body unburied was, similar to the Greek understanding, a major offence and had unpleasant repercussions for the fate of the departed soul as well as for the living (Toynbee 1971: 43). The covering with soil was a crucial requirement, which ensured that the deceased had a respectful resting place. A resting place could even be constructed if there was no body available for burial or as an honorary grave for a person buried elsewhere; such cenotaphs were designed to give the soul a place in which to dwell.

A very interesting interface between cremation and inhumation is the rite of *os resectum* (literally cut bone, mentioned by Cicero, Festus and Varro), in which a small part of the corpse, usually a joint from a finger, was severed from the body and retained from cremation to be buried separately (Graham 2011). Archaeological evidence for this rite exists in the form of 300 small inscribed vessels containing bone fragments from San Cesareo on the Via Appia (Graham 2009: 55–57) and the recent discovery of a vessel underneath the altar dedicated to M. Nonius Balbus at Herculaneum, which very likely contains his phalanx (Graham 2009: 57–67, Pappalardo 1997). The purpose of this practice is little understood (Hope 2007: 108, Toynbee 1971: 49). It might reflect a residual of the earlier custom of inhumation, and it helped to legitimise cremation by providing an interment in earth, which was at times believed to be necessary to create a proper resting place for the soul. This brings up the question of the ontological status of cremated bones as they may not have been conceptualised in the same way as unburnt bones. Furthermore, the practice of *os resectum* has recently been re-interpreted as a significant part of the funerary ritual of purification and remembrance to ensure spiritual well-being of both the community of the dead and that of the living. In the process of these rituals, the *os resectum* might have been subject to a subsequent, second cremation, albeit on a smaller fire employed to purify the household of the deceased (Graham 2009).

Cremation was, however, not practiced everywhere in the Roman Empire; in fact only the western part of the Empire primarily cremated their dead (Morris 1992: 68), while the eastern part of the Empire primarily buried their dead, which caused the writer Petronius to call inhumation a 'Greek Custom' in the first century AD (Morris 1992: 52). A seminal change in burial customs occurred in the second century AD, when inhumation again became the dominant burial rite throughout the Roman Empire. The speed in which this change took place mirrors some class-specific dynamics as well as core-periphery effects in the provinces of the Empire. Rich families seemed to spearhead the new fashion, gradually affecting imperial styles first, then through to the lower classes and out into the remote areas of the empire (Morris 1992: 54). Jewish, Christian or influences from Eastern mystery religions may have contributed to this change, although in the mid-second century AD these influences were probably not very strong and the idea of a bodily resurrection had not yet fallen on fertile ground. According to Toynbee, inhumation was felt to be gentler and more respectful as the belief in a blissful, individual afterlife became stronger (1971: 41). The change, however, might not have been more than a change of 'fashion' (Morris 1992: 33, Nock 1932: 331) as it cannot be clearly connected to a fundamental shift of ideas. As most of the other funerary procedures and rituals to which much attention was paid remained the same, the difference between inhumation and cremation was simply not considered to be of much significance to the Romans.

Case study 3: burial practices in Bronze Age Central Europe

The third and final case study in this chapter is set at the transition from inhumation to cremation practices in Central Europe during the course of the Middle to Late Bronze Age. In the absence of written sources, the relationship between burial practices and beliefs are difficult to tackle and need to be inferred through traces of ritual practices (Sørensen and Rebay-Salisbury forthcoming).[1] As we have seen, no simple and single cause tends to trigger changes in burial practices in the historically well-documented situations; likewise, we should not expect this in prehistory. All we can observe is how the essential changes unfold and, through this, attempt to engage with how people – through action, interpretations, negotiation, and discursive experimentations with forms and reasons – make changes rather than being merely subjected to change. In this, we might be able to highlight some trends regarding how attitudes to and perceptions of dead bodies changed over time.

In the course of the Middle to Late Bronze Age transition cremations gradually replaced inhumation graves over most of Europe, and this change resulted in urn burials becoming the dominant burial form. This change was associated with other developments: the rise of large cemeteries with hundreds, if not thousands of individuals, and an increasing reluctance to bury objects with the remains of the cremated body. It is interesting to explore people's actions and practices when confronted with having to chose between inhumation and cremation: the range of variations and similarities in this case studies speaks to the formulation of local responses to mega-trends and how overarching tendencies and structural changes are worked through at individual sites and on a regional level.[2]

This deliberate transformation of bodies into other substances at death indicates a radical shift in beliefs about what constituted the body and how its parts 'belong' together after death (Sørensen and Rebay 2008c). Earlier research has often neglected the question of how this transformation happens in detail and what people actually did in practice and instead primarily focused on associating the spread of cremation with movements of peoples (Böhm 1937, Childe 1950, Kraft 1926) or saw cremation as the expression of new religious beliefs, for instance as a result of the development of various kinds of soul beliefs. Since the 19th century it has been argued that cremation took place to facilitate liberation of the soul from the prison of the physical body, allowing the continued existence of one aspect of a person beyond death, albeit in a different realm (Hertz 1907, Müller 1897). Whereas these views might not be wrong, such ideas can be associated with inhumation practices as well, as we have just seen in the previous case study above.

Despite the fact that a robust concept of the Urnfield culture is used in the literature to demarcate a cultural phase marked by the introduction of cremation (Sørensen and Rebay 2008a: 57), it is clear that there is no fixed point in time when cremation is introduced. We can identify a period of transition when burial practices undergo gradual but seminal changes, during which inhumations and cremations are often treated very similarly in a number of aspects. This is, for instance, exemplified by the graves from Streda nad Bodrogrom in Slovakia (Polla 1960: 353, Sørensen and Rebay-Salisbury 2008: 56–57). Cremation grave 24 and inhumation grave 35 were both placed in a rectangular pit of similar size and orientation, and a very similar set of pottery was placed at the feet of the inhumation or at the end of the space the scattered cremation occupied (Fig. 3.4). Both bodies have been treated in the same manner, with the same care, regardless of whether they were inhumed or cremated. Building from this and other examples, our wider analyses suggest that it usually takes a couple of generations before the burial practices within regions were fully changed from an inhumation ritual, focussed on the wholeness

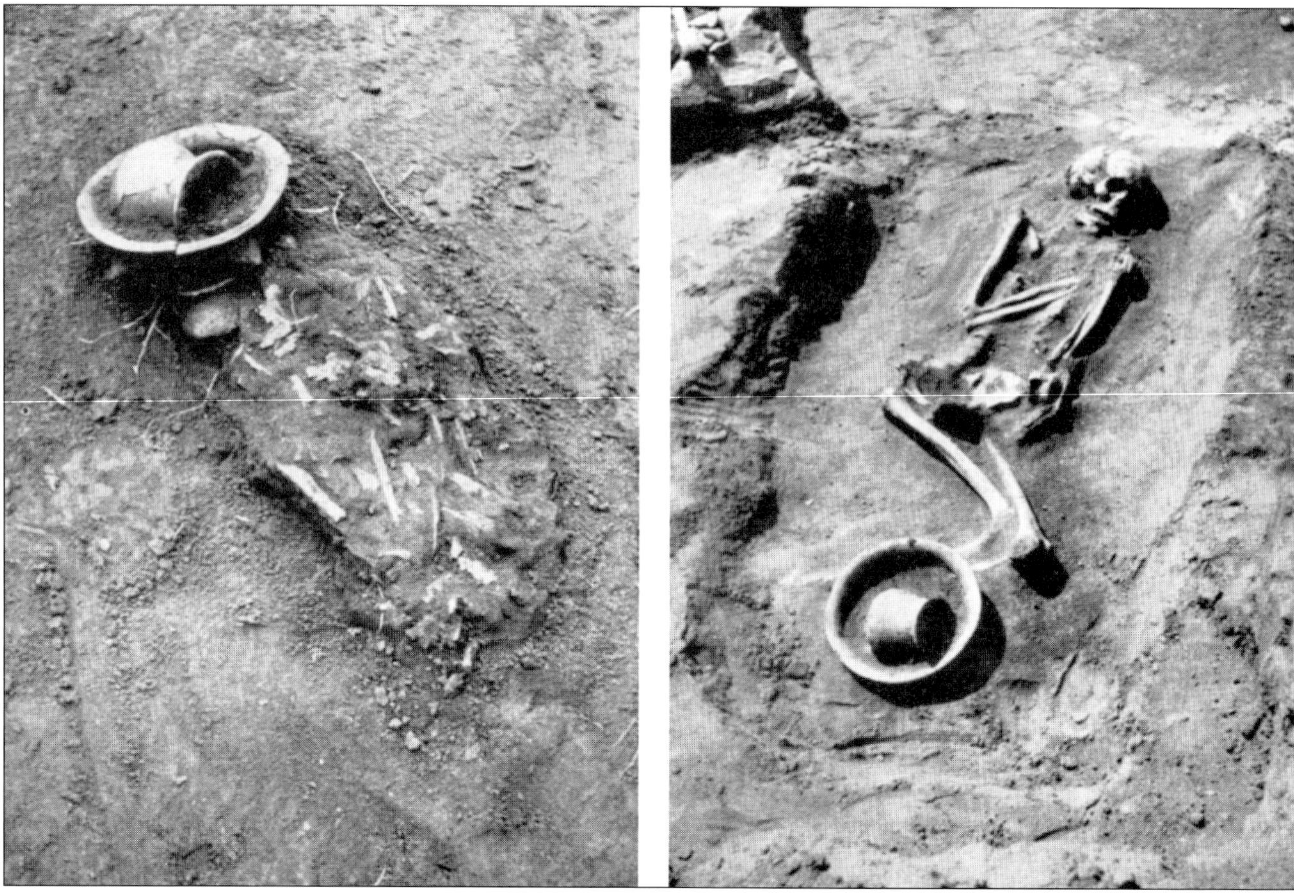

Fig. 3.4: Cremation and inhumation at Streda nad Bodrogom, Slovakia (after Polla 1960: 353).

of the body, to a treatment of the body that deals with it as fragmented and contained in an urn. The transition period, therefore, takes different forms in different regions, and does not happen simultaneously over the whole continent. Close neighbours often change different aspects of their practices or develop different innovative elements or variations.

The range of activities and types of manipulation that take place after the cremation show clearly that the cremation – the burning of the body itself – is not the final stage of this ritual and the handling of the dead body. Rather, this stage is followed by a number of actions that involve manipulating the remains. The choice of one or the other way of doing things may pre-determine some of the outcomes of further decisions along the way, enabling us to think of this process in terms of a *chaîne opératoire*. Despite the fact that Hindu cremations (Parry 1994) are a much beloved analogy often used in archaeology (Fahlander and Oestigaard 2008, Kaliff and Oestigaard 2004, Oestigaard 2000), Bronze Age cremations during the transitional period probably had a very different internal logic and rationale. In particular, in Bronze Age cremations there is a pronounced and continuous focus on the remains of the physical body, and even if this weakened over time, the cremated bones never cease to be a matter of importance to which attention has to be given. In addition, the wide range of variation and 'experimentation' suggests that the beliefs or ontologies affecting the change in burial practices are not fixed, but in the process of becoming formulated. Only at the end of the transitional period can we argue that a normative understanding of cremation burials has emerged.

During the transitional period, much attention is given to making the body whole again: the cremated bones are treated in a manner that creates a likeness to the skeletal or fleshed body. The physicality of the cremated bones is explored in various ways as the link to the re-imaged body, which then becomes the starting point for the reconstitution of the body as a kind of whole. The cremated bones are typically reassembled or annotated so as to regain similarities or similar properties to the lived body. Through these practices and concerns, the body as a materiality is remade after its cremation. The bones themselves are used and re-organised to shape a new kind of spatial presence, either as long and slender and in a realistic proportion, or three-dimensional within an urn; in some cases emphasis is added to particular body parts such as the head. It is also common to see objects used, in particular elements of

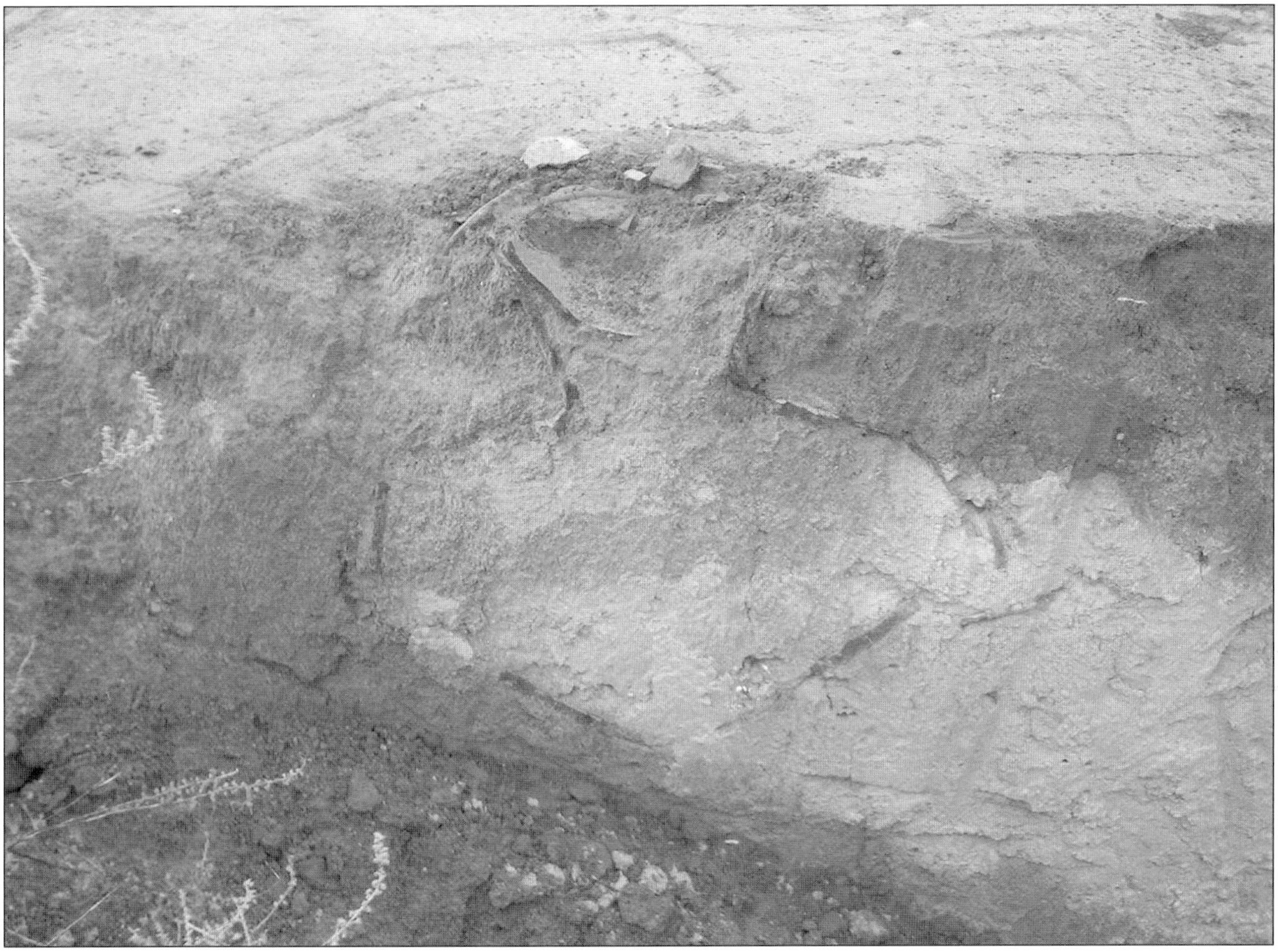

Fig. 3.5: Cut through a Vatya Culture storage vessel used as an urn (photo: K. Rebay-Salisbury).

dress, to annotate the body, for instance when pins are placed where the chest would be, or armrings are placed at the side (for examples from the cemetery of Pitten see Sørensen and Rebay 2008b: 168). Objects, in particular pottery, may also be used to outline the space of the body or even to indicate its extreme points, such as the head and feet (for examples from Hungary see Sørensen and Rebay-Salisbury 2008: 56, 66). These and other characteristics suggest that the concern during the transitional period is the wholeness of the body, despite the fragmentation introduced by cremation (Rebay-Salisbury 2010).

The place of burial is also part of the discursive engagement about how burials should look and what they are about. The various forms of coffins or chamber constructions that had been in use for inhumation burials continued to be used almost unchanged when cremation was first introduced, despite the apparent differences in the physical realities of decomposing and cremated bodies, and only gradually changed and transformed. Their traditional form was useful for insisting on the wholeness of the body, both in terms of the graves' traditional connotations and through the ability to literately mark out the body shape. In the gradual reshaping of the graves, which usually first affected the size and dimensions, we see enormous local variation and the use of familiar local materials and construction techniques. These materials and familiar ways of doing things may have brought associations to mind and gave meanings to particular forms of practices, for instance when storage pits and vessels are utilised in burial practices to evoke notions of closure and secure containment (Fig. 3.5, Sørensen and Rebay-Salisbury 2008: 65). The reflection on the shape and form of graves became a central and dynamic discursive structure through which new ways of treating and thinking about the dead body were being forged.

During the change from inhumation to cremation the roles of pottery and the manner of its use within the grave changed in a number of ways, and they are useful in terms of tracking changes in the meanings surrounding the buried body. During the Early Bronze Age, pottery became a standard element of inhumation graves in most areas. Whereas pottery also remained in use as part of the furnishing in cremation graves for a while, new roles and understandings of pottery were added, culminating in the development of the use of pots as burial urns. In some areas pottery was employed as a kind of construction material used to line and define the grave, while in others special pottery sets may have been used to annotate the body (Sørensen and Rebay-Salisbury 2008). Most importantly, the vessels also began to replace the coffin or grave chamber as the container for the bones. Initially, urns were often placed within various stone constructions, but with time the urns themselves became the grave (the cemetery of Vollmarshausen provides a nice example of this development, see Bergmann 1975). At that stage, the burial practices substantially changed and the body was treated in a radically different manner than in the earlier cremation burials. In some rare cases, urns not only contained the body, but literally took on properties of the body by adding on body parts of clay or using body proportions (Kneisel 2006, Kovács 1992, La Baume 1963, Poroszlai 1992).

Continuous engagement with the cremated and buried body even after the funeral can often be observed in Bronze Age burials. Again there seems to be a common, inter-regional theme, but post-funerary engagement was particularly varied in its local implementation and translation. Bronze Age funerary rituals were not complete after the interment and ongoing engagement with the grave and the bodily remains was very common. This engagement can be broken down in two phases: the phase in which the dead body seemed to still be perceived of as a kind of living body, with needs that had to be catered for; and a second phase in which the physical body became irrelevant and can be deprived of grave goods, robbed and disturbed. Creating and maintaining physical access to the body can be one objective. In the example from Pitten, access was facilitated through the grave architecture, as a doorway was built in the grave construction (Hampel, Kerchler, and Benkovsky-Pivovarová 1981, Sørensen and Rebay 2008b). Furthermore, providing for the needs of the person can be an ongoing concern rather than a one-off event at the funeral and requires close physical contact with the remains. Graves from Királyszentistván (Bóna 1975), for example, contained animal bones on serving platters and cups placed upside down directly on the cremation, suggesting that liquids were poured out onto the cremated bones. In another case from Vollmarshausen, holes were punched into the sides of urns repeatedly after the funeral to offer food and drink directly onto the cremation (Bergmann 1982).

What all the practices described above suggest is that the transformation of the body through cremation was not a complete destruction of the person, nor was there – at least at first – a radical change in what was viewed as being necessary practices. Physical needs were catered for through the creation of space, shelter, food and even community, as if the person was believed to still be living in some way. But even when urn burials became relatively simple affairs towards the end of the Bronze Age, the bodies never became totally irrelevant or meaningless – evidenced by the respectful treatment afforded to the physical remains of the dead.

Discussion and conclusion

In all three examples in this chapter, the practices of cremation and inhumation are entangled in a diverse range of practices that are associated with beliefs, with inhumation and cremation becoming part of a discourse about what is the moral thing to do. The treatment of a dead body and the performance of rituals are actions that build on customary practices informed by conscious and unconscious beliefs about the body. Practical, embodied knowledge and techniques contribute to the material expressions of funerary rituals, from the building of associated structures such as crematoria to the creation of resting places in the form of graves, and furthermore, to ideas about which goods are needed during and after burial.

Cremation, as the deliberate transformation, fragmentation and destruction of the body, might seem like the most radical way to deal with the body after death, but remains only one of the multiple variables that make up burial traditions. As such, the view of cremation as a radical deviation from what is considered 'normal' depends on the broader context. In the example of the re-introduction of cremation after the Enlightenment in Central Europe, cremation is utilised to make a political statement of progress and modernity, while simultaneously expressing disbelief in the traditional doctrines of the Roman Catholic Church. In the Ancient Greek and Roman world burial rites were mainly family affairs, but also subject to changing fashion. A range of beliefs existed about the proper and respectful way to pay respect to the deceased, but these views could be applied to the practices of both inhumation and cremation. In Bronze Age Central Europe, where only archaeological data is available to shed light on issues of knowledge and belief, the idea of a material body with physical needs survived cremation, and initially the remains of a cremation were treated in a very similar way to inhumations. Gradual, small adjustments to existing practices were made to accommodate the changed properties of a cremated body; these small changes then accumulated to become a significant change with a new and different understanding of the body after death.

Burial traditions consist of discursive as well as non-discursive traits, and whereas parts of the burial ritual can be used as a showground for social negotiation, other practices continue despite lost meanings and are justified purely by referencing tradition. Multilayered motives and conflicting beliefs may feed into the way burial practices are carried out. There is, however, no single and straightforward explanation of what cremation means in itself; transferring analogies about associated beliefs from one context to the other is therefore very problematic. Cremation does not have a meaning in itself but can become part of what is perceived to be the right thing to do. It is worthwhile to look at the specific practices associated with cremation burial rites; it is these practices around the body that give us further indications of what was believed about people and their bodies in life and death.

Notes

1 This research was carried out together with Marie Louise Stig Sørensen between 2005 and 2009 at the University of Cambridge, as part of the Leverhulme Trust funded 'Changing Beliefs of the Human body' project. I would like to thank her and all other members of this project for a fruitful cooperation and many stimulating discussions.

2 As case studies, we first looked at the cemetery of Pitten (Hampel, Kerchler, and Benkovsky-Pivovarová 1981) and Vollmarshausen (Bergmann 1982), as the transition can be best observed during the course of occupation in these cemeteries. We then moved on to some regional studies: Hungary (Sørensen and Rebay-Salisbury 2008) because it has often been suspected that the origin of cremation can be found there, the Lüneburg area and Denmark for their marginal position, the Marburg Area (Dobiat 1994) because of the clear continuity to barrow groups, and lastly, we selected two southern German urn fields (Schütz 2007, Ullrich 2004) as representatives of the core of the Central European Urnfield Culture.

References

Bauer, W. T. 2004. *Wiener Friedhofsführer. Genaue Beschreibung sämtlicher Begräbnisstätten nebst einer Geschichte des Wiener Bestattungswesens*. Wien: Falter Verlag.

Bergmann, J. 1975. Ein Brandgräberfeld der jüngeren Bronzezeit von Vollmarshausen im Ldkr. Kassel. *Monographien RGZM* 1, 1: 134–154.

Bergmann, J. 1982. *Ein Gräberfeld der jüngeren Bronze- und Eisenzeit bei Vollmarshausen, Kr. Kassel. Kasseler Beiträge zur Vor- und Frühgeschichte 5*. Marburg: Elwert.

Boardman, J. 1998. *Early Greek vase painting (11th–6th centuries BC): a handbook*. London: Thames and Hudson.

Böhm, J. 1937. *Základy hallstattské periody v Čechách. Obzor prehistorický 10*. Praha.

Bóna, I. 1975. *Die mittlere Bronzezeit Ungarns und ihre südöstlichen Beziehungen. Archaeologia Hungarica 49*. Budapest: Akadémiai Kiadó.

Buklijas, T. 2007. Surgery and national identity in late nineteenth-century Vienna. *Studies in History and Philosophy of Biological and Biomedical Sciences* 38: 756–774.

Carr, C. 1995. Mortuary practices: Their social, philosophical-religious, circumstantial, and physical determinants. *Journal of Archaeological Method and Theory* 2, 2: 105–200.

Cherryson, A. 2010. "In the pursuit of knowledge: Dissection, post-mortem surgery and the retention of body parts in 18th- and 19th-century Britain," in K. Rebay-Salisbury, M. L. S. Sørensen, and J. Hughes (eds) *Body Parts and Bodies Whole: Changing relations and meanings*. 135–148. Oxford: Oxbow.

Childe, V. G. 1944. *Progress and Archaeology*. London: Watts.

Childe, V. G. 1950. *Prehistoric Migrations in Europe*. Oslo: Aschehoug.

Clark, G. 1960. *Archaeology and Society. Reconstructing the prehistoric past*. London: Methuen.

Davies, D. J., and L. H. Mates (eds) 2005. *Encyclopedia of Cremation*. Ashgate: Aldershot.

Davies, P. J. 2001. *Beethoven in Person: His Deafness, Illnesses, and Death*. Westport, Conn.: Greenwood Press.

Dobiat, C. 1994. *Forschungen zu Grabhügelgruppen der Urnenfelderzeit im Marburger Raum. Marburger Studien zur Vor- und Frühgeschichte 17*. Marburg: Hitzeroth.

Ebner, P. 1989. *Der Streit um die Feuerbestattung zwischen Katholischer Kirche und Sozialdemokratie: eine Studie zum Kulturkampf in der 1*. Republik, Universität Wien.

Fagerstöm, K. 1993. Wealth Destruction as a Sign of Iron Age Political Strife: the Greek Example. *Current Swedish Archaeology* 1: 49–57.

Fahlander, F., and T. Oestigaard. 2008. "The Materiality of Death: Bodies, Burials, Beliefs," in F. Fahlander and T. Oestigaard (eds) *The Materiality of Death: Bodies, Burials, Beliefs, British Archaeological Reports International Series 1768*. Oxford: Archaeopress.

Graham, E.-J. 2009. Becoming persons, becoming ancestors. Personhood, memory and the corpse in Roman rituals of social remembrance. *Archaeological Dialogues* 16, 1: 51–74.

Graham, E.-J. 2011. "From fragments to ancestors: re-defining the role of os resectum in rituals of purification and commemoration in Republican Rome," in M. Carroll and J. Rempel (eds) *Living through the Dead: Burial and Commemoration in the Classical World*. 91–109. Oxford: Oxbow.

Gruber, R. H., and R. Bouchal. 2005. *Der Stephansdom. Monument des Glaubens – Stein gewordene Geschichte*. Wien: Pichler.

Hampel, F., H. Kerchler, and Z. Benkovsky-Pivovarová. 1981. *Das mittelbronzezeitliche Gräberfeld von Pitten in Niederösterreich. Ergebnisse der Ausgrabungen des Niederösterreichischen Landesmuseums in den Jahren 1967 bis 1973 mit Beiträgen über Funde aus anderen Perioden, Band 1: Fundbericht und Tafeln. Mitt. Prähist. Komm. Österr. Akad. 19/20*.

Hauf, H. 1996. Die Feuerbestattung in Wien. Diplomarbeit, Universität Wien.

Hertz, R. 1907. Contribution à une étude sur la représentation collective de la mort. *Année Sociologique* 10: 48–137.

Hope, V. 2007. *Death in Ancient Rome. A Sourcebook*. Oxon, New York: Routledge.

Hope, V. M., and E. Marshall (eds) 2000. *Death and Disease in the Ancient City*. London: Routledge.

Jupp, P. C. 1990. *From Dust to Ashes: The Replacement of Burial by Cremation in England 1840–1967*. London: Congregational Memorial Hall Trust.

Kaliff, A., and T. Oestigaard. 2004. Cultivating Corpses. A Comparative Approach to Disembodied Mortuary Remains. *Current Swedish Archaeology* 12: 83–104.

Kneisel, J. 2006. *Anthropomorphe Gefäße in Nord- und Mitteleuropa während der Bronze- und Eisenzeit. Studien zu den Gesichtsurnen – Kontaktzonen und sozialer Kontext*. PhD thesis, Freie Universität Berlin.

Koubsky, K. 2010. *A short history of the ossuary in Kutná Hora – Sedlec.* Retrieved 15.1.2010, from http://www.kostnice.cz/.

Kovács, T. 1992. "Glaubenswelt und Kunst," in W. Meier-Arendt (ed.) *Bronzezeit in Ungarn. Forschungen in Tell-Siedlungen an Donau und Theiss.* 76–82. Frankfurt: Pytheas.

Kraft, G. 1926. Beiträge zur Kenntnis der Urnenfelderkultur in Süddeutschland. *Bonner Jahrbücher* 131: 153–212.

La Baume, W. 1963. *Die Pommerellischen Gesichtsurnen. Kataloge Vor- und Frühgeschichtlicher Altertümer 17.* Mainz: Römisch-Germanisches Zentralmuseum.

Lehr, R. 1979. *Hallstatt. Geschichte und Gegenwart.* Linz: Oberösterreichischer Landesverlag.

McKinley, J. I. 2006. 'Cremation...the cheap option?' in R. Growland and C. Knüsel (eds) *Social archaeology of funerary remains.* 81–88. Oxford: Oxbow.

Meredith, W. 2005. The History of Beethoven's Skull Fragments. *The Beethoven Journal* 20: 3–46.

Mims, C. A. 1998. *When We Die.* London: Robinson.

Momigliano, A. 1963. An Interim Report on the Origins of Rome. *Journal of Roman Studies* 53: 95–121.

Morris, I. 1987. *Burial and Ancient Society: The rise of the Greek city state.* Cambridge: Cambridge University Press.

Morris, I. 1992. *Death Ritual and Social Structure in Classical Antiquity.* Cambridge: Cambridge University Press.

Müller, S. 1897. *Vor Oldtid.* København: Det nordiske Forlag.

Nock, A. D. 1932. Cremation and burial in the Roman Empire. *Harvard Theological Review* 25, 4: 321–359.

Oestigaard, T. 2000. *The Deceased's Life Cycle Rituals in Nepal: Present Cremation Burials for the Interpretations of the Past. British Archaeological Reports International Series 853.* Oxford: Archaeopress.

Pappalardo, U. 1997. Nuove testimonianze su Marco Nonio Balbo ad Ercolaneo. *Mitteilungen des Deutschen Archäologischen Instituts, Römische Abteilung* 104: 417–428.

Parry, J. P. 1994. *Death in Benares.* Cambridge: Cambridge University Press.

Parsons, B. 2005. *Committed to the Cleansing Flame: The Development of Cremation in Nineteenth Century England.* Reading: Spire Books.

Piggott, S. P. 1965. *Ancient Europe from the Beginnings of Agriculture to Classical Antiquity. A survey.* Edinburgh: Edinburgh University Press.

Polla, B. 1960. "Birituelle Füzesabonyer Begräbnisstätte in Streda nad Bodrogom," in B. Chropovský, M. Dušek, and B. Polla (eds) *Pohrebiská zo staršej doby bronzovej na Slovensku I.* 299–386. Bratislava.

Poroszlai, I. 1992. "Százhalombatta-Földvár," in W. Meier-Arendt (ed.) *Bronzezeit in Ungarn. Forschungen in Tell-Siedlungen an Donau und Theiss.* 153–155. Frankfurt: Pytheas.

Prothero, S. R. 2001. *Purified by fire: a history of cremation in America.* Berkeley: University of California Press.

Rebay-Salisbury, K. 2010. "Cremations: Fragmented Bodies in the Bronze and Iron Ages," in K. Rebay-Salisbury, M. L. S. Sørensen, and J. Hughes (eds) *Body Parts and Bodies Whole: Changing Relations and Meanings.* 64–71. Oxford: Oxbow.

Richardson, N. 1985. "Early Greek views about life after death," in P. Easterling and J. Muir (eds) *Greek religion and society.* 50–66. Cambridge: Cambridge University Press.

Schütz, C. 2007. *Das urnenfelderzeitliche Gräberfeld von Zuchering-Ost, Stadt Ingolstadt. Materialhefte zur bayerischen Vorgeschichte 90.* Kallmünz: Lassleben.

Sørensen, M. L. S., and K. Rebay-Salisbury. 2008. Landscapes of the Body: Burials of the Middle Bronze Age in Hungary. *European Journal of Archaeology* 11, 1: 49–74.

Sørensen, M. L. S., and K. Rebay. 2008a. "The impact of 19th century ideas on the construction of 'Urnfield' as chronological and cultural concept: tales from Northern and Central Europe," in A. Lehoërff (ed.) *Construire le temps. Histoire et méthodes des chronologies et calendriers des derniers millénaires avant notre ère en Europe occidentale. Actes du XXXe colloque international de Halma-Ipel, UMR 8164 (CNRS, Lille 3, MCC), 7–9 décembre 2006, Lille, Bibracte 16.* 57–67. Glux-en-Glenne: Bibracte – Centre archéologique européen.

Sørensen, M. L. S., and K. Rebay. 2008b. Interpreting the body: burial practices at the Middle Bronze Age cemetery at Pitten. *Archaeologia Austriaca* 89 (2005): 153–175.

Sørensen, M. L. S., and K. C. Rebay. 2008c. "From substantial bodies to the substance of bodies: analysis of the transition from inhumation to cremation during the Middle Bronze Age in Central Europe," in D. Borić and J. Robb (eds) *Past Bodies. Body-Centred Research in Archaeology.* 59–68. Oxford: Oxbow.

Sørensen, M. L. S., and K. Rebay-Salisbury. forthcoming. *Changing Bodies. The Introduction of Cremation in Bronze Age Europe.* Cambridge: Cambridge University Press.

Sørensen, T. F., and M. Bille. 2008. Flames of transformation: the role of fire in cremation practices. *World Archaeology* 40, 2: 253–267.

Tarlow, S. 1992. Each slow dusk a drawing-down of blinds. *Archaeological Review from Cambridge* 11: 125–140.

Thalmann, R. 1978. *Urne oder Sarg? Auseinandersetzungen um die Einführung der Feuerbestattung im 19. Jh. Bern.* Frankfurt-Las Vegas: Lang.

The Associated Press. 2008. *Greece: Ruling Allows Cremation.* Retrieved 27 September 2008, from http://query.nytimes.com/gst/fullpage.html?res=9B02E0DD103CF934A1575AC0A96E9C8B63.

Toynbee, J. M. C. 1971. *Death and Burial in the Roman World.* London: Thames and Hudson.

Ucko, P. J. 1969. Ethnography and Archaeological Interpretation of Funerary Remains. *World Archaeology* 1, 2: 262–280.

Ullrich, M. 2004. *Das urnenfelderzeitliche Gräberfeld von Grundfeld/Reundorf, Lkr. Lichtenfels, Oberfranken. Materialhefte zur bayerischen Vorgeschichte 86.* Kallmünz: Lassleben.

Westerhoff, W. 1989. *Karner in Österreich und Südtirol.* St. Pölten-Wien: Niederösterreichisches Pressehaus.

Williams, H. 2004. Death Warmed Up. The Agency of Bodies and Bones in Early Anglo-Saxon Cremation Rites. *Journal of Material Culture* 9, 3: 263–291.

Zilkens, S. 1983. *Karner-Kapellen in Deutschland. Untersuchungen zur Baugeschichte und Ikonographie doppelgeschossiger Beinhaus-Kapellen.* Köln: Kunsthistorisches Institut der Universität Köln.

4. Delusion and disclosure: human disposal and the aesthetics of vagueness

Tim Flohr Sørensen

In Lars von Trier's film *Antichrist* (2009, Zentropa) a woman (*She*) and a man (*He*) tragically lose their son, a tragedy that leaves her psychologically devastated. *He* is a therapist, and discontent with the help she is offered upon the death of their son, he wants to treat *She* psychotherapeutically himself, believing that her catatonic trauma is enforced by anxieties more deep-seated than the loss of their son. *She* and *He* retreat to a secluded cabin in a forest, where he wants her to expose herself to her innermost traumas, revolving around the cabin and the area around it. Without going into further detail, his therapeutic project fails utterly, and the deeper the therapist makes *She* dig into her anxiety, the more agonising, brutal and self-destructive she becomes.

Psychological enlightenment and self-unveiling are not portrayed in a flattering way in *Antichrist*, to say the least, and one possible interpretation of the film is to see it as a passionate denunciation of psychotherapy. In the film, achieving knowledge through the excavation of the psyche (Bernfeld 1951) and by facing and verbalising one's fears and repressions is thus essentially mistrusted. This chapter is – in one sense – inspired by *Antichrist*. It seeks to move past the verbalisations of sorrow that occur in relation to death, and it problematises the widespread idea that facing the naked reality of one's loss is always emotionally helpful. Instead, the chapter suggests that strategies for coping with loss and bereavement may be found in the absence of disclosure and in the construction of a vague knowledge of death.

Exploring the themes of the volume, the chapter engages in a critical reading of the ways in which embodiment and knowledge are orchestrated in the material institutions circumscribing human disposal. While many studies of knowledge focus on epistemic acquisition or dissemination, this chapter explores the observation that knowledge can also be absent or unwanted (see e.g. Højer 2010, Hornborg 2006). Similarly, embodiment is frequently examined from the perspective of the bodily modalities (e.g. Barsalou, Barbey, Simmons, and Santos 2005), yet this chapter seeks to complement this view by drawing attention to dissolving and vanished bodies. While abstract, the chapter seeks to connect to a sense of social actions and responses to loss, and it therefore examines the material institutions surrounding the corporality of death in modern and late modern Denmark as a case study. Disembodiment and lack of knowledge are argued to be the central aims of the technology and aesthetics of disposal of dead bodies in contemporary Denmark.

The ways in which knowledge and (dis)embodiment are constructed in relation to death in recent and contemporary Denmark are examined by tracing the material forms of the beliefs that surround death and burial. Belief is in this context defined as the ideas that people create of phenomena located outside of immediate, sensory experience and beyond empirical knowledge. In this context, belief roughly corresponds with imagination in the sense that belief can build on hopes and desires that mitigate cognised knowledge (cf. Casey 2000: 42–43). Moreover, belief and imagination are not seen as solely mental activities, as solipsistic occurrences residing purely within the mind of the individual. Imagining things is shaped and framed by previous experiences in the social realm and hinges on sensory recollection (ibid. 41). Imagination can also spring from one's surroundings in the sense that the material environment creates a framework for imagining the world and one's position therein. The aim of this chapter is thus to explore the intersections of emotions, belief and imagination in contemporary Danish burials, focusing on the material institutions that circumscribe death. It thereby holds that aesthetics (in the classical meaning as the sensory; e.g. Gosden 2001, MacLagan 2001) are central to how ideas and beliefs about the dead are forged.

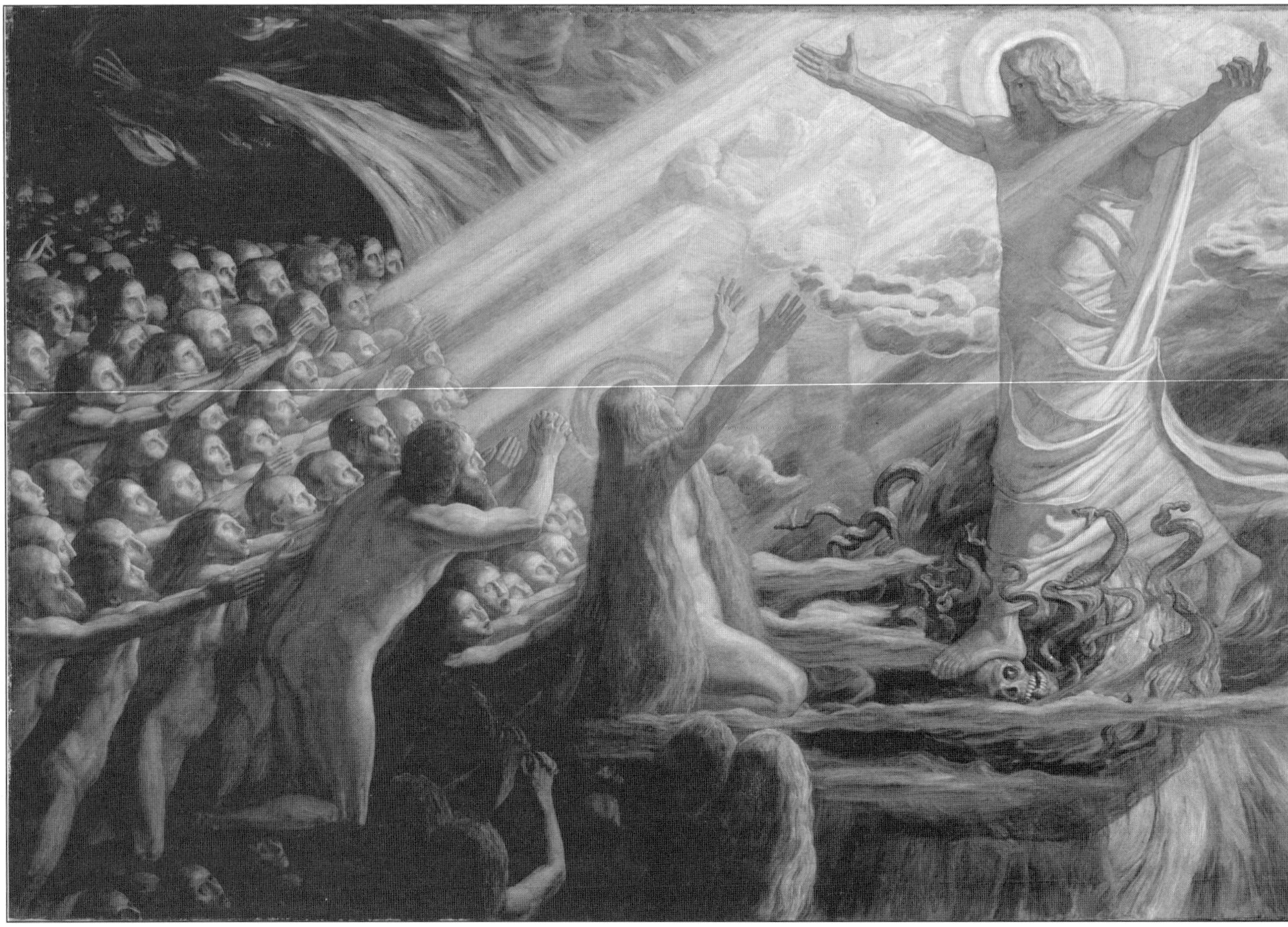

Fig. 4.1: Kristus i de dødes rige ('Christ in the realm of the dead'). Joakim Skovgaard (1891–94), with permission from Statens Museum for Kunst.

Disposal, belief, and imagination

Belief as imagination and conceptualisation may thereby be set apart from religious belief, denoting trust and faith in the presence, power, love, benevolence or justice of a numinous metaphysical or supernatural entity (Otto 1958). Nevertheless, imagination and faith do also intersect in some cases and assume concrete material form. Death and the disposal of dead bodies are often connected with a religious belief in an afterlife and concerns for the destiny of the soul or the like. In much of the Christian world and during most of Christian history, inhumation has been seen as the religiously most agreeable form of disposal in accordance with biblical references and the Christian tradition that the dead will be resurrected at the Second Coming of Christ (Fig. 4.1). Since early medieval times, this tradition – which may rest both on popular beliefs and canonised theology – has been a formative factor in the structural layout of cemeteries in Denmark (Sørensen 2009). Graves have been oriented east-west so as to anticipate Judgment Day, where the dead will sit up in their graves and face the light coming from the east with the Redeemer (Sørensen 2009: 127). Colloquial metaphors for death furthermore envisaged the dead as sleeping or resting until their resurrection (Kragh 1996), offering images of the dead in a peaceful and pleasant – and persistently embodied – condition.

However, the advent and prevalence of cremation during the 20th century has challenged the materially and bodily grounded metaphors that circumscribed the traditional rural cemetery and has also caused a shift in the relationship between religious beliefs surrounding death and ways of imagining death. In part, the ardent debate over cremation in the late 19th and early 20th century revolved around the intersections of religious belief and theology on the one hand and images and visualisation of the dead on the other (see also Rebay-Salisbury in this volume for the similar process in Austria). The cultural and political history of cremation in various parts of the mid-late 19th and early 20th century West can be found elsewhere (e.g. Dennie 2003, Jupp 2006, Nicol 2003, Parsons 2005, Prothero 2001, Sørensen 2009); here the focus is on the figurative – i.e. visually imaginative

– aspects of the schism between inhumation and cremation. The late 19th century debate over cremation offers numerous discussions with very sensuous descriptions of how the dead body decomposes in cremation and inhumation respectively. In 1885, the secretary of the San Francisco Cremation Society, Max Levy, thus described

> ... the bowels of the earth (...), wherein lie deposited the dead of the human race, assigned to them for lasting rest. "For lasting rest." Alas! Some light has been thrown upon that sentiment, sufficient for any one to see the folly of such superstition, except those who *prefer* to carry the veil of superstition, and such who find it in their interest to keep that veil in its place (...). The dead do *not* rest in their graves. It is no profanity, but the bare truth if we assert that they are running a disease factory, and mighty busy they keep it. And when the material gives out, which might be in fifty, or might be in a hundred years, the spade again penetrates the soil and makes an end of that "everlasting rest" (Levy 2003: 5–6).

Advocates of cremation described the organic putrefaction of the dead body, detailing the collapse of the organism with regard to texture, mass, colour, liquids, smell and the animals and bacteria that participate in the consumption and disintegration of the body. And they argued that the unpleasant rotting of the body could be countered by cremating the body, swiftly reducing it to a mass of ash and bone fragments that were sterile, light and dry, maybe even immaterial (e.g. Cobb 1892, Eassie 1878, Hopps 1894, Thompson 1873).

The French doctor, Jules Rochard, was less convinced of the alleged sanitary benefits of cremation and was more concerned about the emotional consequences of the cremation process and its product. Even though the actual burning of the body would be concealed, he argued that the swift transformation of the body exposed its violent destruction, which would have devastating emotional effects on the bereaved. Rochard believed that the efficiency and speediness of the cremation technology eradicated the potential for a lasting and desirable memory of the deceased as spirit *and* body. After the swift cremation process, he concluded, the bereaved would be offered a small vessel with the meek remains of their loved one: 'This is all that is left for us; all delusion is lost' (Rochard 1891: 24; my translation). In the Danish debate, Bishop Bruun Juul Fog further claimed that cremation merely was a means of eradicating corpses rapidly and believed that the dead would be equally quickly erased from the memory of the bereaved, causing an emotional amnesia (Blomquist 1998: 72). Alternatively, the inhumation of the body would offer a chance of what Rochard terms 'delusion', meaning that the bereaved would be allowed to maintain the visual image of their loved one as sleeping peacefully in the subterranean grave. The notion that the deceased was in a state of sleep is still reproduced in the colloquial reference to the grave as the 'final resting place' and in other popular metaphors and paraphrasing of the verb 'to die'.

Architectures of disposal

In Denmark, the debate over cremation continued well into the 20th century, but its use gradually became increasingly widespread, first in the cities and later also in rural parts of the country. The population of Western Denmark has been the most reluctant to adopt cremation, which is commonly ascribed to a larger component of conservative Christianity among this population segment and stronger ties to tradition (see Sørensen 2009). Today, Denmark has one of the highest cremation rates in the Western world, reaching 78.55% of all deaths in 2011.[1] The practice of cremation as it developed in the 19th and 20th century has challenged the traditional burial form in two ways. First, it eliminates the body as the metaphorical reference to resurrection, disentangling the spatial format of the cemetery from its cosmological referent, as the corporeal rising of the dead no longer constitutes a material allusion in the cemetery's spatial structure. Second, it fragments the funerary ceremony as a rite of passage by entailing a hiatus between the ecclesiastical ritual for the deceased and the subsequent deposition of the ashes following cremation.

The disposal procedure thus revolves around three primary institutional spaces in contemporary Denmark: the church, the crematorium and the cemetery. The following will outline the constitutions of each of them, and explore how imagination and belief issue forth in relation to the dead body and the material culture of these institutions.

The church and the crematorium

Despite the prevalence of cremation as the preferred mode of disposal, the church has remained the most common site of burial ceremonies in Denmark throughout the 20th century and into the 21st. Even though early crematoria of the late 19th and early 20th century were constructed architectonically to take over the ceremonial purpose otherwise held in the church, funerals have remained an ecclesiastical affair. In 2010, only 11% of all burials were conducted in the absence of an ordained vicar.[2] As 18.5% of the population were not members of the national church,[3] this means that the vast majority of the population chose to have a church ceremony. This may also be the case even when the deceased was not a member of the church.

At the church ceremony, the deceased forms the pivotal focus of the ceremonial activities. When mourners arrive at the ceremony, the closed casket will already have been placed in the church by funeral directors and cemetery staff. The casket with the deceased is located between the congregation and the altar. The vicar may stand either at the pulpit or at the end of the casket during the various stages of the ceremony, including when s/he addresses the deeds, beliefs and personality of the deceased, and if any of the bereaved gives a eulogy they will typically be standing next to the casket, sometimes touching it. Usually, the congregation then sings a hymn and the casket is carried out of the church and placed in a hearse. Another

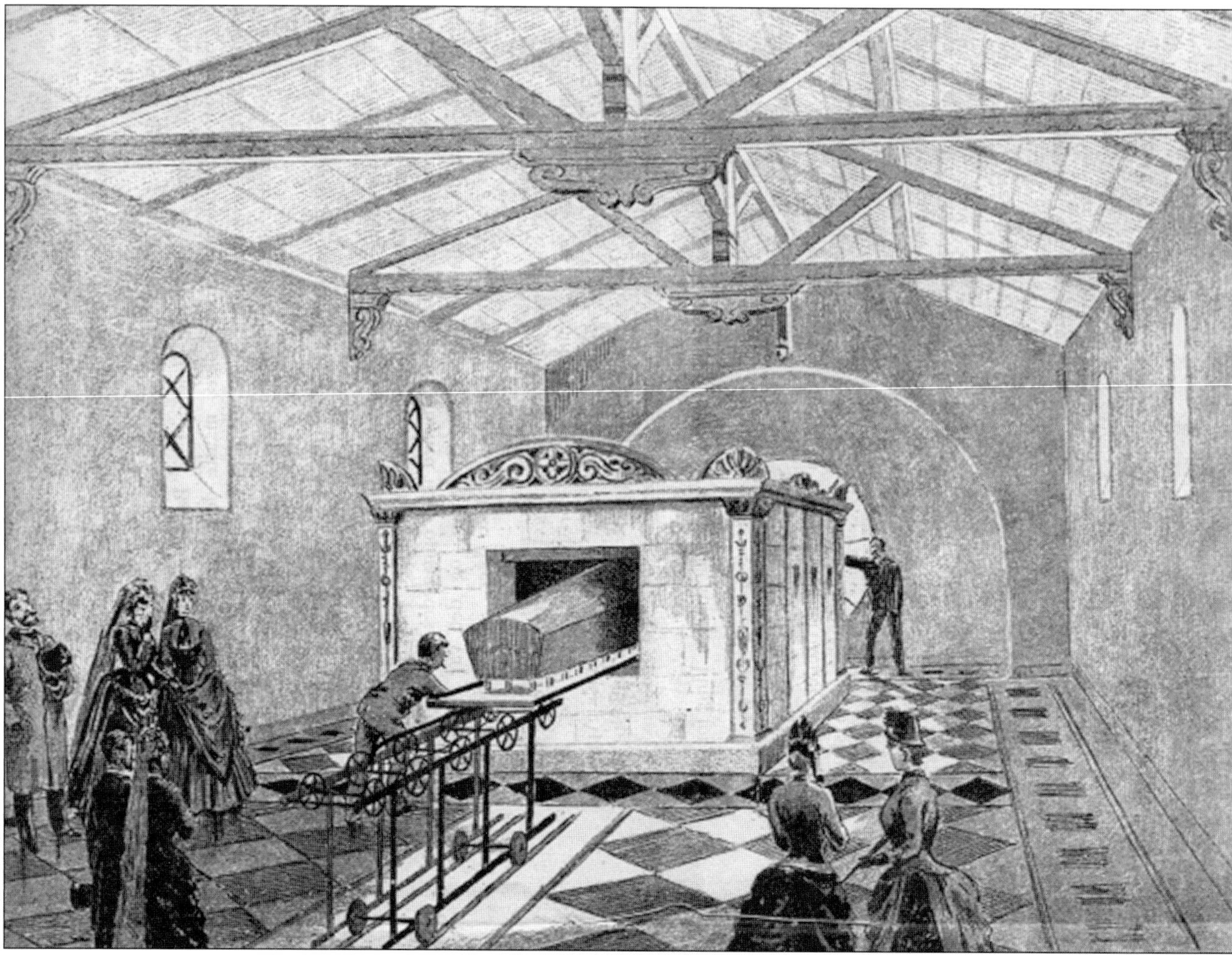

Fig. 4.2: Frederiksberg crematorium (built 1886, discontinued 1908). The illustration shows a hypothetical cremation ceremony with mourners witnessing the insertion of the casket into the incinerator. Drawing dated early 1900s, after Somner 2009: 532.

hymn is then sometimes sung; the hearse is then closed and driven off to the crematorium.

This is normally the last the bereaved will see of the deceased. Or rather, this is the last they will see of the deceased's casket. Seeing the deceased's body used to be customary, either in the week up to the funeral, where the local community would visit the body of the deceased in her or his home, or on the day of the funeral before the casket was closed (Kragh 2003: 40, Møller 1929: 272–274). Today, most of the mourning community is not likely to have seen the dead at any point, since approximately 75% of all deaths in contemporary Denmark occur either in hospitals or in care centres.[4] The custom of seeing the deceased is to some extent still practiced, but usually this takes place in the hospital or in the care centre very shortly after death and only involves the closest of kin. Once death has been established and the nearest of kin have paid their last visit to the deceased, hospital staff, care centre staff or funerary directors wash and dress the deceased and place her or him in a casket. The standard procedure is to then place the casket in a cold-storage room in the hospital or in a chapel of rest by the church, where the bereaved remain out of contact with the deceased, contrary to the older tradition of having the dead lying in wait in her or his home until the funeral (Sørensen 2010a: 221–222).

Even though the church has remained the prevalent site of the funerary ceremonies, early crematoria in Denmark were nevertheless built to offer a place for ceremonial parting with the deceased. Many were deliberately designed to mimic existing sacred architecture or to represent what was perceived as a solemn atmosphere through references to antiquity (Sommer 2006), as has also been the case elsewhere (e.g. Davies 1996, Fischer 2008, Grainger 2000, Winter 2001). This design idiom was in part mixed with technological fascination, and 19th century crematorium architecture celebrated the merits of fire

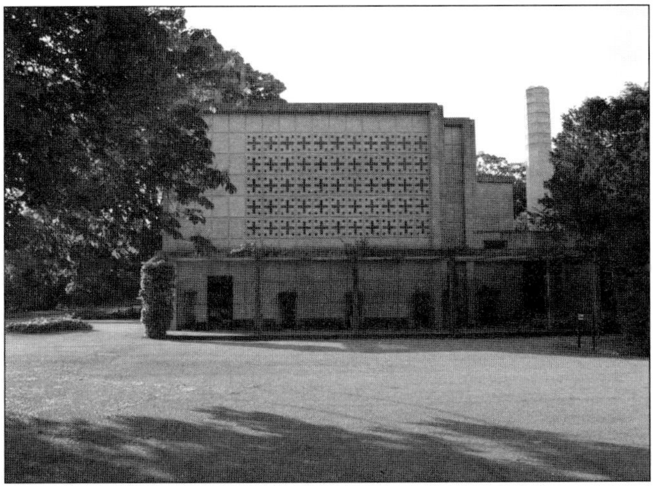

Fig. 4.3: Mariebjerg crematorium (built 1937), north of Copenhagen (photo by Tim Flohr Sørensen, 2011).

disposal by displaying the machinery of the pyro-technology, apart from the actual burning (Sommer 2006). Crematoria, in other words, were built to be visited during cremations and used for funerary ceremonies. Here, the mourner would have been positioned in relative immediacy to what took place in the incinerator (Fig. 4.2).

In 2009, ca. 20% of all ceremonies prior to cremation were conducted at chapels of the crematoria.[5] In many cases, crematoria are located many kilometres from where the deceased and the bereaved live, and they are thus geographically disconnected from what is considered the home of the deceased. If the crematorium is busy when the casket arrives, it is placed in a cold-storage room until cremated. According to Danish law, the deceased is to be cremated within a week after the funeral ceremony, which again can take place up to eight days after the death (Sørensen 2010a: 224). The actual cremation takes one to three hours, after which the ashes and bone fragments are ground to a coarse dust and collected in an urn, all of which ordinarily takes place in the absence of the mourners.[6]

Even though vicars can perform funerary ceremonies in chapels of crematoria, the church remains the preferred funerary scene, constituting a customary and tradition-bound institution close to the home of the deceased. In this light, the limited use of the crematorium as a ceremonial institution may be the result of a general wish of the bereaved for the funeral to take place in a familiar setting, and at a place that is not exclusively associated with death, but also with other social functions and religious perspectives. The early modern attempts to formulate crematoria as semi-religious institutions[7] has therefore never had much impact, and they have largely served as functional rather than ceremonial institutions. Religious belief has thus never become grounded in the crematorium, and burials and the church have remained conjoined institutions. Danish crematoria are rather the places to which the casket and the dead are taken away, and where the actual processing of the dead body is taken out of perceptible immediacy (Fig. 4.3).

Furthermore, Danish crematoria are currently being technologically modified in ways that create an even greater perceptual distance to the processing of the corpse, as the smoke is made invisible and odourless by the use of cooling systems and filters. The smoke from the cremation is often discussed in terms of pollution from CO_2, heavy metals, dust and heat, yet it may also be suggested that making the smoke invisible, heatless and odourless is a way of achieving a less graphically suggestive image of the cremation process. In this sense, the fascination with the technological aspects of cremation so pronounced in the early stages of its modern development have been turned on its head, and the corporeal consequences of cremation are now desired to be moved out of sight and out of touch. The way in which the dead body is recognised and known to the bereaved is thus to remain an elusive and abstract occurrence.

The cemetery

In some other countries, the destination of ashes after cremation is not prescribed in detail by law. This allows individual considerations of how the division, scattering or general indeterminacy of the whereabouts of the ashes affect the dead as a material substance and as a subject, and what emotional consequences this would entail (e.g. Hockey, Kellaher, and Prendergast 2007, Kellaher, Hockey, and Prendergast 2010, Mädler 2008, Prendergast, Hockey, and Kellaher 2007). Danish law, however, strictly binds the ashes to the cemetery and does not permit random scattering of the ashes or dividing ashes between burial places, let alone keeping the urn at home. Upon cremation, the urn can then be collected by the bereaved at the crematorium, but is more commonly taken to the cemetery by the funerary director and handed over to the bereaved at the cemetery. By Danish law, urns are not allowed to be stored anywhere but in consecrated ground, apart from a few exceptions such as scattering over the sea or deposition in private soil, which is only practiced in about two percent of all cases (for further details see Sørensen 2009, Sørensen 2010a: 225–226).

The urn can be deposited in any type of grave at the cemetery, including traditional casket grave plots. The most typical destinations of urns are specific urn gardens (especially at urban cemeteries) and lawn sections (at urban and rural cemeteries alike). Urn gardens assume the form of downscaled casket sections, structured around a grid of pebble paths, where low hedges divide or enclose the individual grave plots. A more common form of urn burial in contemporary Denmark (acknowledging local exceptions) is the deposition of urns in lawn sections. These sections can either be in the form of a lawn with individual grave plots under simple, rectangular, horizontal gravestones, or a communal grave field with unmarked graves (Fig. 4.4, 4.5 see also Sørensen 2007, Sørensen 2009). Danish lawn cemeteries are not to be confused

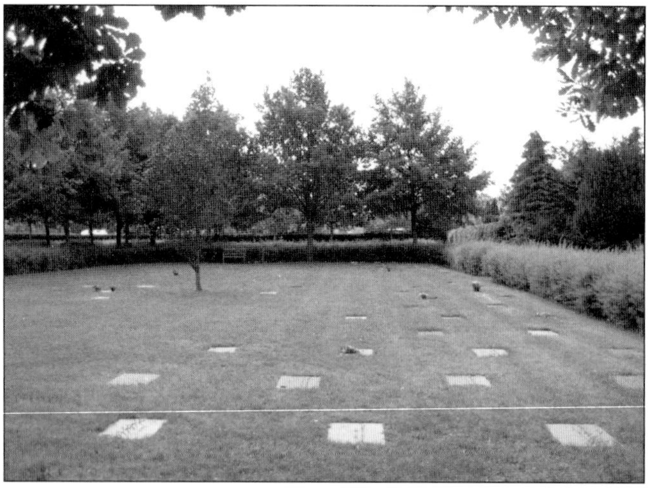

Fig. 4.4: Nykøbing Sjælland cemetery, lawn section, laid out 1975 (photo by Tim Flohr Sørensen, 2007).

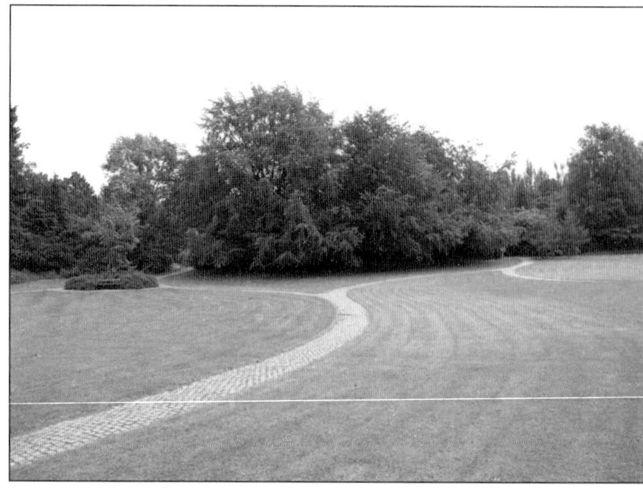

Fig. 4.5: Unmarked communal grave section at Bispebjerg cemetery, laid out 1960 (photo by Tim Flohr Sørensen, 2011).

with the lawns of cemeteries in Britain or in the USA, which have their origin in a different cultural and architectonic tradition (compare Kath 1994, Rugg 2006, Sørensen 2009). In Scandinavia they have their origin in the Modernism and aesthetics of the early 20th century; they are expressions of a minimalist, functionalist design idiom and are in part allusions to notions of 'nature' – disregarding how enculturated and tamed this image of nature may be.

Unmarked communal graves have become increasingly widespread and a recent report showed that 43% of all urn burials at eleven randomly selected cemeteries were burials in unmarked communal sections (Guldager, Elkjær, and Kjøller 2008: 16). In the unmarked section, the burial of the urn is not witnessed by the bereaved, the point being that the exact location of the ashes of the deceased is insignificant as the burial is to be part of the larger collective and not individualised. In fact, even at sections with demarcated graves, only few people bury the urns of their relatives themselves. In the course of the fieldwork for my thesis (Sørensen 2010a), conversations with cemetery staff revealed that they were often tasked with depositing urns in the absence of bereaved families. No records are kept of how many urns are buried in the presence or absence of mourners, but cemetery staff at the 12 cemeteries in my fieldwork area estimated that between half and two thirds of all urns were deposited by the cemetery personnel without the bereaved family attending the deposition. In addition, cemetery staff related that in cases where the bereaved family did bury the urn themselves, the burial was for the most part only attended by the closest of kin (e.g. the spouse, adult children and grandchildren of the deceased), while the church ceremony prior to the cremation would often be attended by a larger number of people.

We cannot only consider the practices and statistics of the use of the cemeteries in order to understand disposal in this case. Appreciating the particular aesthetics of the cemeteries is just as crucial, because the sensory engagement with the cemetery is of the utmost importance to the kinds of ties people establish with the dead. I have explored this theme in detail elsewhere (Sørensen 2009, Sørensen 2010a, Sørensen 2010b, Sørensen and Bille 2008), and argued that the lawn sections in contemporary Danish cemetery culture represent a general move towards the subjectification of the image of the dead. This means that places for the dead are characterised by fewer obligations of embodied practices than previously, in the sense that the corporeal friction between mourner and cemetery has been reduced. Earlier, grave plots were maintained by the bereaved and required attention and care, because the grave functioned as a small garden plot displaying the social propriety of the deceased's family towards the local community. Today, cemetery staff have largely taken over the maintenance of grave plots (Sørensen 2010a), and urn burial plots and lawn sections mean that material commemoration requires less physical labour and in general issue forth as a more subjective and un-embodied exercise (Sørensen 2009, but see also Sørensen 2011). Memories and memorials of the dead are generally not as public and communal as they were some 30 years ago, where the social merits and identity of the deceased were central.

Another consequence of the changed cemetery idiom – and key to the argument in this chapter – is that the minimalist and smooth material expression of burial furthermore produces more abstract and vague contours of the material reality of the deceased individuals. At the casket grave, the size and layout of the grave plot is suggestive of the subterranean resting place of a human body, and in the urn garden, the miniature grave plot expresses the physically limited product of the cremation. The lawn section, on the other hand, is unbounded by physical demarcations that might suggest any kind of corporeal existence, and instead offers a continuous and compliant surface, suggesting an entirely disembodied image of the dead, unburdened by their carnal or cremated remains (Fig. 4.6).

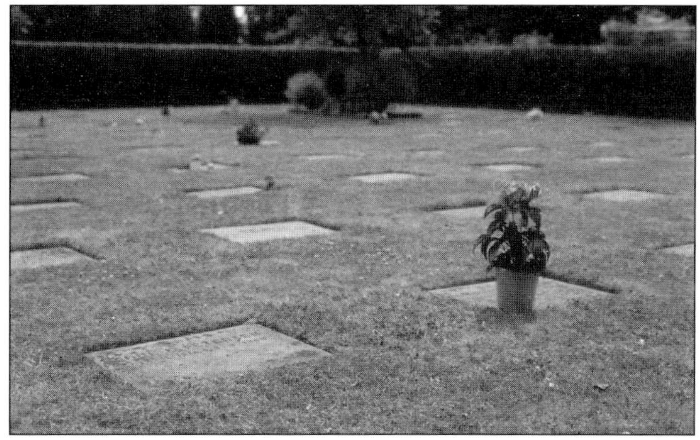

Fig. 4.6: Vig cemetery, lawn section, laid out 1977 (photo by Tim Flohr Sørensen, 2009).

The etherealisation of death

In summary, if we divide the mortuary process in the case of cremation into four stages, beginning with the death of a person and ending with the burial of her or his urn, we may construct the following epistemological sequence as seen from the point of view of the bereaved. At the first stage, the dead body is mainly kept in storage at a hospital and in most cases handled by hospital personnel and funeral directors, and is thus only encountered physically by the bereaved to a limited degree. At the second stage, i.e. at the funeral, the body is concealed in a casket and there is no or only a very limited tradition for viewing the deceased; there is no sensory contact with the deceased. At the third stage, the body is processed at a crematorium, which is a part of the funerary process in which the bereaved have traditionally not taken part, as mentioned earlier. As an epistemic occurrence it is thus void for the bereaved. At the fourth stage, the urn is handed over to either the bereaved or a funeral director, who will take it to the cemetery for burial.

Altogether, we may observe how the architecture of mortuary institutions in Denmark builds a sensory screen between the deceased and the bereaved: the hospitalisation of the dying, the cold storage of the corpse, the opaque casket, the sterilised technology of the crematorium, the opaque urn, and the minimalism and smoothness of the lawn cemetery. In short, knowledge of the corporeal consequences of death thereby becomes reduced to a number of agreeable abstractions. Following the theory of objectification (Miller 1987, Miller 1998), objects – and architecture – are not finalised materialisations of ideas, but may feed back on and re-create ideas. If this is true, the ideas surrounding death must be affected by the places and things in which they play out. This suggests that the material form of mortuary institutions have given rise to a desire for not sensing the dead as an immediate experience, meaning that the dead are primarily encountered through proxies, metaphors, imagination and memory.

The abstraction and inconclusiveness identified above are clearly exemplified in the ideals of the lawn sections. They are characterised by transparency and the desire to collapse visually the boundary between interior and exterior, abiding to the hallmarks of Modernist design and architecture (Lefas 2009, Overy 2008, Quetglas 2001, Rowe and Slutzky 1997, Treib 1992, Worpole 2003). We may thus argue that there are deep culture-historical issues of mentality at play in the material construction of the institutions surrounding the dead in Denmark. Such issues offer alternatives to the explanations of changes in mortuary customs and cemetery designs that simply refer to shifting practical challenges and changing commemorative ideals and traditions. If we consider the historical organisation of mortuary institutions in Denmark as circumscribed by an intimate recursiveness of matter and ideal (i.e. 'objectification'), we may observe how the transparency and visual continuity at the lawn sections give rise to certain ethereal or unbounded qualities that dismantle the barrier between the material and the immaterial, the present and the absent, the seen and the unseen. And more importantly, the collapsing boundaries further result in a breakdown of the sharp distinction between the natural and the cultural, as the transparency at the lawn sections are characterised by loss of the vertical barriers that stand out as distinctly constructed by humans. Instead, the lawn *emerges* as a seamless expanse with little or no friction between natural processes (such as decomposition) and cultural choices (such as urn burial). These qualities, in turn, can be seen as creative of associative understandings of the dead as being made equally ethereal or abstract. In this relationship, it is also important not to confuse transparency with exposure of facts, because the visually penetrable character of a cemetery constitutes a medium for imagination and suggestion rather than factual disclosure. Or, as architects Colin Rowe and Robert Slutzky argue, 'transparency ceases to be that which is perfectly clear and becomes instead that which is clearly ambiguous' (Rowe and Slutzky 1997: 23). Taking this line of reasoning to the next level, we may consider how material culture and aesthetics have the capacity to affect the ways in which people think about the dead in terms of the psychology of grief work and bereavement.

Leaving-while-keeping the dead

Recognising things for what they really are – facing reality – forms a tenet for what defines sanity in most popular and much professional psychology. The contention that failing to admit to one's feelings and realise the loss of a loved one is a fundamental maladaption to reality comes from Freudian psychology's belief in the benevolence of disclosure (Freud 1966). To have things exposed facilitates a cathartic desensitisation, relieving the bereaved of painful memories and pathological delusion. Sigmund Freud saw love as attachment, which meant that a person who continues to remain attached to the one that has passed away will not be able to form new attachments and thus does not adapt to life without the deceased (ibid.).

This psychological model is widespread in the Western world, encouraging people to face their emotional reaction to death in order to be able to cope with the reality of their loss (Stroebe *et al.* 2002: 169). Avoiding doing so and instead ignoring or repressing one's feelings, or retreating into denial or embellished imaginations, it is argued, may stir even greater emotional distress and/or prolong grief (Bonanno, Keltner, Holen, and Horowitz 1995: 975–976). This line of reasoning forms an undercurrent in much professional and popular psychology on loss, trauma and anxiety, leading to the common contention that it is psychologically indispensable to 'work things through' and 'get things out in the open' (see also Lichtenthal and Cruess 2010, Watkin 2003).

The notion that unveiling is emotionally constructive is, for example, pronounced in advice to people who have experienced pre- and neonatal loss, where it is argued that the bereaved should have physical contact with the dead child, caress it and take photos of it. This is based on the belief that refusing to see the child will potentially lead to speculations about the child's actual appearance and that these thoughts result in traumatic visualisations that will continue to linger with the bereaved parents (see Sørensen 2011).

However, the notion that traumas and grief need to be processed through exposure has been challenged in a number of psychological studies (Bonanno *et al.* 1995, Stroebe *et al.* 2002, Stroebe, Schut, and Stroebe 2005, Wortman and Silver 1989). It has been suggested that emotional equilibrium during bereavement may be maintained for some people by inducing positive memories of the deceased, avoiding potentially painful aspects (Hofer 1984). Directing attention towards the fact of the death of the loved one, it is suggested, conversely makes it more difficult to achieve these pleasant memories. In other words, coping with death in a balanced way is not necessarily achieved through the disclosure of subjective emotional experiences of loss. Studies have instead indicated that bereaved individuals may find it easier to navigate between grieving and ordinary, everyday responsibilities by invoking positive imagery of their loss (Bonanno *et al.* 1995: 985).

While many studies of bereavement have focused on the ways in which people part with the deceased and have emphasised the finality of death, recent studies have turned towards the bonds that may persist or be manufactured between the bereaved and the deceased (e.g. Howarth 2000, Howarth 2007, Jensen 2010, Klass, Silverman, and Nickman 1996, Parrott 2010, Woodthorpe 2010). The contentions propelled by these studies are that the ways in which such bonds are forged and maintained build on the negotiation and navigation of a viable route between abandoning the deceased (in the abovementioned Freudian sense) and staying in touch with her or him. The challenge is thereby not so much to find strategies for abandoning the dead or of maintaining their presence, but rather to find forms of leaving them while at the same time keeping them close (Sørensen 2010b: 115). I suggest that we may conceptualise this strategy as *leaving-while-keeping* the dead. The capacity for accomplishing this relationship needs to reach beyond the psychological and achieve material forms. Elizabeth Hallam and Jenny Hockey (2001: 18) argue that objects may 'preserve a material presence in the face of embodied absence' (see also Bille, Hastrup, and Sørensen 2010). This turns our attention towards the qualities of material culture in the constitution of bonds with the dead, and we may further suggest that material culture is inherently, and necessarily, part of the formulations and ritualisations of memorial acts in the private as well as in the collective sphere (Parrott 2010, Sørensen 2010a).

Ritual and the limits of meaning

Critiques of Freud's conception of grief work as a subjective process have also shed light on the role of the social milieu in bereavement, arguing that sharing one's grief is a way of exposing oneself to painful emotions. It is argued that such endeavours may build a higher level of tolerance towards one's trauma, gradually facilitating a desensitisation towards the loss. More importantly, the social sharing of grief is argued to be a way of making sense of loss (Stroebe, Schut, and Stroebe 2005: 399). From the perspective of embodiment and material culture studies, this again calls attention to the material frameworks surrounding bereaved individuals, because all sociality is a performance or practice that needs a spatial, corporeal and material grounding. We may also ponder the kinds of stimuli the bereaved encounter when they carry out the performances that are associated with disposing of the deceased. To what extent do the material framework and the embodied practices surrounding the dead make it possible to retain positive images of the deceased and make sense of her or his death?

Much work has addressed the ritual meaning of funerals and mortuary ceremonies. The ritual foundation of many mortuary practices has been studied in terms of symbolic representations of a larger, transcendental meaning, clarifying or translating otherwise incomprehensible phenomena and feelings. Nicole Boivin (2009), for instance, argues that rituals can serve to facilitate comprehension of the elusive and unknown. This resonates with studies that see rituals and ritualisation as a way of finding or establishing meaning, including the argument that rituals can form a medium for finding meaning in death, thereby easing the mourning process (Kobler, Limbo, and

Kavanaugh 2007, Kratz 1994, LeVine 1982, Metcalf 1981, Moules 1998, but see also Nilsson Stutz 2006).

On the other hand, Matthew Engelke has recently suggested that uncertainty of meaning may actually be a resource in ritual practices. He observes that 'ritual makes things clear; it also makes them unclear. Ambiguity and clarity are mutually constitutive' (Engelke 2007: 79). The power of a ritual act or the significance of a ceremonial form is thus not necessarily reducible to their meaning-content, but may just as well reside in their capacity to produce mystification or ecstasy, or to bring the participants into an affective presence with the sublime. This draws attention to a potential difference between *finding meaning* as opposed to *making sense*. While the former is here taken to signify the projection of cognised knowledge, the latter denotes the ways in which sensory stimuli and emotions correlate and create balance in the psyche. This is why contradictory, confusing and undisclosed elements in a ritual can make perfect sense, even though they may not be meaningful. Thus, in the elusive ritual, the emotionally straining facticity (or banality) of death may thus be mediated and made tolerable; thoughts and emotions may be distracted or enchanted and inherently allowed to remain inconsistent. Aesthetics thereby become central to the process of grief (*contra* Sturken 2004), because they allow the mourner to get a sensory and emotional feel of simultaneous contact with and veiling of their loss.

Vague knowledge and imagination

Suggesting that death may be experienced as potentially inconsistent in contemporary Denmark has been pursued elsewhere (Sørensen 2010a, Sørensen 2010b), showing that death in a more or less banal way hinges on a conflation of presence and absence. Here, I wish to expand on the observation that absence gives rise to a creativity that builds material and mental images of the dead. The void that the dead create offers an inconsistency in the narratives and images that surround them, both in terms of blanks they leave in the lives of the bereaved, but also – and more importantly for my argument here – in terms of the corporeal ambiguities they foster. As should be clear from the chapter so far, the dead are to a large extent taken out of sight in contemporary Denmark, and bereaved are not automatically in sensuous contact with the deceased as part of the mortuary process. Being in touch (*sensu* Runia 2006) with the dead is thus a technique that has to be acquired, and I argue that people learn and develop new ways of constructing images and imaginings of the dead through the material culture and the physical institutions that now surround death. It is this relationship between mortuary institutions and the knowledge of the dead that I hold to be recursive and mutually engendering, yet not in terms of how death is constituted meaningfully or how loss is being made meaningful, but instead in terms of the ways in which people are in touch with the deceased. I am thus arguing that inconsistency and vagueness of knowledge are resourceful ways of being in touch with the dead in contemporary Denmark and ascribe large parts of the reason for this condition to the architecture and aesthetics of death.

Edward Casey suggests that imagination does not necessarily issue forth as clear images of the imagined object, but can be characterised by an indeterminacy and the lack of a 'strictly specifiable form or content' (Casey 2000: 36). The material infrastructure around death is suggestive of precisely such a lack of specifiable forms, as the minimalism and opaqueness of mortuary institutions render the embodiment of death and the deceased invisible, vague and abstract. The vagueness that characterised the sensible knowledge of the dead gives rise to images and conceptualisations of the deceased that hinge just as much on how the deceased is imagined as how s/he is known. This is why lack of abstraction and clarity may be considered a resource in the mortuary institutions; the bereaved are allowed to imagine the state of their loved ones. Being in touch with the dead thus issues forth through daydreaming and imagining as much as through strict knowledge of the corporeal destiny of the deceased. As Gaston Bachelard states, daydreaming – or, in his terminology *reverie* – is 'where consciousness becomes clouded' (1971: 5). He further claims that the poetics of reverie 'listens to the polyphony of the senses' (ibid.), meaning that daydreaming is never divorced from the experience of the tactile world, and hence that daydreaming is not synonymous with illusion.

These contentions allow us to return to Jules Rochard's desire for delusion, and to suggest a re-evaluation of the request for the possibility for imagining death, the dead and her or his bodily destiny in ways other than those offered by the disclosure of biological facts or desensitisation through emotional exposure of the experience of loss. Rochard's desire for delusion is thereby not a pejorative term for the failure of seeing things for what they really are, as living in an illusion or under the misconception of facts. Delusion may instead be a creative and fertile product of imagination and daydreaming. Daydreaming – imagining what hypothetically could be real – is thus not simply a means of passing time (cf. Ehn and Löfgren 2010), but can work as a way of handling reality, as a creative negotiation of the 'real' world's sometimes dull, bitter or saddening facts.

The aporetics of bereavement

In this light, it may be suggested that rituals, aesthetics and daydreaming can be means of translating the potentially disturbing facticity of death, decay and loss into manageable and opaque forms, protecting the bereaved by a layer of vague knowledge. Daydreaming and delusion thus have the capacity to shroud realities that are undesired, where material elements play a central role in the orchestration of this perceptual vagueness. The blurry boundary is important to the construction of vagueness, and the perceptual screen between the subject and her or his object of imagination is essential in this regard. The object could be disclosed, but the resourceful vagueness rests on the potential for creating more compliant images of the object. Vagueness is thus not a failure to face reality, but a strategy for navigating in reality. This suggests a potential distinction between a desire for *epistemic uncertainty* (believing that lack of specificity can be better than facing facts) as opposed to *ontological ambiguity* (that conceptual categories are confused or conflated).

Just as paying attention is a skilled craft (Ingold 2001), the desire and ability to be distracted in a certain way also require skill. Daydreaming is a technique for which people can be trained to establish vague knowledge of undesired facts. For example, in rhetoric and philosophy an inconsistent object is captured in the notion of 'apory', denoting the useful and generative bewilderment of the encounter with the unclear and potentially meaningless. Hence, *aporetics* is a strategy for mobilising reflection or dialogue (Rescher 2009: 1–2). Accordingly, we may argue that the simultaneous transparency and opacity that characterise mortuary institutions can capture an *aporetics of bereavement*, stirring helpful yet not entirely truthful images and realisations of the dead.

It is important here to note that the mortuary ideal in contemporary Denmark is not the total ignorance or rejection of the dead, nor of entire disclosure and knowledge. By comparison, Julie Rugg refers to the British debate over cremation in the 1960s, wherein the superintendent of the cemeteries in Wolverhampton stated that it would be better for the bereaved not to see any tangible evidence of their loved ones, and suggested that cremation would be even more successful if it was technologically possible to entirely eradicate the ashes so that there would be nothing left of the body (Rugg 2006: 230). The Danish case exemplifies a need for a proximity to and contact with the dead, yet at the same time for distance and vagueness. This allows us to suggest that delusion and disclosure thereby always co-exist, because total exposure can be just as hurtful and damaging as living in entire illusion.

It is probably not so much the desire for vague knowledge of death that has changed in the course of the shift from inhumation towards cremation over the past 120 years or so. Instead, what have changed are the material institutions that give rise to different images of the dead. The beliefs that surround death are thus culturally and historically contingent on the aesthetics of mortuary institutions, and effectively Denmark has witnessed a gradual shift from what is considered respectively 'exposure' and 'concealment' from the late 19th century to the beginning of the 21st (compare Fig. 4.2 and 4.7). The cultural perceptions of technology have changed accordingly, and the particular technology that was once considered new and almost futuristic is now commonplace and the default option for most people facing loss. This has taken effect in a way so that the agency that fire and other materials may exercise in the context of pyre cremation (Sørensen and Bille 2008: 256, Williams 2004) is entirely avoided in modern incineration cremation. The technology of disposal plays a central role in the change in perception of the dead during the 20th century. Rochard argued that cremation would be too swift, yet this was a property of cremation disposal taking place entirely at the crematorium, not of the practices where cremation is a distinct part of but not the end of the funeral practice. As contemporary cremation in Denmark is divided into several stages, cremation actually takes up to a week longer than inhumation. It thus allows time to enter and leave various modes of imagining the deceased before s/he is finally disposed.

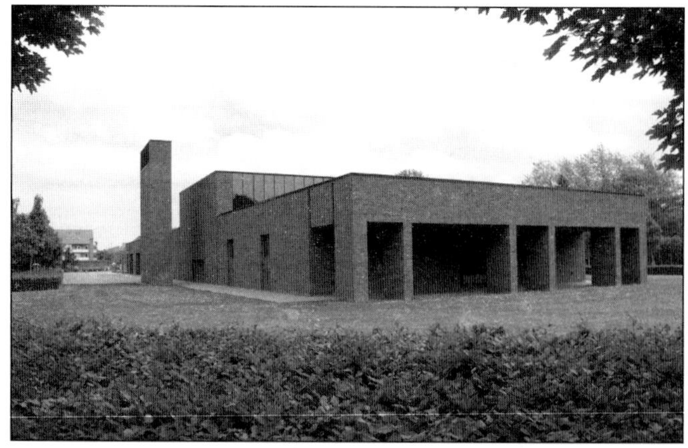

Fig. 4.7: Bispebjerg new crematorium, built 2002 (photo by Tim Flohr Sørensen, 2012).

While the dead previously were envisioned as sleeping in their graves, current conceptions of the dead seem to have been dissolved to a largely abstract phenomenon. To some extent, cremation can thus be seen as a means of immaterialising the dead, turning her or him into an ethereal memory, unburdened by the decaying process of the organism. Cremation in contemporary Denmark is thereby not to be seen as the destruction of the dead body, but instead as the body's immaterialisation. By being cremated, the body is taken beyond sensory proximity to the bereaved; the cremation relieves the body of an undesired corporeal reality.

Conclusion

In *Antichrist*, the bereaved parents hardly spend any time at the cemetery, and the mortuary institutions apparently only serve functional disposal purposes. Instead, the parents retreat and chose social seclusion, digging into their loss through words and emotional disclosure. This chapter suggests that the material scaffolds that people build around the dead and

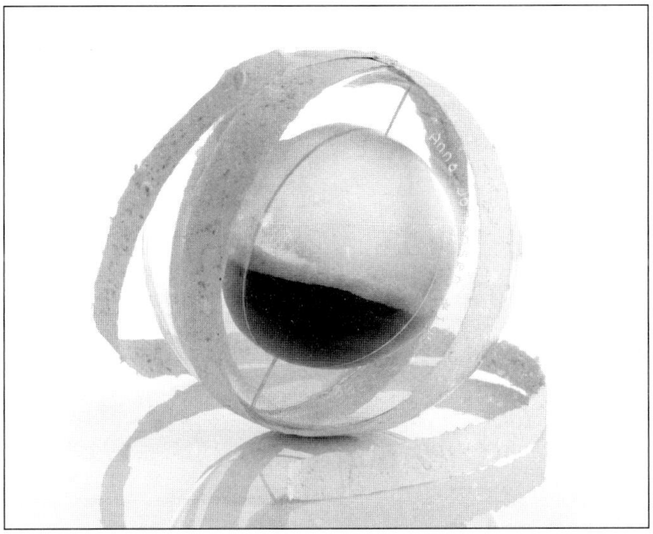

Fig. 4.8: Beyond, bio-degradable urn (concept and product by Pia Galschiøt, 2007. Reproduced with permission from the designer).

the mortuary institutions in their own right are means of maintaining a constructive and healthy screen of vagueness between themselves and the ones they have lost. The screen is a boundary, a bridge and a prism, establishing connection and separation at the same time. The screen is a tool for *leaving-while-keeping* the dead and for revisiting them through manageable images and daydreams.

Just like memory can be reconciled with forgetting (Connerton 2008, Connerton 2009, Forty 1999, Rowlands 1999), so, we may suggest, can knowledge be negotiated with unawareness, which means that we cannot speak of 'total knowledge'. Within the cultural perception of what needs doing when someone dies, knowledge of the processing of the dead cannot be avoided, and I do not wish to suggest that people desire complete ignorance in the face of death. Rather, the contact with the material and visual reality of the deceased is circumscribed by a protecting layer of vagueness, filtering out the most unwanted details of the physical processing of the dead. The emotive agency of material forms is hence minimised, and the ideal seems to be to avoid affecting people emotionally through the sensuous contact with the transformation of the dead body.

Technological development continues to affect the ways in which the dead are handled and deeply influences ways of imagining them and their aesthetics. In 2007, industrial designer Pia Galschiøt created a biodegradable urn for woodland cremation, intended for non-religious ceremonies.[8] The product – called *Beyond* – consists of a decomposable spheric urn, made from maize-based PLA plastic. The urn has two layers: an outer transparent ball and an inner semi-transparent ball, which allows the bereaved to dimly see the actual ashes inside the urn. In addition, a ribbon set with flower seeds is attached to the urn for lowering it into the grave, and the seeds are then to germinate and grow on the grave plot. The designer's intention is that the bereaved can later pick flowers grown from the ashes of their loved one. Such a material format for burial obviously offers an alternative to the veiling of the product of cremation, yet it still plays on and exploits the relative vagueness that allows *leaving-while-keeping* the dead.

Acknowledgements

I owe my thanks to Katharina Rebay-Salisbury and Marie Louise Stig Sørensen for inviting me to contribute to this volume, which has offered me a chance to rethink previous work. They have also made valuable suggestions for improving an earlier draft of the paper. I also take the opportunity to thank Juliane Wammen for suggesting *Antichrist* as a framework for thinking about emotional verbalisation and self-therapy, and Hans Henrik Lohfert Jørgensen for stimulating conversations about vagueness, ambiguity and para-visuality.

Notes

1 Statistics were kindly provided by The National Association of Danish Crematoria.

2 54,368 persons died in Denmark in 2010 of whom 48,079 were buried from a church. The numbers represent people of all ages, geographical locations and religious beliefs. See http://www.km.dk/folkekirken/statistik-og-oekonomi/kirkestatistik/kirkelige-begravelser.html (accessed 2/5/2011) and http://www.statistikbanken.dk/statbank5a/default.asp?w=1280 (accessed 2/5/2011)

3 http://www.km.dk/folkekirken/statistik-og-oekonomi/kirkestatistik/folkekirkens-medlemstal.html (accessed 2/5/2011)

4 http://www.sst.dk/publ/tidsskrifter/nyetal/pdf/2005/05_05.pdf (accessed 2/5/2011)

5 Statistics based on figures kindly provided by the 25 of the 28 crematoria in Denmark.

6 To be fair, media have recently mentioned some persons' interest in being present at the actual cremation (e.g. Schnabel 2010), yet this does not represent a common trend and I therefore choose to leave it aside here.

7 The architecture of crematoria were designed with reference to religious architecture and not to industrial architecture in late 19th and early 20th century Denmark. Their design made space for formalised imitations of church procedures. I thus use the term 'semi-religious' to indicate that they offered the inclusion of religious perspectives and sentiments (and ordained vicars), but were not religious institutions *per se*.

8 www.beyondlife.dk (accessed 2/5/2011)

References

Bachelard, G. 1971. *The Poetics of Reverie: Childhood, Language, and the Cosmos*. Boston, MA: Beacon Press.

Barsalou, L., A. K. Barbey, W. K. Simmons, and A. Santos. 2005. Embodiment in religious knowledge. *Journal of Cognition and Culture* 5, 1–2: 14–57.

Bernfeld, S. C. 1951. Freud and archaeology. *American Imago* VIII: 107–128.

Bille, M., F. Hastrup, and T. F. Sørensen. 2010. "Introduction: An Anthropology of Absence," in M. Bille, F. Hastrup, and T. F. Sørensen (eds) An Anthropology of Absence: Materializations of Transcendence and Loss. 3–22. New York: Springer.

Blomquist, H. 1998. Aktivister for ligbrænding. En [sic] studie i ligbrændingslovgivningens tilblivelse. *Historie* 1998: 67–90.

Boivin, N. 2009. Grasping the Elusive and Unknowable: Material Culture in Ritual Practice. *Material Religion* 5, 3: 266–287.

Bonanno, G. A., D. Keltner, A. Holen, and M. J. Horowitz. 1995. When avoiding unpleasant emotions might not be such a bad thing: Verbal-autonomic response dissociation and midlife conjugal bereavement. *Journal of Personality and Social Psychology* 46, 5: 975–989.

Casey, E. S. 2000. *Imagining: A Phenomenological Study*, 2nd edition. Bloomington: Indiana University Press.

Cobb, A. C. 1892. *Earth-Burial and Cremation. The History of Earth-Burial and its Attendant Evils, and the Advantages Offered by Cremation*. New York: G.P. Putnam's Sons.

Connerton, P. 2008. Seven types of forgetting. *Memory Studies* 1, 1: 59–71.

Connerton, P. 2009. *How Modernity Forgets*. Cambridge: Cambridge University Press.

Davies, D. J. 1996. The Sacred Crematorium. *Mortality* 1, 1: 83–94.

Dennie, G. M. 2003. Flames of Race, Ashes of Death: Re-inventing Cremation in Johannesburg, 1910–1945. *Journal of Southern African Studies* 29, 1: 177–192.

Eassie, W. 1878. The Systems of Cremation Now in Use upon the Continent. *Journal of the Society of the Arts* 26: 230–237.

Ehn, B., and O. Löfgren. 2010. *The Secret World of Doing Nothing*. Berkeley: University of California Press.

Engelke, M. 2007. "Clarity and Charisma: On the Uses of Ambiguity in Ritual Life," in M. Engelke and M. Tomlinson (eds) *The Limits of Meaning: Case Studies in the Anthropology of Christianity*. 63–83. Oxford: Berghahn Books.

Fischer, N. 2008. "Schauplatz Krematorium: Zur Aktualität und Geschichte des verborgenen Todes," in T. Klie (ed.) *Performanzen des Todes: Neue Bestattungskultur und kirchliche Wahrnehmung*. 44–55. Stuttgart: W. Kohlhammer.

Forty, A. 1999. "Introduction," in A. Forty and S. Küchler (eds) *The Art of Forgetting*. 1–18. Oxford: Berg.

Freud, S. 1966. "Mourning and Melancholia," in *The Standard Edition of the Complete Psychological Works of Sigmund Freud*, Volume XIV. 243–258. London: The Hogarth Press and the Institute of Psychoanalysis. First published 1917.

Gosden, C. 2001. Making sense: archaeology and aesthetics. *World Archaeology* 33, 2: 163–167.

Grainger, H. J. 2000. Golders Green Crematorium and the architectural expression of cremation. *Mortality* 5, 1: 53–73.

Guldager, S., J. R. Elkjær, and C. P. Kjøller. 2008. *Driftsstyring på kirkegårde – nøgletal til belysning af kirkegårdenes forvaltning*. Arbejdsrapport Skov & Landskab. Hørsholm: Skov & Landskab.

Hallam, E., and J. Hockey. 2001. *Death, Memory and Material Culture*. New York: Berg.

Hockey, J., L. Kellaher, and D. Prendergast. 2007. "Sustaining Kinship: Ritualization and the Disposal of Human Ashes in the United Kingdom," in M. Mitchell (ed.) *Remember Me: Constructing Immortality – Beliefs on Immortality, Life, and Death*. London: Routledge.

Hofer, M. A. 1984. Relationships as regulators: A psychobiologic perspective on bereavement. *Psychosomatic Medicine* 46, 3: 183–197.

Højer, L. 2010. "Absent Powers: Magic and Loss in Post-Socialist Mongolia," in M. Bille, F. Hastrup, and T. F. Sørensen (eds) *An Anthropology of Absence: Materializations of Transcendence and Loss*. 149–166. New York: Springer.

Hopps, J. P. 1894. *The Etherealisation of the Body. An address delivered at the cremation of Alice Dunn*. London: Williams & Norgate.

Hornborg, A. 2006. Animism, fetishism and objectivism as strategies for knowing (or not knowing) the world. *Ethnos* 71, 1: 21–32.

Howarth, G. 2000. Dismantling the boundaries between life and death. *Mortality* 5, 2: 127–138.

Howarth, G. 2007. "The Rebirth of Death: Continuing Relationships With the Dead," in M. Mitchell (ed.) *Remember Me: Constructing Immortality – Beliefs on Immortality, Life, and Death*. 19–34. London: Routledge.

Ingold, T. 2001. "From the Transmission of Representation to the Education of Attention," in H. Whitehouse (ed.) *The Debated Mind: Evolutionary Psychology versus Ethnography*. 113–153. Oxford: Berg.

Jensen, A. M. B. 2010. "A Sense of Absence: The Staging of Heroic Deaths and Ongoing Lives among American Organ Donor Families," in M. Bille, F. Hastrup, and T.F. Sørensen (eds) *An Anthropology of Absence: Materializations of Transcendence and Loss*. 63–80. New York: Springer.

Jupp, P. C. 2006. *From Dust to Ashes: Cremation and the British Way of Death*. Basingstoke: Palgrave Macmillan.

Kath, L. 1994. *Forest Lawn Memorial-Parks. A Place for the Living*. Glendale, CA: Forest Lawn Memorial Park.

Kellaher, L., J. Hockey, and D. Prendergast. 2010. "Wandering Lines and Cul-de-sacs: Trajectories of Ashes in the United Kingdom," in J. Hockey, C. Komaromy, and K. Woodthorpe (eds) *The Matter of Death: Space, place and materiality*. 133–147. Basingstoke: Palgrave Macmillan.

Klass, D., P. R. Silverman, and S. L. Nickman (eds) 1996. *Continuing bonds: new understandings of grief*. London: Taylor and Francis.

Kobler, K., R. Limbo, and K. Kavanaugh. 2007. Meaningful Moments: The Use of Ritual in Perinatal and Pediatric Death. *MCN: The American Journal of Maternal/Child Nursing* 32, 5: 288–295.

Kragh, B. 1996. Romantik på kirkegården. *Humaniora* 3: 8–12.

Kragh, B. 2003. *Til jord skal du blive... Dødens og begravelsens kulturhistorie i Danmark 1780–1990*. Sønderborg: Aabenraa Museum/Skrifter fra Museumsrådet for Sønderjyllands Amt.

Kratz, C. A. 1994. *Affecting Performance: Meaning, Movement and Experience in Okiek Women's Initiation*. Washington DC: Smithsonian Institution Press.

Lefas, P. 2009. *Dwelling and Architecture: From Heidegger to Koolhas*. Berlin: jovis Verlag.

LeVine, R. A. 1982. Gusii Funerals: Meanings of Life and Death in an African Community. *Ethos* 10, 1: 26–65.

Levy, M. 2003. *Why Modern Cremation Should Replace Earth Burial: an exposé of the dangers caused by inhumation and an explanation of the superior merits of incineration upon new and scientific principles*. San Fransisco: Bacon & Company. First published 1885.

Lichtenthal, W. G., and D. G. Cruess. 2010. Effects of Directed Written Disclosure on Grief and Distress Symptoms Among Bereaved Individuals. *Death Studies* 34, 6: 475–499.

MacLagan, D. 2001. *Psychological Aesthetics: Painting, Feeling and Making Sense*. London: Jessica Kingsley Publishers.

Mädler, I. 2008. "Die Urne als 'Mobilie': Überlegungen zur gegenwärtigen Bestattungskultur," in T. Klie (ed.) *Performanzen des Todes: Neue Bestattungskultur und kirchliche Wahrnehmung*. 57–75. Stuttgart: W. Kohlhammer.

Metcalf, P. 1981. *Meaning and Materialism: The Ritual Economy of Death*. Man 16: 563–578.

Miller, D. 1987. *Material Culture and Mass Consumption*. Oxford: Blackwell.

Miller, D. 1998. "Why some things matter," in D. Miller (ed.) *Material cultures: Why some things matter*. 3–20. Chicago: University of Chicago Press.

Møller, J. S. 1929. *Fester og Højtider i gamle Dage. Skildringer fra Nordvestsjælland med Forsøg paa Tydninger*. 2den Levering: Død og Begravelse. Holbæk: P. Haase & Søn.

Moules, N. J. 1998. Legitimizing Grief: Challenging Beliefs That Constrain. *Journal of Family Nursing* 4, 2: 142–167.

Nicol, R. 2003. *This Grave and Burning Question: A Centenary History of Cremation in Australia*. Adelaide: Adelaide Cemeteries Authorities.

Nilsson Stutz, L. 2006. "Escaping the Allure of Meaning. Toward new paradigms in the study of ritual in prehistory," in A. Anders, K. Jennbert, and C. Raudvere (eds) *Old Norse religion in long term perspectives*. 89–92. Lund: Nordic Academic Press.

Otto, R. 1958. *The Idea of the Holy: an inquiry into the non-rational factor in the idea of the divine, and its relation to the rational*. Oxford: Oxford University Press. First published 1923.

Overy, P. 2008. *Light, Air and Openness: Modern Architecture Between the Wars*. London: Thames & Hudson.

Parrott, F. R. 2010. "Bringing Home the Dead: Photographs, family imaginaries and moral remains," in M. Bille, F. Hastrup, and T. F. Sørensen (eds) *An Anthropology of Absence: Materializations of Transcendence and Loss*. 131–146. New York: Springer.

Parsons, B. 2005. *Committed to the Cleansing Flame: The Development of Cremation in Nineteenth Century England*. Reading: Spire Books.

Prendergast, D., J. Hockey, and L. Kellaher. 2007. Blowing in the wind? Identity, materiality, and the destinations of human ashes. *Journal of the Royal Anthropological Institute* (N.S.) 12, 4: 881–898.

Prothero, S. 2001. *Purified by Fire: A History of Cremation in America*. Berkeley: University of California Press.

Quetglas, J. 2001. *Fear of Glass: Mies van der Rohe's Pavilion in Barcelona*. Basel: Birkhäuser.

Rescher, N. 2009. *Aporetics: Rational Deliberation in the Face of Inconsistency*. Pittsburgh, PA: University of Pittsburgh Press.

Rochard, J. 1891. *Ligbrænding*. Copenhagen: Karl Schønbergs Forlag.

Rowe, C., and R. Slutzky. 1997. *Transparency*. Basel: Birkhäuser. First published in 1964.

Rowlands, M. 1999. "Remembering to Forget: Sublimation as Sacrifice to War Memorials," in A. Forty and S. Küchler (eds) *The Art of Forgetting*. 129–145. Oxford: Berg.

Rugg, J. 2006. Lawn cemeteries: the emergence of a new landscape of death. Urban *History* 33, 2: 213–233.

Runia, E. 2006. Presence. *History and Theory* 45, February: 1–29.

Schnabel, L. E. 2010. "Pårørende vil se afdøde, før kisten lukkes," in *Kristeligt Dagblad*. 20 January.

Sommer, A.-L. 2006. Den moderne død: etik og æstetik. *Kirkegårdskultur* 2006: 50–64.

Sommer, A.-L. 2009. *Den danske arkitektur*. København: Gyldendal.

Sørensen, T. F. 2007. Nærvær og distance – Ligbrændingens placering i dansk kirkegårdskultur. *Kirkegårdskultur* 2007–08: 6–17.

Sørensen, T. F. 2009. The Presence of the Dead: Cemeteries, cremation and the staging of non-place. *Journal of Social Archaeology* 9, 1: 110–135.

Sørensen, T. F. 2010a. *An Archaeology of Movement: Materiality, affects and cemeteries in prehistoric and contemporary Odsherred, Denmark*. Unpublished doctoral thesis, University of Aarhus.

Sørensen, T. F. 2010b. "A Saturated Void: Anticipating and Preparing Presence in Contemporary Danish Cemetery Culture," in M. Bille, F. Hastrup, and T.F. Sørensen (eds) *An Anthropology of Absence: Materializations of Transcendence and Loss*. 115–130. New York: Springer.

Sørensen, T. F. 2011. Sweet Dreams: Biographical blanks and the commemoration of children. *Mortality* 16, 2: 161–175.

Sørensen, T. F., and M. Bille. 2008. Flames of transformation: the role of fire in cremation practices. *World Archaeology* 40, 2: 253–267.

Stroebe, M. S., W. Stroebe, H. A. W. Schut, E. Zech, and J. van den Bout. 2002. Does disclosure of emotions facilitate recovery from bereavement? Evidence from two prospective studies. *Journal of Consulting and Clinical Psychology* 70, 1: 169–178.

Stroebe, W., H. Schut, and M. S. Stroebe. 2005. Grief work, disclosure and counseling: Do they help the bereaved? *Clinical Psychology Review* 25, 4: 395–414.

Sturken, M. 2004. The aesthetics of absence: Rebuilding Ground Zero. *American Ethnologist* 31, 3: 311–325.

Thompson, H. 1873. The Treatment of the Body after Death. *Contemporary Review* 23: 319–328.

Treib, M. 1992. *Modern Landscape Architecture: A Critical Review*. Cambridge, MA: MIT Press.

Watkin, W. 2003. Melancholia, Revolution and Materiality in the Work of Julia Kristeva. *Paragraph* 26, 3: 86–107.

Williams, H. 2004. Death Warmed Up: The Agency of Bodies and Bones in Early Anglo-Saxon Cremation Rites. *Journal of Material Culture* 9, 3: 263–291.

Winter, H. 2001. *Die Architektur der Krematorien im Deutschen Reich 1878–1918*. Dettelbach: J.H. Röll.

Woodthorpe, K. 2010. "Private Grief in Public Spaces: Interpreting Memorialisation in the Contemporary Cemetery," in J. Hockey, C. Komaromy, and K. Woodthorpe (eds) *The Matter of Death: Space, place and materiality*. 117–132. Basingstoke: Palgrave Macmillan.

Worpole, K. 2003. *Last Landscapes: The Architecture of the Cemetery in the West*. London: Reaktion.

Wortman, C., and R. Silver. 1989. The myths of coping with loss. *Journal of Consulting and Clinical Psychology* 57, 3: 349–357.

5. Material culture, embodiment and the construction of religious knowledge

Mads Dengsø Jessen

In recent cognitive science there has been an increased focus on how corporeal understandings influence cognitive operations. In this paper, I shall sketch the theoretical development of a cognitive science leading to the formation of a theory of embodiment as it has come to be understood over the last 30 years. The point of departure is the digital revolution of the late 1940s and onwards, and the associated theoretical standpoints about the mental world of humans, which came to form the more traditional branch of cognitive science. From there I will venture into ideas put forth by the 'corporeal movement' that have advocated for an embodied notion of cognition and how this is seen in recent developments within material culture studies. As a case study, I will analyse the materiality of religious thinking, and in particular the position of portable altars in the transmission of theological concepts, such as the Eucharist, and relate the function of the portable altars to the transition from heathendom to Christendom in Southern Scandinavia.

Embodiment theory and cognitive science

Following the invention and development of the digital computer, beginning in the early 1940s, an avalanche of theories of the mind started to form. Scientists within the field of Artificial Intelligence (AI) saw technological advances as the solution to the mysteries of the workings of the human mind. The computer, even at the earliest primitive stages, had the ability to mimic a wide range of the capacities of the human mind, and could to a great extent even 'outperform' its alleged organic counterpart when it came to tricky mathematical questions or the handling of complex data sets. Alan Turing, if anyone the father of classical AI, even introduced a test – the Turing Test – which was supposed to examine if a computer (or in principle any machine) could give such correct and 'natural' answers to specific questions as to fool a 'real' human being into believing that s/he was having a conversation with a person (Turing 1950). This led to the belief that the digital computer could be used as an electronic version of the human mind, which again spawned the conviction that technological progress and theoretical insight within the field of AI could provide answers to problems and questions which scientists faced in psychology, neurobiology and related disciplines. Consequently, the theoretical idea associated with the earliest forms of AI is for the most part built upon the computer-model of the mind, or mind-as-machine metaphor. The correlation assumed to exist between the neural networks of the brain and the soldered iron and silicon chips of the digital computer not only triggered the hope of creating genuine AI, it also inspired the idea of the brain working as a computer. The digital computer was regarded as providing a fitting analogy to the human mind, and computer-operations on abstract symbolic structures by means of a variety of algorithms were believed to parallel the mental-symbolic content characterising the actions of the brain – hence introducing the mind-as-machine metaphor (Anderson 2003: 92, Haugeland 1993: 3). The main tool employed in analysing the human mind, i.e. the digital computer, hereby transformed the object of analysis into its own image, which gave AI the feel of the possibility of creating exactly that – an artificial intelligence imitating the human mind, freed from the material chains of neurons and synapses (Turing 1950: 451, Pylyshyn 1986: 3).

To carry out true cognitive science in the eyes of the AI movement is therefore to regard cognitive science as 'A science of structure and function divorced from material substance' (Pylyshyn 1984: 68). Within this perspective, the workings of the human mind only have importance in an input-output relation, making the actual template generating the outcome an insignificant part of the structure of the system. Be it silicon or synapses, as long as the functions they perform resemble each other, the actual 'fact of matter', as one might put it, does not hold any significant position in the system. Generally speaking, the objective notions found within traditional cognitivism encompass the ideas that: 1) the mind is functioning as a mirror of nature, meaning that there can be no direct interaction with the world, but only mental, symbolic representations of it, 2) that in nature, which is separated from the experiencer, there exist discrete entities with discrete logical combinations of properties and relations among these entities, and 3) that mental representations and symbols correspond to these entities and their relationally constructed categories (Varela, Thompson and Rosch 1991: 133, Lakoff 1988: 119–120). Cognitivism therefore assumes that meaning is independent of the state of people's minds and does not accept the relevance of human cognition and perception to the construction of meaning.

Building a theory of the body

The suggestions put forth within modern cognitive science and embodiment-theory as it has been formulated in the past 30 years are contrary to the ideas of cognitivism. The profound lack of consideration of the entire bodily sphere in classical AI led researchers to question the traditional ideas present in theories concerned with the workings of the mind. The intentions, emotions, experiences and actions in the world that so obviously play a significant role in the lives of human beings had to a large extent been left out. The theory of embodiment and the idea of the human mind as constituted in a series of relations to bodily and social experiences of the organism (as a whole) took up this challenge. This has, over the last couple of decades, evolved into a real and growing alternative to the formerly dominant cognitivist approach. The embodied standpoint is largely built upon a phenomenological tradition that recognises the organism as placed firmly within the environment or life-world in which it is living, and regards the body as a central contributor to semantic understanding. It thereby takes a profound non-objectivist stance towards human cognitive life. This basic tenet implies that through its place within the environment, the organism is at all times coupled with its surroundings. Therefore, there is no apparent separation between the mental phenomenon within the skull of the perceiver and the environment to which the perceiver directs its attention (Dreyfus and Spinoza 2002: 249, Edelman 1992: 161–164). Being human means at all times to be placed within a certain environment, which we inevitably incorporate into our cognitive system through our perceptual capacities (Slingerland 2004: 4, Varela, Thompson and Rosch 1991: 172). The being or organism doing the cognising is therefore a generative and necessary constituting factor for thinking and information retrieval, and consequently cannot be left out of a theory of mind. Likewise, the environment plays an important part in the way we think, as it is where we receive our input. The consequence of such a view upon human ontology is the renunciation of any restricted separation of subject and object. The mind and body organisation can be seen as consisting of both outer and inner structures between which a constant circulation and mutual elaboration unfolds; a dialectic system in which one cannot exist without the other.

The distributed mind

One of the more recent advances in the cognitive study of embodiment has been to include the manipulated environment produced and used by the human body. Situated studies of how the human body functions in space, and how such physical positioning influences how the human mind relates to, for example, objects, artefacts, buildings or landscapes, have received increased attention (Gallagher 2005, Jeannerod 1997, Suchman 1987). Of particular interest are the ideas concerned with the distributed mind, mainly advocated by Edwin Hutchins. In his seminal book 'Cognition in the Wild' (Hutchins 1995, see also Hutchins 2005) Hutchins demonstrates how humans with the aid of 'material anchors' are able to prolong and stabilise the often complex conceptual content (such as mathematical questions, systems of exchange or social hierarchies) we try to produce between one another in different social situations. In particular, Hutchins has brought attention to the extended functional properties of the human mind by showing that humans, when processing complex conceptual knowledge, can distribute it among several individuals and that such cognitive processes always are mediated through some sort of artefact (linguistic or other) which anchors the mental processes to a particular environment (Hutchins 1995: 283–285). In his analysis, Hutchins demonstrates, for example, that when navigating a large ship the chief officer must necessarily rely on information provided by the other crewmembers and on the readings presented by the instruments placed on the bridge. A simple and recurring task, such as determining the ship's position, must be distributed over several individuals and instruments, as aligning the ship's position requires abstract conceptual knowledge of intersecting longitudinal and latitudinal lines, visual evaluation of landmarks, drawing positional lines on the charts, getting auditory information from other crew members about instrument readings, etc. (ibid. 264–279). In other words, the collective endeavour of navigating a ship safely around the globe entails that every person is deeply involved with other humans as well as extra-somatic entities – each individual *thinks* through other persons as well as through things – and a systemic division into separate minds, bodies, and external objects therefore becomes implausible. The cognitive processes are distributed throughout the extensive physical environment on the bridge of the ship and involve several embodied tasks in order to achieve a functional outcome, so that an entire range of embodied interaction with other persons and cultural artefacts supports the abstract notion of navigation (Hutchins 2010).

David de Léon, clearly influenced by the works of Hutchins, has likewise made a very convincing analysis of the importance of 'artefactual intelligence', as he terms it (de Léon 2003: 7–8), regarding the evolution of different artefact designs. He believes a sort of intelligence (not to be confused with artificial intelligence) can be held by the physical qualities of the object itself. That is, the sensori-motor performances afforded by certain artefacts have been promoted by their alterations through time, as exemplified by the evolution of the modern rifle. Such 'simple' innovations as the introduction of a slow burning fuse, instead of a piece of red hot iron, to ignite the powder, and eventually the composite bullet with a primer, have not only greatly increased the mobility of the shooter, but also improved aiming and thus precision (ibid. 111–132). As a result, modern rifles have 'distributed' the traditional corporeal process of igniting the powder into the artefact (or later cartridge), which has freed other physical manoeuvres, such as movement and taking aim. Clearly, the gunman and his firearm constitute an inherently distributed organism, which transgresses the conventional division of mind and matter (see also Bateson 1979: 195 for a related discussion about the embodied nature of firearms).

Similar ideas have been presented by other scholars, who advocate the notion of the extended mind,[1] which likewise seeks to redefine the traditional borders of cognition (Clark and Chalmers 1998, Haugeland 1993, Sharifian 2003, Sterelny 2004). One of the main principles in the extended mind paradigm is that cognition necessitates a fundamental engagement with the environment, comprised of the relation between mind, body and world (social as well as instrumental), and that cognitive processes cannot function without being related to some sort of 'external' reality (Malafouris 2009, 2010: 265–266). Andy Clark, in particular, underlines the hybrid structure of cognition, emphasising that we are biologically attuned to extra-somatic influences. This 'externalised' nature of mind affects the computational organisation of the individual as well as the group. As Clark observes 'external structures function as to complement our individual cognitive profiles and to diffuse human reason across wider and wider social and physical networks whose collective computations exhibit their own special and dynamic properties' (Clark 1997: 179).

This way we minimise the effort we have to put into complex or lengthy mental operations – we 'extend' or 'distribute' our cognitive capacities. Another gain in displacing cognition into the cultural dimension is at the level of complexity. When several individuals, through distributed cognition, become part of a wider network of communication (which also includes artefacts), there is a potential increase in the level of complexity that can be handled, to a level far greater than what is possible for a single human being. Further, the sheer quality of the mental operations seems to attain new levels, such as in navigation (Hutchins 1995: 284–285). That means we utilise material structures to scaffold particular understandings of the world, and thereby establish collective concepts that can perhaps be transferred into more abstract forms, such as economic, aesthetic or religious concepts (Clark 1997: 59–63, Clark 1999).

Even though the field of distributed cognition often concentrates on more practical and goal oriented actions and may seem far from, for instance, Catholic rituals, it nevertheless illustrates the necessity of regarding human social performances, which involve abstract (or theological) knowledge, as tightly connected to specific physical environments. This is particularly so when the abstract content is supposed to generate a commonality of conceptual understandings among the involved individuals, be it reading a marine chart or consuming the blood and flesh of a divine being (see below for a further elaboration). Furthermore, the cultural artefacts that are located in such an environment are not just passive paraphernalia, but afford particular types of embodied behaviour, and must be recognised as constituting elements in the creation of the different abstract concepts.

When referring to abstract concepts, I do not differentiate between a religious mode and a secular mode. The reason for this is that, from a cognitive perspective, there seemingly is no absolute dividing line between the two categories. The problem of defining religious concepts as exceptional is a topic that has sparked a series of investigations and intense debate, especially concerning the widespread presence of counterintuitive elements in religious belief systems (Barrett 2008, Barrett and Nyhof 2001). However, concepts such as a 'wealthy rodent' or a 'healing rodent' each contain counterintuitive elements, because they violate the normal (i.e. ontologically intuitive) characteristics of a rodent, and could therefore potentially both be categorised as religious concepts. But why is it then, that we do not revere Mickey Mouse as a deity (Atran 2002), but only regard him as an unusual and funny character? Seemingly, the situations where counterintuitive concepts are in use and affect people the most relate directly to the contextual setting, and how they are presented (Norenzayan and Atran 2004, Tweney *et al.* 2006). Furthermore, in order to be perceived as religiously significant, the abstract concepts need to contain some sort of attention-grabbing elements in order to settle and be memorised by the believers. Such attention could be generated by counterintuitive elements, but in order to generate a religious salient concept, the abstract concepts need to be accompanied by recognisable narratives, performed by authoritative persons and related to a familiar environment. In other words, since concepts are not transmitted as unitary vehicles of information all by themselves, the saliency of abstract concepts seems to be generated by the cultural context (Bulbulia 2009: 166–167, Gervais and Henrich 2010, Jessen in press).

Therefore, the generation of abstract concepts, to which religious concepts per definition belong, require the incorporation of the material environment, which can provide the thinking subject with a tangible experience and at the same time focus attention. For this reason, the dialectic relationship between the *physical context* and the *abstract content* will figure prominently in this chapter. Of particular interest is how religious concepts might, consequently, be understood as an intermingled constellation of cultural instruments and abstract thinking, and how the cognitive operations might be distributed across brain, body and extra-somatic entities. Hence, investigating the different behaviours and the particular material environment in which ritual activities are situated will benefit from special attention to the broader interactions between the persons involved in the ritual acts, as well as the interactions between the mind and the particular types of material culture framing the situation. On this basis, in the following I shall interpret the importance of the portable altar during missionary travels in Viking Age and Early Medieval Scandinavia, where it can be recognised as an indispensable instrument in the spreading of religious concepts. In particular, I will investigate the performance of the Eucharist sacrifice and the idea about the consummation of a consecrated food as an associative notion of an adaptation of the Christian faith, with the view to show that abstract religious concepts are generated on the basis of distributed and embodied behaviours.

The religious environment

There is little doubt that within the ritual setting of Christian cosmology the church building plays a prominent role. This is especially so after the recognition of Christianity as a state religion under the reign of Constantine the Great, who also promoted a series of prominent church constructions (Krautheimer 1965: 17–18, Christie 2006: 268–269). The buildings were created to house the liturgical and faithful life of living communities. These communities consciously built their religious environment in order to partake in liturgical events and to follow prescribed religious processions and practices. It was a place invested with devotion and affection, but also a place fostering social ties among believers in the community. The importance of the church as a place of religious assembly is reflected in the early Christian writings, where its metaphorical extensions meander between the congregation and the building itself. For example, in the Apostle Paul's letter to the Ephesians he writes 'Christ himself being the cornerstone, in whom the whole structure is joined together and grows into a holy temple in the Lord; in whom you are also built into it for a dwelling place of God in the Spirit' (Ephesians 2: 20–22). Even at that time, when there were no regulated Christian cult buildings, the importance of a place of worship clearly shines through, and underlines the intermingled conception of believer and place of worship.

An example more directly related to the Nordic situation regarding the introduction of Christianity is the description by Rimbert telling the story of Ansgar's endeavour, in the early 9th century, to baptise the Danish King Horic in Haithabu (present day Schleswig, Germany). 'Following this wish our Holy Father (which is Ansgar) advised Horic to prove to Christ our Lord his willingness, that which would be most dear to him, to allow a church be built in his Kingdom' (Robinson 1921: chapter XXIV). Here again, the church figures prominently as an important actor in the true way of Christian worship, as well as providing a token for the true believer to acknowledge. Of course, the opportunity for Ansgar to get an actual plot of land and approval to build a church upon it would surely ease his missionary enterprise, as he would then have a locality from where the priests could spread the Christian word further north. Furthermore, the church would function as the centre of the major Christian rituals and as the focal point of the community's attention during mass. Because this place of worship is the scene of the more prominent *rites de passage*, such as communion, burial requiems, and later in time also baptisms and weddings, a slight paradox arises when the priests are required to perform their rituals in places outside the church. This is, for example, the case when people are in too poor health to come to the church or when the priests are carrying out missionary activities. During the missionary travels, when there is no church, no environmental setting supports the performance of the most essential ceremonies. In missionary activities, the priests have to do without the physical space in which the theological concepts normally would unfold. The constant and substantial ritual anchor which the religious building – the church – normally conveys is nonexistent during missionary work.

The missionaries journeyed to distant locations and would therefore necessarily have to travel light, but at the same time, they had to be able to perform the most essential of the Catholic ritual acts. Of course, the Catholic Church found a way around this problem, and they provided the travelling priests with a sort of minimum requirement for a church-like environment. By equipping the missionaries with a 'miniature church', the necessary extra-somatic support for the ritual activities were in place. This was done by means of the so-called 'portable altars' which habitually accompanied the missionaries in their spreading of 'true ideology' into pagan societies, and it must be presumed that the altars were regular components in their travel gear (Gustafsson 2008: 162–163). This religious equipment basically consisted of a small, and thus portable, altar and usually also other religious items, such as miniature chalices to contain the wine and patens to hold the bread to be used in the Eucharist (Braun 1924: 73–74). The portable altars epitomised the essence of the normal sized churches in that they reproduced the most essential architectural part of the church, namely the altar. Further, such small altars might regularly have functioned as small shrines with reliquaries inside them and they were consecrated through particular ceremonies just like the

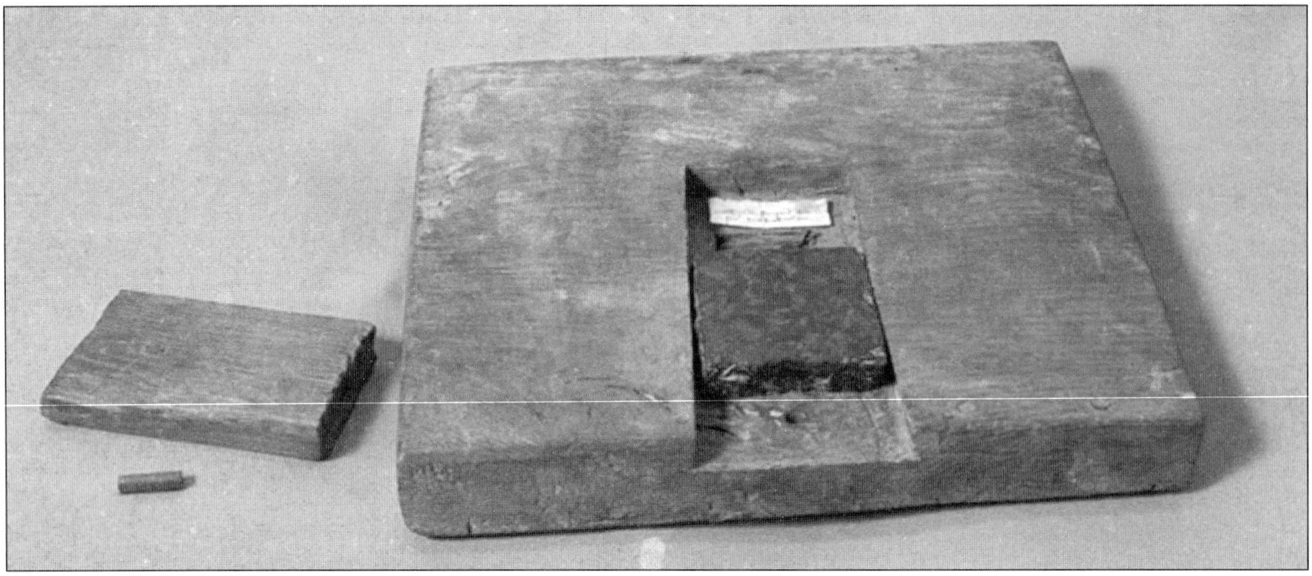

Fig. 5.1: Portable altar from Gosmer Church, Jutland. This particular piece is dated to the 11–13th century AD, and was found inside the main altar during restoration (Licht, Michelsen, and Poulsen 1983–87: 2764). The porphyry stone framed by an oak board measures 7x6 cm and underneath it, in the small sepulcrum, the reliquaries are still clearly visible (photo reproduced with permission from the National Museum of Denmark).

bigger, stationary altars were (Stolt 2001: 31). Therefore, from a theological point of view there were no obvious differences between the two types of altars, as both make clear reference to the tomb of Christ as well as to the table used during the Last Supper. The portable altars were designed for the same purposes, and they clearly eased the missionaries' endeavour; they did not have to make compromises regarding the liturgical traditions because all essential Catholic rituals could be carried out by the use of the portable altars (Braun 1924: 71, Stolt 2001: 28). Both the stationary and the portable altars functioned as the focal point during the Catholic rituals, inside as well as outside the church. Usually, when the portable altars were not accompanying the priest on his travels, they would be kept in the depository of a church or cloister, or they could even function as a small side-altar in the actual church building (see Wangsgaard 2010 for an overview of alternative positions of altars inside Late Medieval Churches).

From the perspective of the present chapter, the portable altars present an interesting object to investigate because they embody the distributed character of religious thinking. Theologically, these altars pass on *abstract content*, such as the process of transubstantiation taking place during the Eucharist, while at the same time the liturgical practices demand that they have a specific and concrete physical appearance in order for this to happen. Secondly, as the altars are movable, they themselves can be used to generate the required *physical context* in places outside the church that are not normally regarded as theologically viable locations to perform the Catholic mass. For these reasons, a more thorough evaluation of the portable altars is required.

The portable altar

In order to get a more precise idea about the ritual position of the portable altars I will take a closer look at the components comprising the altars, and why these components are used. In addition, I will examine the more concrete frame of use, and define the ritual situations in which the altars participated. Such an exposition should provide a platform for discussing the historical development of the use of portable altars, which types of authoritative persons would administer their use, and the theological reasoning behind the design of the portable altars. In effect, this should lead to an outline of the distributed characters of the portable altars.

One of the more famous examples of such missionary equipment is the portable altar of St Cuthbert (see Fig. 5.1). His reputation of enduring missionary activities, particularly in northern England and Scotland, gave him a special position and led to the development of a devout cult during the Middle Ages (Marner 2007: 11). His portable altar, most likely used during his travels around northern Britain, is presumed to be the oldest existing example and consists of a silver-coated, wooden altar, approximately the size of a small shoebox, originally found together with a chalice and a paten; the latter two are now missing. The altar appears to have been refurbished in at least two instances, but there seems to be a general agreement that the original altar consisted of a rather plain wooden box, incised with crosses and a Latin inscription, which can be dated to the latter half of the 7th century (Coatsworth 1989: 300–301, Oman 1957: 63–65, Radford 1956: 326–329).

What is missing from St Cuthbert's altar is the very prominent piece of stone placed in the upper surface – the *mensa* – of the altar, which is a common Anglo-Saxon feature (Okasha and O'Reilly 1984: 35). This stone becomes a recurring feature of the portable altars in later versions from continental Europe. For example, a stone is incorporated in the lavishly decorated examples from Paderborn Cathedral, belonging to Bishop Heinrich II of Werl (Deckert 1931). It is also, however, found on more plain types, for example the ones from Hedared church in Västergötland, Sweden, and Lom church in Oppland, Norway, which consist of a hollowed out wooden plank with a small piece of stone in the middle (Sanmark 2004: 270–271). A tendency towards privately owned altars being more sumptuously ornamented with gold, silver, precious stones or ivory in comparison with the more plain ones often discovered in churches, which have a simple oak or pine frame around the altar stone, seems clear, but the centrally placed stone is a common feature. Inserting an altar stone in the miniature altar, usually covering a small chamber with a relic, as the ecclesiastical doctrines prescribe, must therefore be regarded as the principal feature and the minimum premise if the altar is to be recognised as theologically functional. In fact, fixing a stone in the top of the mensa (or the whole mensa being manufactured in stone) is a specific requirement for the altar to be approved for consecration, according to the 'Discipline of the Church'. However, it is assumed that the earliest Christian churches used wooden altars, and that stationary stone altars only later became the preferred version (Schulte 1907). Eventually, during the Synod of Epaon in 517 a decree was adopted which stated that 'In order to be consecrated as a true altar to be used in the performance of true sacrifices the altar, that is part of the mensa, must contain a stone' (Kieckhefer 2004: 76). Later, in 857, we find the announcement from the chapter of Archbishop Hincmar of Reims giving additional restrictions to the design of altars, in that the altar stone should be of marble or another dark, unbreakable stone (Way 1847: 241). Therefore, from the 6th century onwards a recurring feature of consecrated altars, including portable altars, was a piece of marble or porphyrite mounted in the centre of the mensa. In liturgical terms, the idea was that the stone should be sufficiently large to hold the sacred host and the better part of the chalice during the Eucharist sacrifice. With regard to the portable versions, the size of the altar stone is more variable, and a more lenient attitude seems to have been endorsed (Stolt 2001: 25). Nevertheless, the similarities between stationary and portable altars tell us that there existed no theological discrepancy between the two, and both types followed the same development from wooden to stone inlaid versions. In effect, the validity of rituals performed around the small, portable altars was equal to those performed around stationary altars. Furthermore, the specific design features pertaining to the altars give evidence to the ecclesiastical investments that were put into insuring that the material context framing the Catholic mass was infallible. Material form clearly mattered in the transmission of theological ideals, and from this, a distributed definition appears of how abstract religious concepts were administered.

Portable altars during conversion in Scandinavia

An interesting situation can be recognised in the transitional phase from heathendom to Christianity in southern Scandinavia during the late Viking Age. For this period there are only a low and presumably insufficient number of churches registered, in which regular sermons for the greater part of the population, or even just for the upper part of society, could be held (Ottosen 2006: 207, Nygaard 1983: 9). A gradual increase in church buildings seemingly commenced around 1050 AD (Sanmark 2004: 270, Tesch 2001, 2007: 60–61, Thastrup-Leth 2004: 209, Fig. 2), but before then, alternatives to going to church must have existed in order to participate in an appropriate Catholic ritual, and perhaps the portable altars partly explain the absence of churches in the early phase of Christianisation. A number of scholars have suggested that the significant Viking Age halls, which figured prominently in the pre-Christian cult, must also have functioned as places of Christian veneration, or at least as places of missionary interest, until the proprietor found it necessary to build a more suitable place for worship (Andrén 2002: 320–321, Jessen 2010, Roesdahl 2004: 205, Skree 1995). One possible way of transposing a Viking Age hall into a Christian space would be through the introduction of a portable altar, because, as mentioned above, such an altar would generate the needed physical context for the transmission of abstract Catholic concepts.

Some of the recovered fragments of porphyrite from Sigtuna suggest that they are found in contexts belonging to non-ecclesiastical buildings, and more likely are connected with the hall-buildings in different plots within the city.[2] Furthermore, the culture layers the stones belong to predate the intense phase of church construction (from around 1050 AD), which strengthens the hypotheses that the traditional hall buildings were used as Christian ritual buildings (Stolt 2001: 34, Tesch 2001). Even though the traditional ritual activities performed in the halls would be considered barbarian and uncivilised by the advocates of Christianity, the place itself could, without any further complications, be incorporated into a pseudo-hallowed space if the priest brought his miniature altar here. This moveable place of worship would thereby change the Viking Age halls into a suitable Christian environment and the former heathen house would function as a Christian locality. The main concern was therefore not to be positioned inside a church, but to ensure the divine communication between the priest and God through the consecrated and portable altar.

It is therefore no coincidence that localities such as Sigtuna (Tesch 2007, containing an excellent overview of Scandinavian altar stones), Haithabu and Schleswig (Bracker-

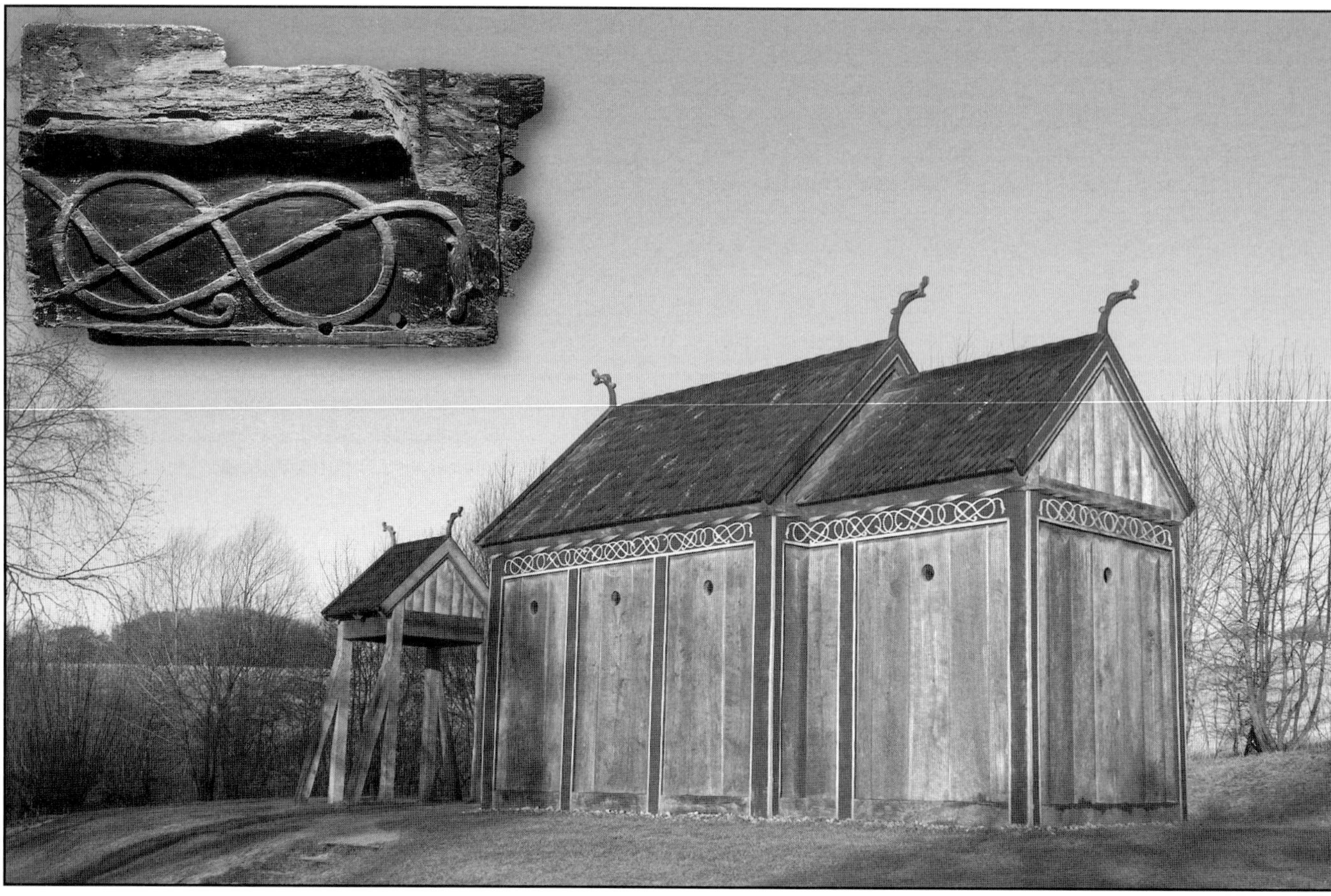

Fig. 5.2: Reconstruction at Moesgaard of the stave-church from Hørning, Jutland. One part of the church, the so-called 'hammer-plank' (top left, is situated just under the roof and is dendrochronologically dated to c. 1060 AD (Krogh and Voss 1961). The layout of the church as it appeared during the excavations have formed the basis for the reconstruction (photo: Per Kirkeby, Moesgaard).

Wester 1989), Roskilde (Andersen, Høj and Sørensen 1989), Varnhem (Vretemark 2007), Björkö/Birka (Gustafsson 2008), and most recently the royal estate at Skjævesland in Norway (Stylegar 2010), all of which figure prominently in the early Christian church in Scandinavia, are the main find spots for worked porphyrite and marble. These localities are considered to have been influential in the consolidation of Christianity in Scandinavia, and have all at some point in their history functioned as a mainstay in the continuous spread of Christianity further north. Other contemporary finds also give witness to the special position of the portable altars. For example, the Icelandic finds, which likewise have a close relation to important early Christian locations, seem to follow the same pattern as the southern Scandinavian examples – they are found at localities that can be regarded as central to the spreading of the Catholic ideology (Kristján Mimisson, Reykjavik University, personal communication). In fact, notations in the inventories of the churches, which eventually were constructed at these places, show that the number of portable altars or altar stones is surprisingly high, with up to five or six portable altar stones in each church (Stolt 2001: 29).

Religious concepts, rituals and their material anchoring

Ethnographical and historical accounts of ritual behaviour are abundant, with many examples of people interacting with cultural artefacts as well as natural objects. Stones, weapons, fluids, feathers, dead bodies, living bodies, buildings, boats and a host of other physical entities turn up in descriptions of people's commerce with the supernatural powers. There is a universal tendency to organise ritual activities around the material environment. Manipulability and reification appear to take a principal position in ritual behaviour. But what might be the reasons for this profound inclination to employ objects into religious thinking? Should not pure mental operation (i.e. 'conventional' thinking) involving the otherworldly suffice? Apparently it does not. Religious cognition almost seems to *require* some sort of material scaffolding in order to function properly. Analysing this propensity to organise ritual behaviour around material structures might therefore constitute a fitting point of departure for a better understanding of the significance of portable altars in Catholic liturgy. Within the cognitive study of religion, only little effort has been put into describing the position of material culture in rituals as well as in the generation of religious concepts (see Liénard 2006, Sørensen, Liénard and Feeny 2006 for exceptions). In general, the material side of ritual behaviour has been regarded as mere curiosities and paraphernalia or as detached representations of inner, mental concepts, without any relevance for the cognitive study of religion. This problematic omission has been specifically addressed by Matthew Day (2004), where he appeals for 'a way of thinking about religions that would move the cultural, historical, and social aspects of cognition from the margins of thought to the core of reflection' (ibid. 102), or perhaps even extend the locus of thought into the extra-somatic environment. He further proposes that the human capacity to have abstract thoughts is firmly grounded in the extra-somatic environment, because off-line, decoupled thoughts entail 'the strategic use of non-biological artefacts to scaffold and augment biological modes of computation' (ibid. 111). Basically, his claim is that by filling the environment with cultural artefacts, humans have become able to transgress the native limitations of the physical brain – perhaps even to a degree which allows abstract concepts to form (see Donald 1991 for an evolutionary account of the same idea). Within archaeology a similar hypothesis has been promoted by Steven Mithen, who recognises that 'networks of minds, disembodying minds and exponentially increasing the range of conceptual spaces available for [mindful] exploration' is caused by material culture, and might even have instigated non-adaptive cognitive domains such as religious thinking – totemism, animism, anthropomorphism, etc. (Mithen 1998: 181). The Swedish philosopher Johan Modée (2005) has even argued that any kind of religious concept functions solely through the notion of some kind of materialisation. Even linguistic utterances should, in his view, be recognised as material concepts, because they have proper functions, which are constituted in social contexts, just as 'traditional' artefacts have (ibid. 112–119). However, there is more to ritual behaviour than language, and I will therefore return to the position of the portable altar and the generation of religious thinking.

The ritual body

The design of the portable altar, as well as of altars more generally, is adapted to one important Catholic ritual in particular – the Eucharist Sacrifice. All the above-mentioned descriptions of how to construct a true altar result from the specific pontifical requirements of a consecration. Such a consecration is necessary if profane objects are to be transferred into a sacred category by the rituals performed by the members of the clergy. The Eucharist aims to make Christ a real presence in the form of the bread and wine on the altar at the time of the consecration during the mass; the Catholic doctrine defines this as a case of transubstantiation. The theological reasoning behind this kind of presence is that in this way, Christ can become present in the life of humans; through their intake of the bread and wine, people relive his physical sacrifice on Golgatha.[3] From an external presence in the sacred foodstuff, the supernatural presence is transferred and embodied during the Eucharist. The body of Christ becomes the body of the believer. Conclusively, the congregation as a whole will present a real divine presence of Christ. The bodies of the congregation become the body of Christ, and a two-fold manifestation of a divine concept is achieved – an understanding which is also echoed in the quote from the Ephesian letter about the church as a constellation of Christ and the congregation in unison.

What the Eucharist in essence prompts is a sense-perceptible version of a theological concept. It is an extra-somatic manifestation of an abstract and non-experienceable phenomenon (i.e. the mystical body of Christ), in which is staged both the veneration of Christ and the intake of his body. In the Eucharist, the transformation of bread into flesh and wine into blood takes place during the *anaphora*, a quiet, almost silent prayer articulated by the priest over the paten and the chalice containing the items to be sanctified. Both must be placed on the consecrated altar, i.e. on the altar stone, if the transformation is to be initiated in the first place. This makes the portable altar an indispensable channel between the worldly and supernatural realms when mass is performed outside the church. During the unfolding of the ritual, the display of the now consecrated host to the assembly usually indicates that a transformation has occurred and the physical substances can be identified as a manifestation of Christ in our world (Seasoltz 2005: 27–29). Strictly speaking, however, the sacrifice cannot be recognised as such before a 'real' change of the visible gift has taken place (for example by burning, crushing, pouring out, slaying, blood shedding or through other means). The changes produced by the *anaphora* are not exactly visible changes, as the bread and wine still exhibit their normal characteristics such as colour, taste or smell, and are influenced by the forces of nature such as gravity and can be manipulated by force. Rather, the transformation should be regarded a magical transfer of *essence* as Jesper Sørensen has illustrated (Sørensen 2007). He has demonstrated how the projection of magical agency to certain elements can be obtained through the performance of certain rituals, and how the ontological assumptions of particular physical elements change accordingly and become endowed with a sacred essence. In the Eucharist, such a transformation is accomplished by means of a metonymic extension of the flesh and blood of Christ into bread and wine due to the transference of the essence of Christ (ibid. 97–102, Jessen 2010). The breaking of the bread and the distribution of the wine to the congregation is thus an abstract extension of the death of Christ, in which he sacrificed his own flesh and blood to deliver mankind from sin.

I would, furthermore, advocate that the layperson's experience of the ritual alteration of the sacramental wine and bread is underpinned by an additional, more subconscious and corporeal process. This observation relates to the concrete physical interactions that circumscribe the sacred food during the Eucharist and the corporeal experience people have attained from interacting in the non-ritual sphere with similar physical articles. Because of the consumption, the bread and wine are no longer available to our sensuous modalities – the ability to see, smell, touch or taste the bread and wine disappears – and the ritual items therefore no longer bestow the same insistent ritual presence. The moment of consumption thus presents an obvious perceptual change. Moreover, any person partaking in the Eucharist would know that the intake and further digestion of the foodstuff would make it dissolve and change inside his or her own body. For that reason, the only observable transformation of the body of Christ is directly connected to one's own physical performance in the Eucharist and one's own bodily functions. In this way, a very personal and intimate relation with the 'blood and flesh' is activated on the basis of an associative relation between everyday activities and ritual activities. Such a sacrificial transformation, furthermore, fully corresponds with the theologically accepted version of the body of the assembly presenting the body of Christ described above.

Eating and digesting are, of course, normally considered profane activities, and the comparison with a ritual setting might seem naïve, but from a cognitive point of view this might actually be the reason why a ritual such as the Eucharist could be especially useful for the transmission of an abstract theological concept. Bringing the abstract concept into contact, so to speak, with the perceptual modalities of the theological patient might have a quite potent influence upon the reception of the concept; if the ritual practice can be associated with an embodied understanding of the abstract religious concept it is more likely to obtain a more *efficient* process of internalisation. That is, if the compatibility between the bodily state and the cognitive state of a person is high, the performance effectiveness increases. As several scholars investigating the interaction between the body and higher cognition have shown (Barsalou *et al.* 2003: 55), there is a significant correlation between bodily actions and emotional condition. Chen and Bargh (1999), for example, have shown that subjects understand the act of drawing closer as positive and pushing away as negative. Therefore, in their experiment when subjects were told to pull a lever towards them when a word appeared before them, they performed the motion more quickly when exposed to positive words (e.g. love) than to words with a negative valence (e.g. hate). Compatibility between the motor performance and the cognitive processing gave better results, indicating that embodied versions of knowledge tend to influence more abstract concepts. Likewise, in the Eucharist when the believer has completed the ritual s/he has actually undergone a physical process that has a very similar structure to the religious concept the theological principles describe. The intake of foodstuff is a fitting echo of the intake of the spirit of 'Our Lord the Saviour.' In other words, the ritual provides an embodied, performative analogy for the desired mental state and most likely helps the believer understand the concepts and reasons behind the actions (Barsalou *et al.* 2005: 44). As was the case in the word recognition study mentioned above, it is very likely that the familiar embodied operations connected with eating increase the chance that the participants will internalise the desired 'theological message' that is conveyed in the ritual, and that they will actually adopt it.

A closer examination of the ritual activities behind the Eucharist reveals that material culture features prominently in the whole process. The notion of a transfer of a divine essence, a non-physical, abstract concept, into the human corpus is anchored by different extra-somatic entities from the initial phase until the believer has finally embodied it. Firstly, a consecrated stone that guarantees a channel of exchange with the divine realm generates the sacralisation of the foodstuff, and a priest, who equally seems to be an agent, administers the channel.[4] Secondly, as the food has been endowed with an otherworldly essence, the priest now seems to have only a secondary function as the significant agency (the capacity to transfer essence in the physical world) belongs to the consecrated 'flesh' and 'blood'. That is, a physical object is guaranteeing the effect the ritual is supposed to bring about. Thirdly, the physical body of the believer becomes the object of the transference, and the dissolving of the bread and wine will effectuate a spread of the essence of Christ. The believer is thus converted into a 'container' for a divine essence, and, in effect, an abstract theological concept has found an embodied expression. Importantly, every step taken during the performance of the ritual is defined by a material entity, which, one way or the other, provides the ritual gathering with the possibility of manipulating a non-existing, abstract concept. The process takes advantage of the anchoring possibilities produced by the distributed organisation of mind and matter in the ritual. I believe this embodied kind of sacrificial change of the consecrated offer has been greatly overlooked in ritual studies and presents a very potent means of transferring an abstract theological concept into a very tangible and corporeal frame of conceptualisation.

The collective body

Incorporating the physical environment into the ritual sphere does not only effect the personal understanding each individual has about the abstract, theological concepts in play, but will influence the broader social uptake as well. Very early in the scientific study of religious behaviour the importance of the collectively experienced ritual environment was in fact acknowledged, and as Emile Durkheim explains:

> It is true that since collective sentiments can become conscious of themselves only by fixing themselves upon external objects, they have not been able to take form without adopting some of their characteristics from other things: they have thus acquired a sort of physical nature; in this way they have come to mix themselves with the life of the material world, and then have considered themselves capable of explaining what passes there (Durkheim 1912: 466).

The point is that the physical environment and material structures have evident qualities – they can support and scaffold non-experiential entities such as divine beings simply by providing the believer and the collective as a whole with *similar qualitative experiences*. Thus, the relationship between the environment and a commonality of experience is absolutely central. By generating an assembly of recurring and stable ritual materialities, a common conceptual understanding is most likely to occur within the wider group of believers. It is by virtue of its physicality that the environment provides an efficient basis for achieving collective conceptual integration and a relative stability of shared meaning. The cognitive gain of having an environmental anchoring to support non-experienceable entities seems a powerful encouragement to invest in the materialisation of abstract concepts. Material culture creates the sense of real, 'actual' experience, even in cases where the meaning and the content is very complex, or hard to comprehend. Still, material structures have evident qualities. By implementing the environment into difficult, intellectual tasks, we create a sense of conceptual stability that cannot exist in purely mental, ephemeral processes (Shore 1991). Material scaffolding will produce the possibility of materialising ephemeral concepts and at the same time exploit a very stable unit of experience. By placing the cultural artefacts as an actual part of the environment, it becomes possible to experience and interact with even the most unbelievable creatures, much better than linguistic symbolisation would. Abstract, theological knowledge becomes embodied in the artefact, and acknowledged by the beholder through tacit means of experience such as vision, touch, smell, taste and sound. Of course, the Eucharist is an individually oriented ritual, but at the same time also involves several believers in its execution – they witness the same mass from the beginning and the structure of the individual ritual is the same from person to person, where they interact with the same materials in the same religious environment. Not only is the ritual structure recognisable between individuals, but also between each time it is performed, and the structural familiarity, perhaps with minor alteration, even spans several generations. Obviously, the material environment is the main reason for this production of similar qualitative experiences. No other part of the ritual organisation contributes with a comparable amount of permanence, and the extra-somatisation in itself is responsible for the generation of a collective conceptual understanding around the ritual, because each individual can relate to a common experience. In other words, the most efficient way of establishing a concept shared by a larger body of people will have to be through an external media, which is able to solidify the concept in question. Material culture fulfils exactly these demands – concepts with material anchoring or embodied associations are concepts that are easy to think with and interact with, and these embodiments should not be perceived as mere appendages or epiphenomenal to the transference of abstract information, but rather be regarded as constituting elements in the formation of concepts (see Jessen in press for a further elaboration of these points).

This is also the reason why it was so important for the missionaries to carry the portable altars with them during their missionary activities. What these items embody is a *movable place of worship*, which mirrors the essence of the church. The altars provide the travelling priest, who is going into uncharted territory or preaches away from his customary ritual environment, with the opportunity to bring the church to the needy. Carrying the portable altar allows the priest to physically bring the most essential part of the church building with him, and the priest can freely choose where to 'situate his church.' Through the management of a physical entity the proper place to perform the Catholic rituals is defined by a non-fixed space, and the miniature altar, paten and chalice will function as a concretisation of the ordinary stone-built church, and therefore also as a concretisation of the divine exchange for the collective to behold. This way, in combination with a particular kind of consecrated item, the ritual environment becomes an extension of the priest's physical position in the world. Where the priest is situated, the church is situated, and the symbiotic relationship between agent and item will generate a genuine ritual environment regardless of the perhaps contradictory setting they might be located within – such as the Viking Age halls. Any environmental backdrop can be transformed into a consecrated space by the use of the portable altar and the priest's presence. Importantly, the entity permitting this plasticity of ritual space is not the ritual agent (the priest) but an extra-somatic entity (the altar). This means that the priest's capacity to communicate between the divine powers and the collective of believers is dependent upon an artefact, and the constellation of the ritual environment must be regarded as a phenomenon distributed across different media – both mental and material.

Concluding remarks

The specific point I have tried to make in this chapter is that human beings are capable of reconciling the abstract and non-experienceable with intuitive experiences by anchoring religion in the material world. The fact that human beings exploit their familiarity with a wealth of material structures and practices in attempts to make sense of other features of the world has an important implication: cognitive saliency is established through the practical and embodied comportments within the world and the physical relations humans have with things. In this fashion, religious thinking assimilates extra-somatic materials with constructs inside the human skull. Physical engagement with the world therefore seems to take centre stage in contextualising concept formation. This point leads me, again, to recognise a striking feature of human religious life, namely the integration of the material world in order to ground the ideological frame used as an explanatory tool in religious concept building. In this fashion, material culture functions as conceptual anchor-points that synthesise abstract knowledge with embodied experience. Evidently, when studying human cognition and ritual behaviour, material culture deserves closer attention: without it, we would not think the same.

Notes

1 All of the different concepts used, be they artefactual intelligence, distributed cognition or the extended mind, have as their main prerogative a desire to discard any type of Cartesian dualism, and the different denominations often seem to have more to do with the wish to formulate original thoughts, than with profound diversities in the ontological understanding of the symbiotic nature of mind and matter.

2 It has to be kept in mind that the evidence for the use of portable altars is scarce, and to my knowledge only one other locality suggests a connection between hall and portable altar. In the recent excavations at Raisio Ihala in south-western Finland a fragment of a polished green porphyrite stone was found. The context of the stone is dated to the early 11th century, which coincides with several finds from the very active Christian centre at Sigtuna in mainland Sweden (Juha-Matti Vuorinen, University of Turku, presentation in Athens, Greece, March 2007).

3 The concept of the transubstantiation emerged sometime in the 8th or 9th century and had its final specifications defined during the 4th Lateran Council (Nilsson 1987: 139).

4 The priest himself can be understood as a physical entity containing a special quality or essence that has been passed on to him from Jesus Christ through the unbroken line of the imposition of hands.

References

Anderson, M. L. 2003. Embodied Cognition: A Field Guide. *Artificial Intelligence* 149: 91–130.

Anderssen, M., M. Høj, and S. A. Søresen. 1986. Et vikingetidshus fra Bredgade i Roskilde. *ROMU Årsskrift for Roskilde Museum* 1986: 33–50.

Barsalou, L., A. K. Barbey, W. K. Simmon, and A. Santos. 2005. Embodiment in Religious Knowledge. *Journal of Cognition and Culture* 5: 14–57.

Barsalou, L. W., P. M. Niedenthal, A. K. Barbey, and J. M. Rupert. 2003. "Social Embodiment," in B. H. Ross (ed.) *The Psychology of Learning and Motivation*. 43–92. Amsterdam: Academic Press.

Bateson, G. 1979. *Mind and Nature – A Necessary Unity*. London: Wildwood House.

Bracker-Wester, U. 1989. Porphyrfunde aus Haithabu und Sleswig. *Ausgrabungen in Schleswig: Berichte und Studien* 7: 9–18.

Braun, J. 1924. *Der christliche Altar in seiner geschichtlichen Entwicklung. Band I: Alte Meister*. München: Guenther Koch and Co.

Chen, M., and J. A. Bargh. 1999. Consequences of Automatic Evaluation: Immediate Behavioral Predispositions to Approach or Avoid the Stimulus. *Personality and Social Psychology Bulletin* 25, 2: 215–224.

Clark, A. 1997. *Being There. Putting Brain, Body, and World Together Again*. Cambridge: MIT Press.

Clark, A. 1999. An Embodied Cognitive Science. *Trends in Cognitive Science* 3, 9: 345–351.

Clark, A., and D. Chalmers. 1998. The Extended Mind. *Analysis* 58, 1: 7–19.

Coatsworth, E. 1989. "The Pectoral Cross and Portable Altar from the Tomb," in G. Bonner, D.W. Rollason, and C. Stancliffe (eds) *St. Cuthbert, His Cult and His Community to AD 1200*. 287–302. Woodbridge: Boydell and Brewer.

Day, M. 2004. Religion, Off-Line Cognition and the Extended Mind. *Journal of Cognition and Culture* 4, 1: 101–121.

de Léon, D. 2003. *Artefactual Intelligence: the Development and Use of Cognitively Congenial Artefacts*. Lund: Grahns Tryckeri AB.

Deckert, H. 1931. Opus Interrasile als vorromanische Technik. *Marburger Jahrbuch für Kunstwissenschaft* 6: 137–164.

Donald, M. 1991. *Origins of the Modern Mind: Three Stages in the Evolution of Culture and Cognition*. Cambridge, MA: Harvard University Press.

Dreyfus, H. L., and C. Spinoza. 2002. "Coping with Things-in-themselves: A Practice-Based Phenomenological argument for Realism," in H. L. Dreyfus and M. Wrathall (eds) *Heidegger Reexamined 2: Truth, Realism, and the History of Being*. 249–278. London: Routledge.

Durkheim, E. 1915. *The Elementary Forms of Religious Life*. New York: The Free Press.

Edelman, G. M. 1992. *Bright Air, Brilliant Fire. On the Matter of the Mind*. New York: Basic Books.

Gallagher, S. 2005. *How the body shapes the mind*. Oxford: Clarendon.

Haugeland, J. 1993. "Mind Embodied and Embedded," in *Proceedings of the International Symposium of Mind and Cognition*. 3–37. Taipei: Academia Synica.

Hutchins, E. 1995. *Cognition in the Wild*. Cambridge: MIT Press.

Hutchins, E. 2005. Material Anchors for Conceptual Blends. *Journal of Pragamtics* 37: 1555–1577.

Hutchins, E. in press. "Imagining the Cognitive Life of Things," in L. Malafouri and C. Renfrew (eds) *The Cognitive Life of Things. Recasting the boundaries of the mind*. Cambridge: McDonald Institute for Archaeological Research.

Jeannerod, M. 1997. *The Cognitive Neuroscience of Action*. Oxford: Blackwell.

Jessen, M. D. in press. "Religion and the Extra-Somatics of Conceptual Thought," in A. Geertz (ed.) *Origins of Religion, Cognition and Culture*. London: Equinox.

Jessen, M. D. 2010. "Altars and the sacred space," in G. Dharampal-Frick, R. Langer and N. H. Peterson (eds.) *Ritual 2008, vol. V. Transfer and Space*. Wiesbaden: Harrassowitz Verlag.

Kieckhefer, R. 2004. *Theology in Stone: Church Architecture from Byzantinium to Berkeley*. Oxford: Oxford University Press.

Krogh, K. J., and O. Voss. 1961. Fra hedenskab til kristendom i Hørning. En vikingetids kammergrav og en trækirke fra 1000-tallet under Hørning kirke. *Nationalmuseets Arbejdsmark* 1961: 5–34.

Lakoff, G. 1988. "Cognitive Semantics," in U. Eco, M. Santambrogio, and P. Violi (eds) *Meaning and Mental Representation*. 119–154. Indiana University Press.

Licht, K. von de Fine, V. Michelsen, and N. J. Poulsen. 1983–87. *Danmarks Kirker. Århus Amt Bd. XVI, 5.* Copenhagen: Nationalmuseet.

Liénard, P. 2006. The Making of Peculiar Artifacts: Living Kind, Artifact and Social Order in the Turkana Sacrifice. *Journal of Cognition and Culture* 6, 3–4: 343–373.

Marner, D. 2007. *St. Cuthbert: His Life and Cult in Medieval Durham*. London: The British Library.

Mithen, S. 1998. "A Creative Explosion? Theory of Mind, Language and the Disembodied Mind of the Upper Palaeolithic," in S. Mithen (ed.) *Creativity in Human Evolution and Prehistory*. 165–192. London: Routledge.

Modée, J. 2005. *Artifacts and Supraphysical Worlds. A Conceptual Analysis of Religion*. Lund: Centre for Theology and Religious Studies.

Nygaard, J. J. 1983. *Det begyndte med en trækirke. Kirkebyggeri i Danmark*. København: Forum.

Okasha, E., and J. O'Reilly. 1984. An Anglo-Saxon Portable Altar: Inscriptions and Iconography. *Journal of the Warburg and Courtauld Institutes* 47: 32–51.

Oman, C. 1957. *English Church Plate 597–1830*. London: Oxford University Press.

Ottosen, K. 2006. *Oldtidens og middelalderens kirkehistorie i dansk og nordisk perspektiv*. København: Books on Demand.

Pylyshyn, Z. 1984. *Computation and Cognition: Towards a Foundation for Cognitive Science*. Cambridge: MIT Press.

Pylyshyn, Z. 1986. "Preface," in Z. Pylyshyn (ed.) *Robots Dilemma: The Frame Problem in Artificial Intelligence*. Norwood: Ablex.

Radford, C. A. R. 1956. "The Portable Altar," in C. F. Battiscombe (ed.) *The relics of Saint Cuthbert*. 326–335. Oxford.

Robinson, C. H. 1921. *Anskar, the Apostle of the North (translated from the Vita Anskarii by Bishop Rimbert, his fellow missionary and successor)*. London: Croydon.

Sanmark, A. 2004. *Power and Conversion – a Comparative Study of Christianization in Scandinavia. Occasional Papers in Archaeology 34*. Uppsala: Universitetstryckeriet Ekonomikum.

Schulte, A. J. 1907. "High Altar," in *The Catholic Encyclopaedia*. New York: Robert Appleton Company.

Seasoltz, R. K. 2005. *A Sense of the Sacred. Theological Foundations of Sacred Architecture and Art*. New York: Continuum.

Sharifian, F. 2003. On Cultural Conceptualization. *Journal of Cognition and Culture* 3, 3: 187–207.

Slingerland, E. 2004. Conceptual Metaphor. Theory as Methodology for Comparative Religion. *Journal of the American Academy of Religion* 72, 1: 1–31.

Sørensen, J. 2007. *A Cognitive Theory of Magic*. Lanham: Altamira Press.

Sørensen, J., P. Liénard, and C. Feeny. 2006. Agent and Instrument in Judgements of Ritual Efficacy. *Journal of Cognition and Culture* 6, 3–4: 463–482.

Sterelny, K. 2004. "Externalism, Epistemic Artefacts and the Extended Mind," in R. Schantz (ed.) *The Externalist Challenge. New Studies in Cognition and Intentionality*. 239–254. Berlin: Walter de Gruyter.

Stolt, B. 2001. "Boktyngder och bärbara altarskivor," in B. Stolt (ed.) *Kyrkliga sällsyntheder på Gorland och annorstädes*. 25–42. Visby: Ödins Förlag AB.

Suchman, L. 1987. *Plans and Situated Actions: The Problem of Human-Machine Communication*. Cambridge: Cambridge University Press.

Tesch, S. 2001. Från hall till kyrka. *Populär arkeologi* 2001, 2: 14–16.

Tesch, S. 2007. Tidigmedeltida sepulkralstenar i Sigtuna – heliga stenar från Köln för såväl hallkult som mässa i stenkyrka. *Sigtune Dei* 2007: 45–68.

Turing, A. M. 1950. Computing machinery and intelligence. *Mind* 59 (236): 433–460.

Varela, F., E. Thompson, and E. Rosch. 1991. *The Embodied Mind*. Cambridge: MIT Press.

Vretemark, M. 2007. *Dakboken, September 2007, Utgrävning vid Varnhem KlosterKyrka*. Retrieved 16.09.2008, from http://www.vastarvet.se/kulturvast_templates/Kultur_ArticlePage.aspx?id=8704

Way, A. 1847. Notices of Ancient Ornaments and Appliances of Sacred Use: The Super-Altare, altare viaticum. Itinerarium, Portatile, Gestatorium, Levaticum, Paraticum, or Tabula Itineraria. *The Archaeological Journal* 4: 239–248.

6. Sealed by the cross: protecting the body in Anglo-Saxon England

Helen Foxhall Forbes

> Hec crux consignet Ælfwinum corpore mente
> In qua suspendens Deus traxit omnia secum

This inscription heads a miniature of the crucifixion in Ælfwine's prayerbook (Fig. 6.1), a small book made somewhere between 1023 and 1031 for Ælfwine, monk and then dean (later abbot) of the New Minster, Winchester. The book now survives as two manuscripts in the British Library, Cotton Titus D.xxvii and D.xxvi (Günzel 1993: 1–3; the crucifixion is at Cotton Titus D.xxvii, f. 65v). Ælfwine's name appears in a number of places in the book, and although it is debatable whether he was one of the scribes of the book, he probably gave instructions for its production (Günzel 1993: 3, Karkov 2010: 104–105, Keynes 1996: 111–113). A number of prayers are personal to Ælfwine, like the inscription on the crucifix which names him: 'may this cross sign [or seal] Ælfwine in body and in mind, on which, hanging, God drew everything to himself.' This inscription shows just one of the ways in which Ælfwine sought to protect his own body: the cross and the sign of the cross were believed to have very strong apotropaic powers (Johnson 2006, Karkov 2010: 110–111). This article examines how people sought to protect body and soul from the dangers of the physical and spiritual worlds, and how this illustrates some of the complex links between body and soul.

In the early middle ages, protecting the body was something of a losing battle: life expectancy was short (in some times/places as little as 25–30 years), disease was frequent, and medicine was mostly more optimistic than effectual (Cameron 1993: 5–18, Crawford 2009: 73–4, 179–197). The Old English poem, *Fortunes of Men*, reminds the audience just how many ways a person could lose his life, even if he seemed to be in good health (ed. Krapp and Dobbie 1936: 154–156). Counteracting the numerous visible and invisible threats to the body involved (amongst other things) recourse to medical remedies, prayer, charms, and the sacraments, not all of which were always easily distinguishable from each other. Christian theology traditionally encouraged a focus on the health of the soul rather than the body, but textual and archaeological evidence indicates that there was significant concern to protect the body as well as the soul. Although the link between body and soul was extremely complex, what protected one might also protect the other, revealing much about how life and death were understood in early medieval England.

Invisible beings: angels and demons

Early medieval Christians believed the world around them to consist of both visible and invisible planes: in addition to what they could see, they were surrounded by benevolent angels and malevolent demons, usually invisible to human eyes, but visible at certain times or to certain individuals (Mango 1980). While demons could attack the soul, tempting it to sin, they could also be the cause of bodily sickness, either by possessing the individual, or by affecting the body physically in other ways. The dramatic effects of demonic attacks are described in a letter written at some point between AD 716 and 719 by Boniface, an Anglo-Saxon missionary to the Continent (Ep. 10, ed. Tangl 1916: 7–15). This letter narrates the visions of a man whose spirit was taken out of his body and who returned, apparently from the dead, to describe what he had seen in the otherworld. He was able to see a number of people who were still living on earth and to observe the angels and demons, invisible to them, who filled the air around them. One in particular was considered worthy of note:

> And likewise he testified about Ceolred, king of the Mercians, who at the time that he saw these things, was still in the body [i.e. alive], without a doubt. He said that he saw him defended by a

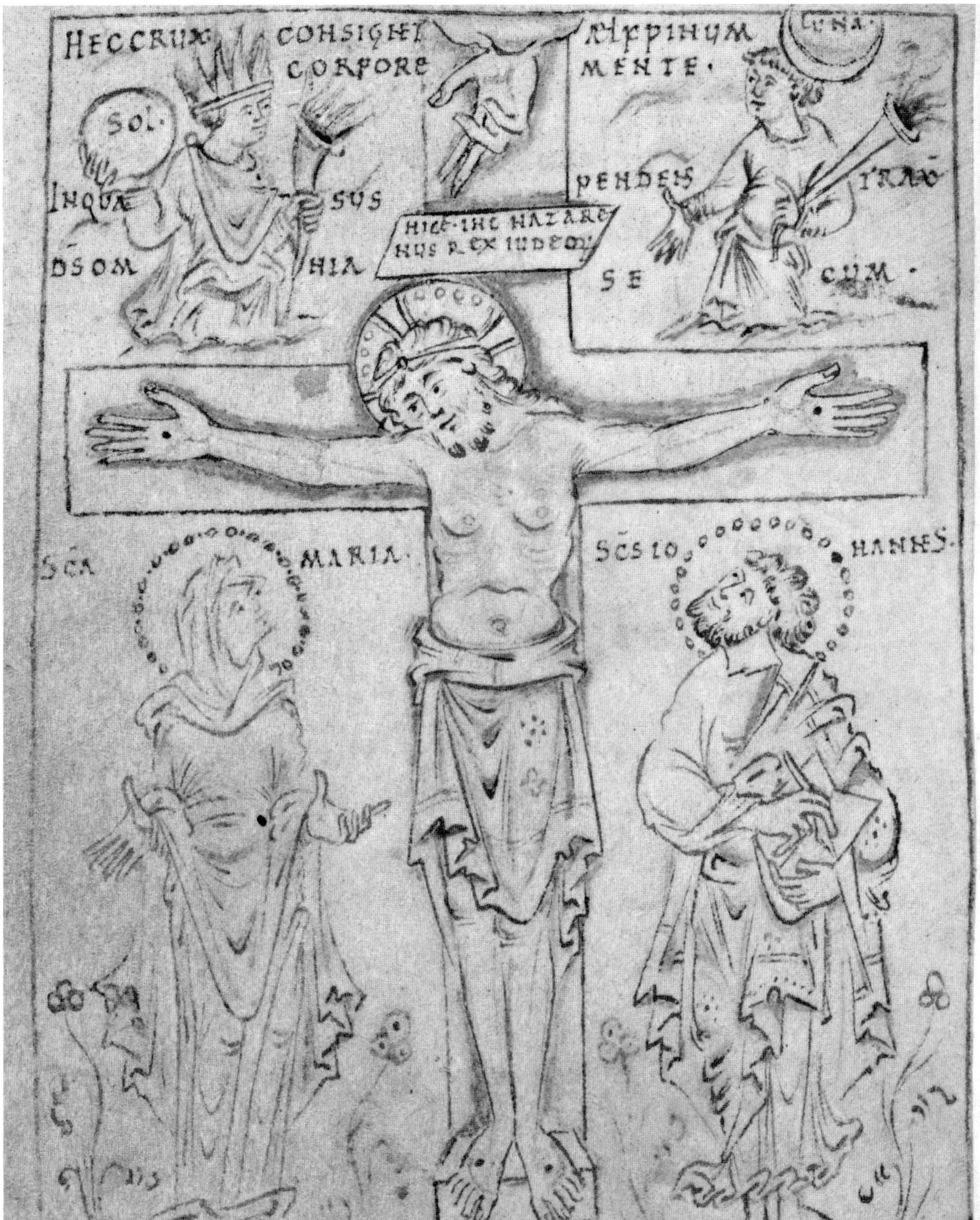

Fig. 6.1: Crucifixion from Ælfwine's prayerbook (London, British Library, Cotton Titus D.xxvii, f. 65v. Reproduced with permission from the British Library).

certain angelic canopy against the onslaught of the demons as if some big book were opened out and placed above him. But the demons asked … that they indeed might be permitted to exercise the wishes of their cruelty upon him with that defence taken away… Then the angels, growing more unhappy, said: 'Alas that this man, a sinner, does not permit us to defend him any longer; and we can offer him no help on account of his own merits. And they took away the protection of the defence placed above him. Then the demons, rejoicing and exulting, came together from all parts of the world in a great multitude…and tore at him with various tortures, tiring him inestimably (Ep. 10, ed. Tangl 1916: 14, ll. 1–19).

Though invisible and immaterial, the devils were able to cause serious bodily harm to Ceolred, and in fact another of Boniface's letters relates that he came to a sticky end at the hand of the devils: 'as those who were present testified, a wicked spirit turned Ceolred, the sinner, to insanity of mind, while he was feasting splendidly with his companions … so that raving mad, babbling frantically and conversing with devils and bad-mouthing the priests of God, he departed from this light without penance and confession and went to the torments of hell, no doubt' (Ep 73, ed. Tangl 1916: 152, l. 29 – 153, l. 7). One of the most important ways of protecting the body was with the sign of the cross, as numerous Anglo-Saxon authors explained: the cross banishes and puts to flight the devil (e.g. Ælfric, On Auguries, ll. 143–147, ed. Johnson 2006, Skeat 1966: I.374–375).

This is one of the ways that Ælfwine's cross could protect him in body and mind, and the prayers which surround the illustration draw attention to the efficacy against demons not only of the cross, but also of devotion to it. Before the image of the crucifixion is a prayer to the cross asking that 'by this sign of the holy cross all my enemies, visible as well as invisible, present as well as absent, and powerful as well as powerless, may be thrown down' (Cotton Titus D.xxvii, f. 64v, ed. Günzel 1993: 122), suggesting that the cross was effective against material as well as immaterial beings. After the image are a series of devotions to the Holy Cross, and four reasons why the cross should be adored, the second of which is that 'if your first work is to the cross, all the demons, if they are around you, will not be able to harm you' (Cotton Titus D.xxvii, f. 70r, ed. Günzel 1993: 126). Further on in the book, a prayer asks for the sign (or seal) of Christ's cross to be upon the supplicant night and day and to defend him from evil works (Cotton Titus D.xxvii, f. 89r–v, ed. Günzel 1993: 139).

In protecting the body, personal piety such as the devotions in Ælfwine's prayerbook could be effective in other ways too. At least as significant as the sign of the cross in protecting body and soul from demonic onslaughts was the presence of benevolent angels, who were believed to be most frequently present around those who devoted themselves especially to God and to divine services (Hom. II.10, ll. 92–103, ed. Hurst 1955: 248–249). Bede, an early eighth-century writer, notes that 'with an invisible presence, angels are frequently at the side of the elect so that they may defend them from the snares of the cunning enemy' (Hom. II.10, ll. 92–4, ed. Hurst 1955: 248). But the angels would only defend those who devoted themselves to God, as the vision described by Boniface illustrates: King Ceolred was protected by angels, until they concluded that he was too sinful to deserve their protection any longer, and it was at this point that the demons were able to cause him serious damage (Ep. 10, ed. Tangl 1916: 14, ll. 14–19).

The sacraments: baptism

The sacraments of the Church included efforts to protect the body as well as the soul from the attacks of demonic powers, and also to invoke the power of the angels. Although the sacraments were directed primarily at the internal and invisible soul, they were performed upon the external and visible body which the soul animated, to indicate outwardly what had taken place within. Through baptism, the soul becomes Christian – but so also does the body. Under the law of the Jews, God's chosen had been marked bodily by circumcision, but now baptism marked out God's chosen, and 'cleaned them from all stains of body and mind'; although baptism was still a ritual of the body, the more important ritual was the spiritual conversion that it indicated (Bede, Hom. I.11, ll. 40–58 ed. Hurst 1955: 74–75, Commentary on the gospel of Luke, 2.4, ll. 305–306, ed. Hurst 1960: 107). The Anglo-Saxon scholar Alcuin, writing to Bishop Arno of Salzburg in 796, stated that in the baptismal rite, 'what the priest performs on the body with water outwardly, the Holy Spirit works invisibly in the soul through faith' (Ep. 113, ll. 39–41, ed. Dümmler 1895: 164). It is the visible body which must receive the action of the sacrament, but this cannot take effect unless the soul within is turned to the Christian faith (Alcuin, Ep. 110, ll. 21–3, ed. Dümmler 1895: 158).

One part of preparing soul and body to become Christian in baptism was the exorcism of devils from the body, and one means of exorcism was exsufflation (Foot 1992: 177). This involved blowing in the child's face, and is described in Alcuin's writings and some later Anglo-Saxon homilies, although it is not found in many Anglo-Saxon liturgical books (Hill 1985: 136). By this exsufflation, immaterial and invisible powers are countered through physical and bodily actions; Wulfstan, Archbishop of York (d. 1023), explains that by God's power, when the priest blows in the face of the baptismal candidate, the devil is terrified and is banished, leaving an entrance for Christ instead (VIIIa, ll. 36–41, VIIIb, ll. 15–36, VIIIc, ll. 29–35, 54–68, 116–142, ed. Bethurum 1957: 170, 172–173, 176–179, 181–182). An eleventh-century sermon notes that a heathen man who is baptised is 'thereby released from devils' (Hom. IX, ll. 82–84, ed. Fadda 1977: 183). Prayers throughout the rite of baptism refer to the banishment of the devil and to the angels who will surround and protect the newly baptised individual (Daniélou 1953: 95–100, see also Kelly 1985).

Baptism also includes the signing of the body with the cross, numerous times and with oil and salt as well as with water, and Ælfwine's inscription above the crucifix may refer to the sign or seal of the cross placed on him at his own baptism (Karkov 2010: 207, Raw 2004: 113–114, Raw 1997: 176–186, esp. 185–186), when his body was marked out as a Christian body. The rite of baptism in the Red Book of Darley (now Cambridge, Corpus Christi College 422) provides a good example from late Anglo-Saxon England, particularly since it includes extensive Old English instructions for the priest (Graham 1993, Page

1978). This book was written in or after 1061, perhaps at the New Minster, Winchester, but was probably used elsewhere, perhaps Sherborne (Pfaff 1995: 21–22). After the priest makes the sign of the cross on the child's forehead, he recites a prayer asking God 'to send your holy angel so that he may guard your servant' (CCCC 422, p. 370), a significant request in keeping the candidate safe in body and soul. The next prayer is directed at the devil, instructing him to 'recognize the judgement of God', to 'recede from this servant of God', and concluding that 'this sign of the cross, which we have placed on the forehead [of the baptised], you will never, wicked devil, dare to violate' (CCCC 422, pp. 370–371).

A number of other prayers ordering Satan, the devil, or unclean spirits to leave the child follow, and then the priest signs the right hand of the child with the cross so that adversities will be kept away, and so that he is marked out and will remain in the Catholic faith forever (CCCC 422, p. 377). The emphasis on the permanent marking of the body with the cross is significant, showing his acceptance by God, and acting as protection against demonic powers; the Old English poem *Andreas* describes that devils attempted to attack St Andrew, but retreated once they recognised that he was sealed with the sign of the cross from baptism (Andreas, ll. 1334–44, ed. Johnson 2006: 87–88, Krapp 1932: 40). The priest also uses his saliva to make the sign of the cross on the baptismal candidate's nose and ears to signify that through both smell and hearing he will receive divine holiness (VIIIb, ll. 29–32 ed. Bethurum 1957: 173), indicating once again the importance of the body – the priest's as well as the candidate's – in a ritual which was primarily aimed at the soul.

The sacraments: anointing of the sick

The link between the physical or material and the spiritual is very strong in the ways that sickness was conceptualised and understood by theologians. Sickness was believed to affect people for a number of different reasons, and just as bodily infirmity might have spiritual causes and cures, the health of the soul could also be improved by rituals performed upon the body, such as the anointing of the sick. Bede observes five reasons from the scriptures for physical infirmity: some sinners become sick so that they can be corrected, others so that they learn while they are alive that an unhappy state awaits them after death. The just may also be afflicted with bodily sickness, either so that they receive a greater reward for their endurance, or so that they may keep the merit of their justice with greater humility. Finally, Bede notes that bodily sickness may be given to humans for the simple reason that the glory and power of the Lord and his saints may be made more widely known when they are healed; bodily sickness can therefore be connected with the greater narrative of mission and the gospel, rather than with the suffering of the individual (Homily I.23, ll. 154–164, ed. Hurst 1955: 165).

For many clerical writers, the first response to sickness was to pray. Bede instructs that remedies for the body should not be sought at all before the inner state has been considered and purified; since it is not immediately obvious whether illness comes on account of sin or whether it is to increase the merits of the just, repentance is the first port of call if bodily infirmity strikes (Hom. I.23, ll. 192–201, ed. Hurst 1955: 166). Priests were likened to doctors and were frequently described as the physicians of souls (Frantzen 1983: 30–31), but their healing work might also apply to the body. The rites for the sick, like baptism, are a sacrament which involves the anointing of the body, and which were intended to cure soul and body, and to prepare the soul for a good death if a bodily cure was impossible. Preparing the soul meant that the sick person was required to confess his sins before his body was anointed, but there are statements in instructional literature and in the rites themselves which command that the priest deny the ritual of anointing to any individual who refused to confess, suggesting that some people wanted bodily anointing without confession (Ælfric, Pastoral Letters, I.91, II.178, ed. Fehr 1914: 21 and 130–131, see also CCCC 422, p. 401).

After the sick person had confessed, the body was anointed with holy oil in a number of places and prayers were said over each one. Surviving liturgical rites from Anglo-Saxon England always include anointing prayers for the eyes, the feet, and the shoulders, but differ as to the inclusion of the head, chest and back, or neck (Keefer 1995: 105). Sins were forgiven partly by the anointing of the parts of the body which represent the five senses, and through which sins are committed (CCCC 422, pp. 415). The prayers focus on the healing of body and soul, emphasising that it is no good to heal the body without the soul: one asks God to keep his servant and to 'grant that he may feel your remedy not only in body but also in soul' (CCCC 422, pp. 408–409). Another prayer asks that the Lord send his angel to visit the sick man, to comfort him and to gladden him (CCCC 422, pp. 409–410). After the anointing the priest asks God to make the holy oil purify the body and mind of the sick person, and to be a defence against the darts of wicked spirits (CCCC 422, p. 414).

The sign of the cross made in oil was intended partly to keep away malign influences from those specific body parts, as the prayers for anointing indicate. In the Red Book of Darley's rite, the sick person's ears are anointed so that spiritual medicine will put to flight what was heard illicitly; he is anointed between the shoulders to provide protection against the stings or darts of the devil; and his chest is anointed to keep away the hostile crowds flying through the air, and the importance of this protection from the holy oil is reiterated in the final anointing prayer (CCCC 422, pp. 411–412, 413). Evidently the crosses made at baptism were not always believed to be as efficacious as might have been hoped, if demonic powers could still cause sickness; but the forgiveness of sins and each Christian's duty to be devout was also important in keeping away demons, and in encouraging the protection of the angels. A number of the rites for the anointing of the sick include prayers for angelic help, often asking specifically for God's angel Raphael to visit the sick man and to cure and comfort him in his illness (Foxhall Forbes 2009: 124, Keck 1998: 63).

Although the rituals for the sick were intended as healing rites, they were also performed for the dying, and perhaps more frequently than for those who were expected to recover (Paxton 1990: 85–87, 128–130). Later Anglo-Saxon texts attest to significant confusion over the rite of the sick and what it was supposed to achieve, and a number of superstitions over its use seem to have existed. Ælfric, writing in the early eleventh-century, complained that some men thought that the anointing would lead them to death, and explained instead that it was simply to forgive sins (Pastoral Letter III.14–15, ed. Fehr 1914: 150–151). Ælfric also had to confirm that if a man recovered after being anointed he was allowed to live with his wife, to eat meat, to be anointed again, and finally (and somewhat bizarrely) that the process of anointing granted forgiveness of sins, but not ordination, suggesting that some people may not have understood the differences between anointing of the sick with oil, and the anointing with oil which was part of the ceremony to change an ordinary man into a deacon or priest (Pastoral Letter II.178–179, 181, ed. Fehr 1914: 130–133). Perhaps because of these misconceptions, Ælfric complained elsewhere that men were reluctant to be anointed and to receive the sacrament when they were sick (Pastoral Letter I.84–6, ed. Fehr 1914: 19–20).

Non-sacramental approaches to sickness and suffering

Ælfric's complaints echo the objections of earlier Continental writers, such as Caesarius of Arles in the sixth century and Eligius of Noyon in the seventh, who instructed their congregations to receive the rites of the sick rather than turning to folk medicine (Paxton 1990: 50, 56). The surviving Anglo-Saxon books of medical remedies are another indication that anointing was not considered to be the only appropriate response to sickness, but a distinction between Church and 'folk healers' should not be overplayed: recent work has emphasised that the frequency of Latin liturgical prayers and book-learning from the classical world in medical remedies, as well as their manuscript contents, point to clerical usage and writing (see especially contributions by Karen Jolly, e.g. 1996, 2005, 2006). Bede evidently had access to some medical texts, and reports that Theodore, Archbishop of Canterbury, had given advice about the correct days for blood-letting; it is possible that in the seventh-century medicine was taught in the school at Canterbury (Cameron 1993: 27–29, Bede, Ecclesiastical History V.3.2, ll. 14–18, ed. Lapidge, Monat, and Robin 2005: III.24). The works of Rabanus Maurus, a Carolingian author, likewise indicate the perceived importance of medical knowledge for priests, although his concern for the soul seems to have outweighed any desire to treat the body (Paxton 1995).

Once again Ælfwine's prayerbook is instructive: it includes several lunaria for blood-letting, one for illness and a recipe for boils (Cotton Titus D.xxvii ff. 2r and 22r–23r, Cotton Titus D.xxvi ff. 6r–v, 8r–9r, 17r, ed. Günzel 1993: 89, 110–111, 146–147, 148–149, 157). The lunaria for blood-letting identify which days are recommended for the practice, or should be avoided, according to the day of the moon; the lunarium for sickness similarly indicates the likely progression of the sick man's illness according to the day of the moon. It is not clear whether Ælfwine was supposed to be responsible for letting blood, or only for the knowledge of the appropriate days on which this could be done: how precisely Ælfwine used his prayerbook in his role as dean is not easy to pin down (Günzel 1993: 3–4). The recipe for boils is interesting too, not least because of the instruction that unconsecrated oil was to be used in its preparation. It might be expected that consecrated oil would be required, since it was powerful in the rite for the healing of the sick; nevertheless the distinction between consecrated and unconsecrated oil suggests a writer and/or user who was able to distinguish between the two, and perhaps had access to both. The extravagance of nine egg yolks (with the whites supposedly thrown away) suggests a context for production where resources were hardly scarce.

Recipes like that in Ælfwine's prayerbook for the treatment of boils approach the problem of disease by treating symptoms in a way which applies a (supposedly efficacious) concoction to various parts of the body, but do not state what the causes of boils were believed to be or precisely how the preparation was supposed to help (Cameron 1993: 40–41, Meaney 1992). Although it is obvious that the Anglo-Saxons did not conceptualise the causes of sickness and disease in a way like that of modern medicine, it is not always easy to match up the many different types of suggested causes in a way which clarifies which types of causes might overlap, and which were distinct. In some cases it seems that the recipes given would have been effective, such as the eye salve for a stye which included onion and garlic (antibiotic properties), bull's gall (detergent properties) and wine (acidic, reacts with the brass vessel specified in the recipe to produce copper salts, which are cyotoxic) (Cameron 1993: 120–121, see also 117–129 on 'rational medicine'). In other cases, such as the remedies for burns which require a selection of herbs (or goat droppings, in one case) boiled in butter or lamb fat, they must have been ineffective and were probably also painful (e.g. Bald's Leechbook, I.60, ed. Cockayne 1864: 130–132).

The rites of the sick assume a demonic cause for disease, and their curative approach thus includes apotropaic signs to keep devils away, as well as asking for angelic protection; they also assume that if the health of the soul is addressed, the health of the body may also be restored. Extra-liturgical efforts to combat bodily troubles also sometimes refer to the agency of devils, such as charms targeting particular devils with specific responsibility for certain diseases, or seeking the help of particular angels in healing. An eleventh-century copy of an Old English translation of Bede's *Ecclesiastical History* (now Cambridge, Corpus Christi College, 41) was used in the mid-eleventh century to record charms and homilies in the margins (Bredehoft 2006: 722–723, Grant 1978: 1), one of which blames 'Lanielum', an evil angel (i.e. a devil), as the cause of stomach aches (Johnson 1998: 67–68). Once again, harnessing the power of the angels was at least as important as repelling demons, and the same charm in CCCC 41 also records that 'Dormielo' was the benevolent angel to whom one should turn for the restoration of health (Johnson 1998: 67–68), while Raphael was another (more orthodox) angel whose help was sought for sickness (as mentioned above).

The physical effects on the body of these immaterial agents of good and evil required appropriate spiritual knowledge so that their powers could be combatted or used effectively. But although the narrative literature and rites of the Church tend to attribute sickness either to demonic influence, or to its imposition as a punishment by God, it is difficult to ascertain whether other approaches to healing similarly assumed that devils were at the root of all or even most sickness. Since very many medical remedies do not mention the causes of the ailments they purport to cure, it is not always easy to understand how the Anglo-Saxons themselves understood disease. Some 'cures' in medical books have a close affinity with rites for the sick, such as the directions for a 'sudden sickness' which require the sign of the cross to be made on the tongue, on the head, and on the chest (Bald's Leechbook, I.47, ed. Cockayne 1864: II.116–117, Jolly 2005: 229–230). Other remedies involve combinations of plant preparations and prayers or snippets from the liturgy or even masses, or the use of church furniture, for example placing herbs under an altar (Jolly 2005: 231–226).

It is possible that some of these cures were intended to combat demons, but there were other supernatural agents which could cause disease, such as elves, who are the focus of some recipes. Elves could also merge with demons or be seen as distinct from them, although they were perhaps not considered to be invisible and ever-present in the same way that demons were (Hall 2007: 67–68, 71–72, 105–106, 121–122, Jolly 1996: 133–145, Jolly 1998); some of the cures for 'elf-ailments' are those which involve extracts from liturgical rituals (Jolly 2005: 231–233). But it seems that there were some ailments which could be caused either by elves or by other (unknown) causes: a remedy for an ailment in horses was supposed to be efficacious whether elves were involved or not (Bald's Leechbook II.65, see Hall 2007: 98–156, esp. 98–102). From the surviving evidence, it is difficult to establish how far the Anglo-Saxons themselves saw elves as responsible for disease, or whether they might occasionally have had a positive role (Hall 2007: 148–155).

Disease was not only understood to be caused by what seem, from a modern perspective, to be supernatural agents. Some remedies attest to a belief that the body was vulnerable to attack from other types of creatures such as worms or insects: tooth decay was believed to be caused by 'tooth worms', and ear ache or ear problems by earwigs or other types of worms (Cameron 1993: 11–13). The modern distinction between natural and supernatural seems not to have been made by the Anglo-Saxons, however, who instead saw a division between themselves and entities which were 'external to humanity' (Neville 1999: 2–3). This division into human and non-human is important, because it suggests that although in some contexts distinctions might be drawn between different types of non-human beings which might attack the body, in some respects creatures like worms were akin to elves or devils. In fact, in Anglo-Saxon texts worms are described as part of (and the word *wyrm* is used to describe) a category of beings which included dragons and serpents, themselves sometimes understood to be agents or images of the devil (Thompson 2004: 132–137). This is significant in understanding how Anglo-Saxons conceptualised the various different types of threat to the body which came in the form of sickness and disease, because it suggests that the differences perceived by modern audiences between these agents of bodily affliction were not necessarily understood in the same way by their early medieval counterparts. The sign of the cross, a prayer, or a preparation against worms might all be effective ways of protecting against or healing a number of different types of ailment or sickness if the causes were considered to be contiguous.

Dying, death and burial

If neither physical remedies nor the spiritual healing of the Church was effective in eliminating bodily sickness, the sick person and his close relatives or companions had to consider the protection of body and soul in death. Bodily sickness before death could actually be beneficial to the soul, taking the place of the suffering which might otherwise apply only to the soul in the interim between death and judgement. Bede relates that St Cuthbert, knowing that the end of his life was near, informed his friend Hereberht of this when they met: Hereberht then asked Cuthbert to pray that they would both leave the earth at the same time and enter heaven together. This happened as Hereberht had requested, but Bede records that Hereberht suffered a long illness before he died, so that in case he was in any way inferior to Cuthbert, he would still be able to enter heaven at the same time because of the chastising pain he had endured (Ecclesiastical History IV.27, ed. Lapidge, Monat, and Robin 2005: II.368–374). Ælfric too noted that some men suffer in this life so that they will not face unending torment in the next (Catholic Homilies, II.19, ll. 259–262, ed. Clemoes 1997: 188).

When performed for those who were near death, the ritual for the anointing of the sick took on a new urgency, and rites for the dying were used in addition to the anointing. All those who were anointed were offered communion, which was especially important if the sick person was not expected to recover. It had been customary for centuries that the dying were offered the viaticum, and it was considered to be a vital part of the death rituals: even those undertaking penance which would normally have prohibited them from receiving the Eucharist were allowed to receive it if their lives were in danger, and statements to this effect were recorded in late Anglo-Saxon books (London, British Library, Cotton Nero A.i, f. 129v, ed. Cross and Hamer 1999: 120–121, Old English Penitential, ed. Frantzen 2008: SXY 41.10.01). But like anointing, the viaticum does not seem always to have been well understood, and Ælfric had to warn that it could only be given to sick men while they are still properly alive, since it had to be eaten (Pastoral Letters, I.84–6, III.11, ed. Fehr 1914: 19–20, 150–151). Such a placement of the viaticum in a dead man's mouth seems to treat it more like a protective talisman than as a sacrament which united the believer with Christ.

Like the rites for the sick, the funerary liturgy alludes to the possibility of demonic attack on the soul: it was widely believed that angels and demons fought over the soul as it left the body, the winners taking it to the place to which it belonged by right. In life these demons attacked the unity of body and soul which made up each individual human being, but in death their threat to the soul alone was of much greater concern, and the body is much less frequently mentioned in written sources as a cause for concern. The role of angels in defeating demons was so important that funeral liturgies contain repeated requests for their help (Keck 1998: 204–205). An anonymous eleventh-century homily depicts these struggles in vivid detail, describing the fights which resulted in a bad soul dragged off to hell by demons after all the sins he had committed were listed by them, and a good soul rescued by angels after they enumerated his good deeds (Hom. XLVI, ed. Napier 1883: 235, ll. 9–13 and 236, ll. 1–6). This struggle could be played out in a number of ways, as an account of a nobleman in Bede's *Ecclesiastical History* indicates (V.13, ed. Lapidge, Monat, and Robin 2005: III.84–90).

On his death-bed, this thegn was visited by angels who showed him his good deeds listed in a very small and beautiful book. After this, the whole house was filled with an army of wicked spirits, and the leader of the troop presented the thegn with a huge, black, horrible-looking book which contained all his wicked deeds. This devil asked the angels why they were present since they knew that the man belonged to the troop of the damned, and agreeing, the angels left. Two of the wicked spirits then struck the man with ploughshares, one on the head and one on the foot, whereupon he felt the strokes penetrating through the innards of his body, knowing that when they met, he would die. Here demonic action apparently resulted in the man's death, but for Bede the knowledge of the soul's eternal death in hell, rather than the bodily death of the individual, was far more important.

But since the body was believed to rise again after the Last Judgement, the fate of the soul was also important in determining the ultimate fate of the body, just as only the souls of the wicked were punished before the Judgement, and after it the wicked would suffer in both soul and body. In a number of depictions of hell, such as those in an illustrated manuscript of Old English poetry, the bodies of the damned are whipped by Satan and his minions (Oxford, Bodleian Library, Junius 11, pp. 3, 20, 36). An image of the Trinity in Ælfwine's prayerbook (Fig. 6.2) includes a depiction of Judas (the betrayer of Christ) and Arrius (a heretic) bound up in hell below the Trinity, with Satan under the feet of God, while a hell-mouth threatens to devour them all (Cotton Titus D.xxvii, f. 75v, for discussion of this illustration see Karkov 2006: 106–107, Karkov 2010: 111–113, Raw 1997: 182–185).

The *Liber Vitae* of the New Minster, Winchester (now London, British Library, Stowe 944), contains a drawing (on f. 7r, Fig. 6.3) which illustrates precisely the fate of the bodies of the damned compared to those of the just: while Peter welcomes the elect into heaven with celebrations, the damned are locked into hell by an angel and their bodies are dragged by wicked spirits into the gaping hell-mouth. The fight between angels and demons is illustrated dramatically in the middle register of the judgement scene, where a boy is tugged from the one side by a devil and from the other by St Peter, who is accompanied by an angel. Both the angel and the devil hold books which presumably record the boy's deeds, and St Peter assists in the spiritual struggle by helpfully smashing the devil in the nose with an extremely large key, his symbol as a saint and the key to heaven which he uses in the upper register of the picture (for detailed discussion of this illustration, see Karkov 2003b).

In these images, the bodies of the wicked are almost all naked, illustrating their vulnerability and emphasising their torment, while the bodies of the just are beautifully clothed (see also Karkov 2003a). Elsewhere when the resurrected bodies of the just were described, writers affirmed that the bodies of the martyrs would retain their injuries, and that these would be signs of glory in heaven. No matter how the body had been destroyed, it would be resurrected whole at Judgement, either for punishment or for reward; the fate of the body at death was theoretically irrelevant (see Bynum 1995; for an in-depth treatment of this topic). Monica, the mother of Augustine of Hippo, seems to have taken this to heart and on her death-bed apparently informed her sons that she did not care where she was buried, asking only to be remembered at the altar, in the Mass (Conf. IX.xi.27, ed. Verheijen 1981: 149). But Monica was unusual in late Antiquity and the early Middle Ages, since from very early on, burial of the body in the vicinity of saints' relics was desirable, and considered to be beneficial in some way to the soul (McLaughlin 1994: 30–31).

This practice continued to be popular at least into the fifteenth century, even though Augustine (amongst others) had concluded that there was no real benefit to the souls

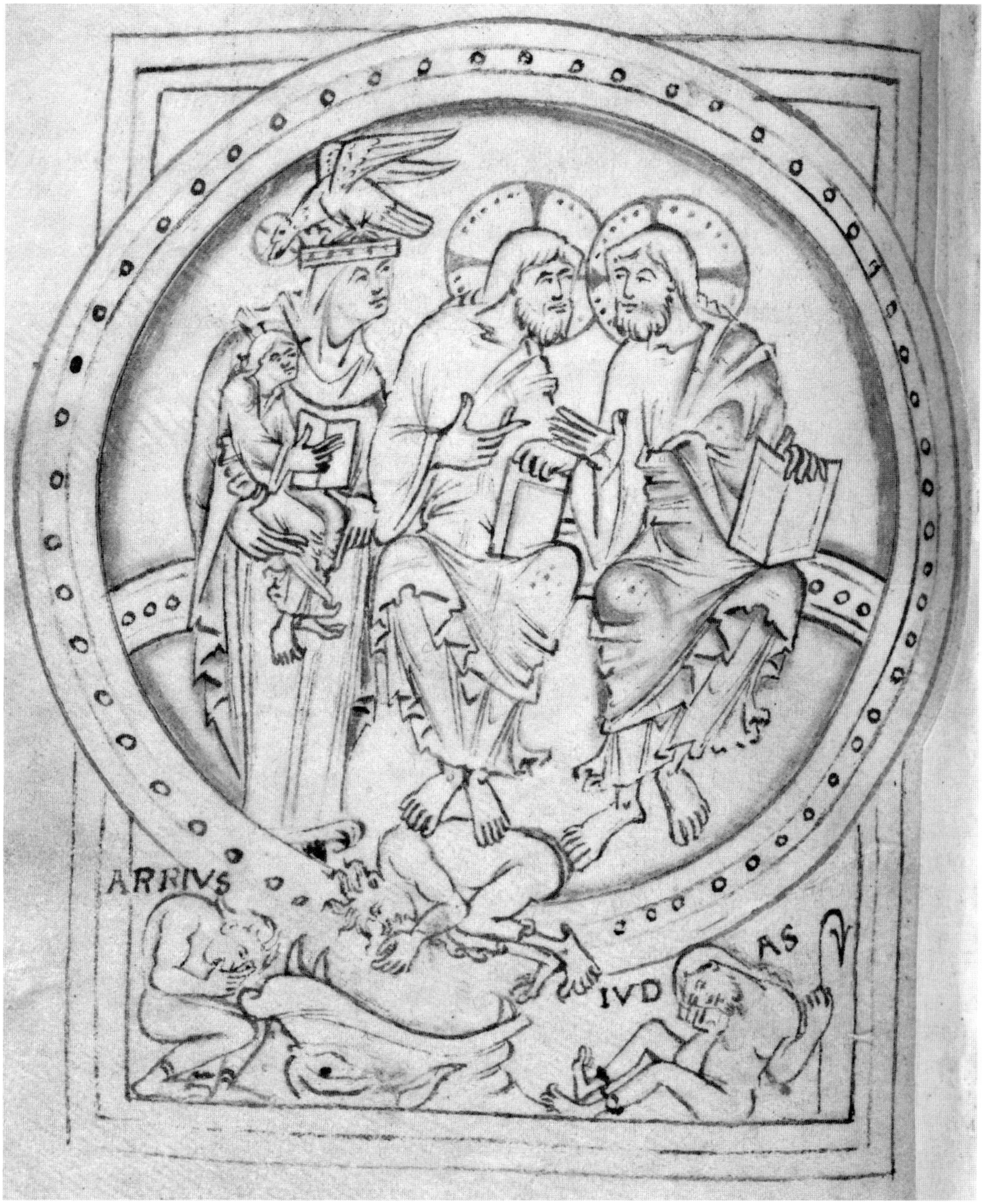

Fig. 6.2: The 'Quinity' (i.e. the Trinity plus Mary and Child) from Ælfwine's prayerbook (London, British Library, Cotton Titus D.xxvii, f. 75v. Reproduced with permission from the British Library).

whose bodies were buried near the saints (De cura pro mortuis gerenda, I.1, ed. Zycha 1900: 621–622, cf. Schmid 2002: 25). The prestige associated with burial near the saints can be seen from Anglo-Saxon and later cemeteries such as that of the Old Minster, in Winchester, where a number of the richest inhumations were in some way associated with the resting place of Swithun, Winchester's most famous saint, or were inside the church and so nearest to the relics of the saints (Biddle 1969: 321–322, Biddle 1970: 320, Kjølbye-Biddle 1992: 223–231). The stress placed on the location of the body hints that this was believed to have a significant bearing on the afterlife, despite Augustine's conclusions. But burial practices are often difficult to interpret, and it should be remembered that not all practices which seem peculiar or inexplicable to modern sensibilities should necessarily be assumed to be salvific or salutory in some way.

A good example is found in the rituals for the sick and dying in the Red Book of Darley, and in a late Anglo-Saxon book of penitential and other liturgical rites (now Oxford, Bodleian Library, Laud misc. 482). These include instructions that linen gloves and socks should be placed on the hands and feet of the sick person when he is anointed, and that if he dies wearing them, they should be worn in the grave (CCCC 422, pp. 412–13; Laud misc. 482, ff. 58b–59a). Victoria Thompson suggests that these gloves and socks were intended as 'an attempt to affect the dead person's prospects of salvation', although it is difficult to find evidence for this (2004: 108). It is difficult to know how to interpret the socks and gloves, although an interesting comparison is found in the anonymous Life of Cuthbert, where the author records that Cuthbert's body was dressed in priestly robes, and wearing shoes in preparation for meeting Christ (IV.13, ed. Colgrave 1940: 130). In his reworking of the Life, Bede omits this detail, although it is not clear why he did so. It may simply be that covering the hands and feet was considered part of a decent burial, since the nails appear to continue to grow in death, and evidence from the monastic customaries also shows a concern to cover hands and feet (Letter to the Monks of Eynsham, 66, ed. Jones 1998: 142, Lanfranc, Constitutiones, 122, ed. Knowles and Brooke 2002: 178–184, Regularis Concordia, 66, ed. Symons 1953: 65).

The careful treatment of the body is evident too in the instructions of funeral liturgies that the dead body is to be guarded and accompanied by prayers and psalm-singing until the burial takes place (e.g. Ordo XLIX, ed. Andrieu 1931–61: IV.529–530, Regularis Concordia, 66, ed. Symons 1953: 65). This may have been partly in order to keep demons away, connected with a belief that the soul might linger near the corpse (cf. Brown 1981: 223 and n.3, see also Paravicini Bagliani 2001). The return of the soul to the body is described in poems and homiletic literature: a good soul praises its body for its good deeds, which will lead to heaven, while a wicked soul chastises its body for its wicked deeds, which can only lead to hell-torment (e.g. Moffat 1990). The body of the wicked soul is described as rotting in graphic and disgusting terms, and is insulted by the wicked soul, who calls him 'foulness of earth', and 'meal for worms' (Moffat 1990: 50, ll. 18, 22). The interpretation of the theology of the poem is complex, and it has been suggested that the disturbing description of decay is indicative of the moral decay of the soul, feared by the living, and in contrast to the incorrupt bodies of saints (Thompson 2002: 234–238, Thompson 2004: 52, 140).

This is difficult though because the body of the good soul also rots, and the good soul consoles its body with the thought of what is to come after judgement (Moffat 1990: 62–4, ll. 127–167). In fact, it is clear that all bodies were expected to rot, regardless of the state of the souls which had inhabited them. Bede notes that when the monks translated the body of Cuthbert, intending to honour him by moving it to a new coffin above ground, they expected to find his bones, with the flesh reduced to dust (Ecclesiastical History, IV.28.1, ll. 11–19, ed. Lapidge, Monat, and Robin 2005: III.374–376). This is precisely why incorrupt bodies were so miraculous: even the bodies of those who were widely acknowledged as good were assumed to be subject to normal processes of decay, and the wide circulation of bones of saints' bodies indicates that although incorruption might be a sign of sanctity, it was not a prerequisite. Similarly, a number of homilies refer to the body waiting for the judgement in the grave, without suggesting fear of the grave, only an expectation that it will house the body until the time of resurrection (e.g. Blickling Homily X, ed. Morris 1874: 109).

This did not mean that people genuinely did not care about how or where the body was buried. From at least the tenth century, cemeteries appear to have been consecrated, and from this time also appear laws which use as a penalty the exclusion of the bodies of criminals from consecrated ground. In the tenth and eleventh centuries criminals were often treated violently, being executed or if allowed to live, their bodies mutilated (Gittos 2002, O'Brien O'Keefe 1998, Reynolds 2002, Reynolds 2009). This indicates precisely how people did *not* want their bodies to be treated in death. The importance of burial in consecrated ground to the relatives of an individual as well as to the individual himself is highlighted by a decree which envisages a situation in which relatives wish to clear a dead man and bury him elsewhere (III Æthelred 7–7.1, ed. Liebermann 1903: 230).The decapitation of criminals and the display of the severed heads which seems to have occurred in some cases served to highlight the bodily effects of misdemeanours (see for example Buckberry and Hadley 2007: 316–325). This is also true of other means of justice in Anglo-Saxon England such as mutilation or ordeals, which required that someone accused of a crime should carry a hot iron bar, take a stone from a cauldron of boiling water, be thrown into deep water to see if he/she floated or sank, or (bizarrely) be required to eat consecrated bread and cheese without vomiting them up; in these ways the body might display signs of guilt, or be used to try to identify them (Bartlett 1986, see also Foxhall Forbes forthcoming, Keefer 1998, O'Brien O'Keefe 1998, Richards 2003). Protecting the body in death as well as in life therefore involved staying within the bounds of the secular law, just as, from the Church's perspective, protecting body and soul required following its commandments and precepts.

Whether or not the dismemberment of criminal bodies was intended (or able) to harm the soul, the exclusion from consecrated ground did not allow their bodies to be placed in the cemeteries of major churches which were considered to be particularly desirable as burial places. It is clear that care was taken in the treatment of dead bodies (Hadley and Buckberry 2005), and one particularly interesting example is found in an elaborate funeral liturgy from the Old Minster in

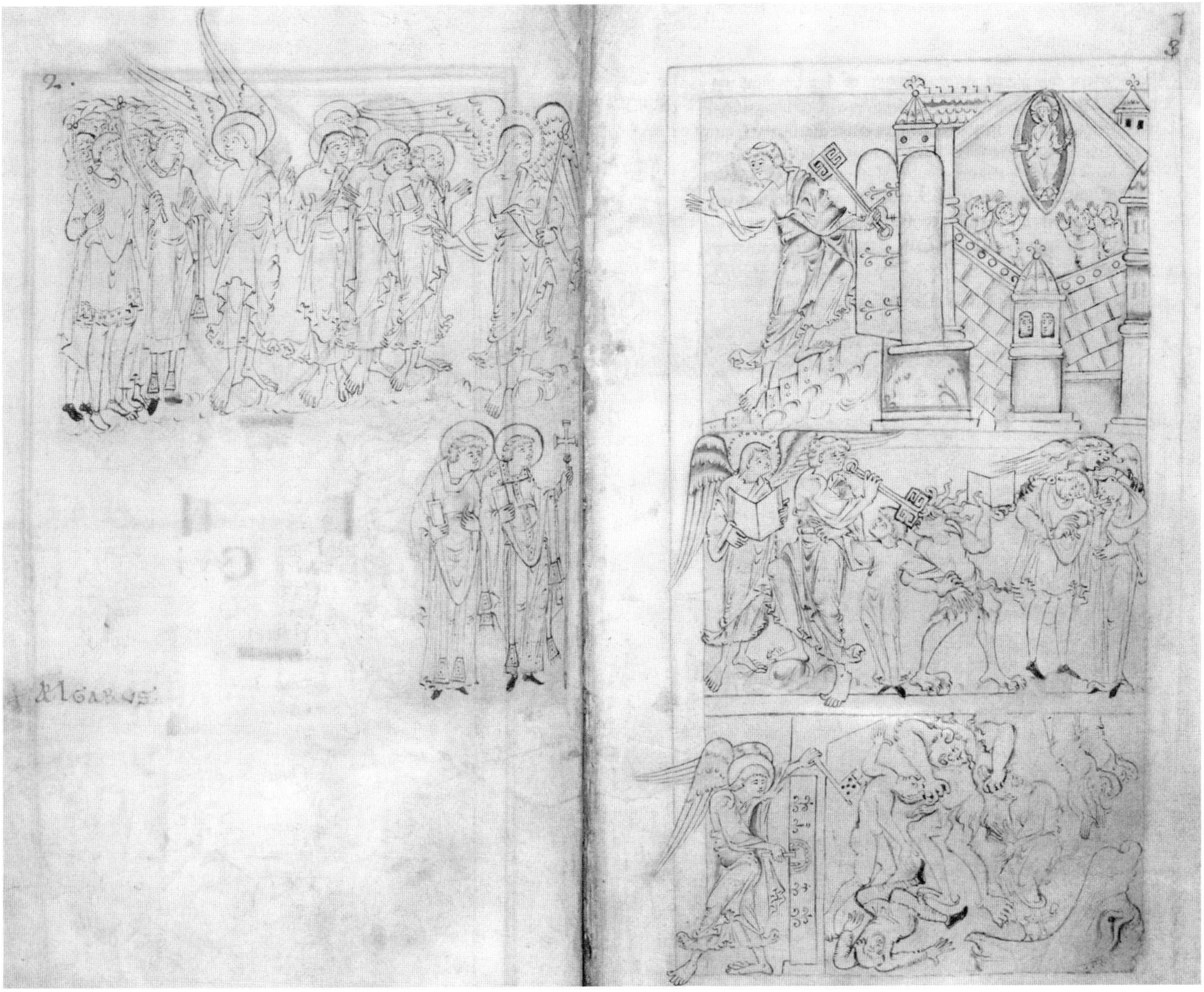

Fig. 6.3: The Last Judgement from the Liber Vitae of the New Minster, Winchester (London, British Library, Stowe 944, ff. 6v–7r). Reproduced with permission from the British Library.

Winchester, which includes instructions that after the funeral mass, the body was to be carried from the Old Minster around a number of churches in the monastic precinct, including the New Minster, before burial in the cemetery (Worcester, Cathedral Library, F.173, f. 21a–b, partially ed. Turner 1916). Throughout the procession, the saints of the churches are called on to intercede for the soul of the person whose office of burial is celebrated. The importance of the saints after death, and the desire to be buried near them, is at least one of the reasons why so many individuals in late Anglo-Saxon England arranged for burial in the cemeteries of major churches, but this is not the only reason, since the prayers and commemorations associated with burial in major cemeteries were evidently also extremely important (for specific examples, see Foxhall Forbes 2009: 189–204). Burial in a major church might speed the soul on its way to heaven, but it would also ensure that the body would be with other christian bodies and hopefully would be resurrected with them.

The personal role that could be played in the protection of the body is clear from Ælfwine's prayerbook. One prayer included in the book asks God to give 'the seat of refreshment to the souls of all the faithful whose bodies rest here' (Cotton Titus D.xxvi, ed. f. 64r, Günzel 1993: 188). Other prayers ask for forgiveness and mercy to be given to 'the souls of all the faithful, of our brothers and sisters and parents and friends, and those whose bodies rest in all the cemeteries', and for the remission of sins 'of my father and mother and of all brothers and sisters' (Cotton Titus D.xxvi, ff. 64r–v and 71v–74r, ed. Günzel 1993: 189, 193–194). These are particularly significant, since Ælfwine's prayerbook records the obits of his parents and siblings, and it can be assumed that he prayed for them, sought to protect their souls and bodies as he did his own, in the hope that they would all be reunited in heaven, guarded by angels and safe from demons, and beautifully adorned and called in by St Peter like those in the New Minster's *Liber Vitae*, a book which Ælfwine himself helped to create.

Conclusion

For an early medieval Christian, especially one so connected with the rites of the Church as was Ælfwine, the health and fate of the soul may in the end have been of more concern than the body, and this is evidently a major concern in his prayerbook (Karkov 2010). But body and soul were not separated so easily as the works of some early medieval authors might suggest, and personal protection of the body was not at all unimportant. One of Ælfwine's devotions to the cross in his prayerbook names his body parts in turn, asking protection from the cross for each one: he lists the head, the eyes, the mouth, the hands and arms, the heart and the innards, the feet and all the limbs – for which he specifically requests protection against the attacks of the devil – and finally, he asks protection for his soul, and prays that God may free it from all his adversaries (Cotton Titus D.xxvii, f. 73r–v, ed. Günzel 1993: 127–128). Catherine Karkov suggests that while praying before the crucifix, 'the supplicant would have traced the body of the Lord with his (or her) mind', and that the body of the Lord becomes united with the body of the supplicant, here Ælfwine, by the focus on the crucified body as well as on the supplicant's own body, for which he asks protection from the cross (Karkov 2010: 108–109). The body could be the medium for sin, as prayers in Ælfwine's book asking to be freed from carnal desires and worldly vanities bear witness (Cotton Titus D.xxvi, ff. 59v–60r, ed. Günzel 1993: 186), but good deeds, including prayer, were also performed bodily and involved the body as a focus for devotion.

Even if prayer comes from the soul, the mind, or the heart, the living Christian still prays bodily, because it is the hands which are needed to hold the book, the eyes which read the prayer, the lips which pronounce the words. Ælfwine's body is represented in his book at the feet of St Peter (Fig. 6.4), where he is shown holding a book, perhaps his own book which he used for prayer (Cotton Titus D.xxvi, f. 19v). Ælfwine's stance here, looking up at St Peter, is reminiscent of a prayer in his book which begins 'I turn my face to you, Lord, I commend my body and soul to your majesty' (Cotton Titus D.xxvi, ff. 71v–72r, ed. Günzel 1993: 193). The psalms too indicate the importance of gesture in prayer, such as the psalm for the beginning of vespers: 'may the lifting up of my hands be an evening sacrifice' (Ps. 140:2, Karkov 2010: 109) Another prayer in Ælfwine's book illustrates the movements in making the sign of the cross which accompany the act of praying: 'save us, Christ ... who suffered on the cross, whose sign we impress on our foreheads and in our hearts' (Cotton Titus D.xxvi, f. 70r, ed. Günzel 1993: 192). The images in Ælfwine's book are intended to draw him towards God and to help him to achieve salvation just as the prayers in the book are, and they illustrate the active ways in which devotion could be expressed, mirroring the mental and invisible activities of prayer (Karkov 2006: 109–111).

The belief in the omnipresence of demons and their powers to cause physical harm or damage meant that protection of the body was considered to be essential even if the soul was ultimately more of a concern. Some bodies were evidently considered particularly vulnerable, such as those of unbaptised children or the sick, but demons could strike at any time, as is illustrated in a vivid account from the *Dialogues* of Gregory the Great, translated into English in the ninth century (I.4, ed. Hecht 1900: 30). A nun, wanting to eat some lettuce, forgot to bless it with the sign of the cross, and a demon immediately possessed her. When an abbot, called to exorcise the demon, ordered him to leave, the demon responded that it was no fault of his, since he had simply been sitting on the lettuce when the nun had bitten him. The sign of the cross with which the priest sealed Ælfwine's body at his baptism was remade by his own fingers probably hundreds of times every day, marking his body with an invisible sign which was visible to supernatural beings, and which served to protect both body and soul from their attacks. The nun's failure to protect her body by making the sign of the cross resulted in demonic possession, which in turn endangered her soul, housed inside the body.

Protecting the body required recourse to the sacraments of the Church such as baptism and the anointing of the sick, as well as to charms, medical recipes and prayers, and in death to careful location of the body and continued prayer. Bad bodies might be isolated from good bodies in death as in life, by exclusion from consecrated cemeteries, but good and bad bodies alike were subject to the processes of decay, just as they were subject to the attacks of demons or other agents, or struck by disease, even if this was visited upon them for different reasons. Recent work on the Holy Cross in Anglo-Saxon England has emphasised the importance of the cross in keeping away demons and malign influences (Jolly, Karkov, and Keefer 2008, see the essays in Karkov, Keefer, and Jolly 2006, Keefer, Jolly, and Karkov 2010), but the opposing powers of angels and how these were harnessed has been much less explored (for treatment of this subject see Foxhall Forbes forthcoming). It is clear, however, that the spiritual knowledge of how to appropriate benevolent powers was as important as the knowledge of how to defend against malevolent ones in the protection of body and soul.

As the centuries went on, guardian angels which stayed near the person and watched over him or her became increasingly important in the fight to protect body and soul (Keck 1998: 64, 163–164, 209), as a prayer to a guardian angel written into Ælfwine's prayerbook by a woman in the twelfth-century

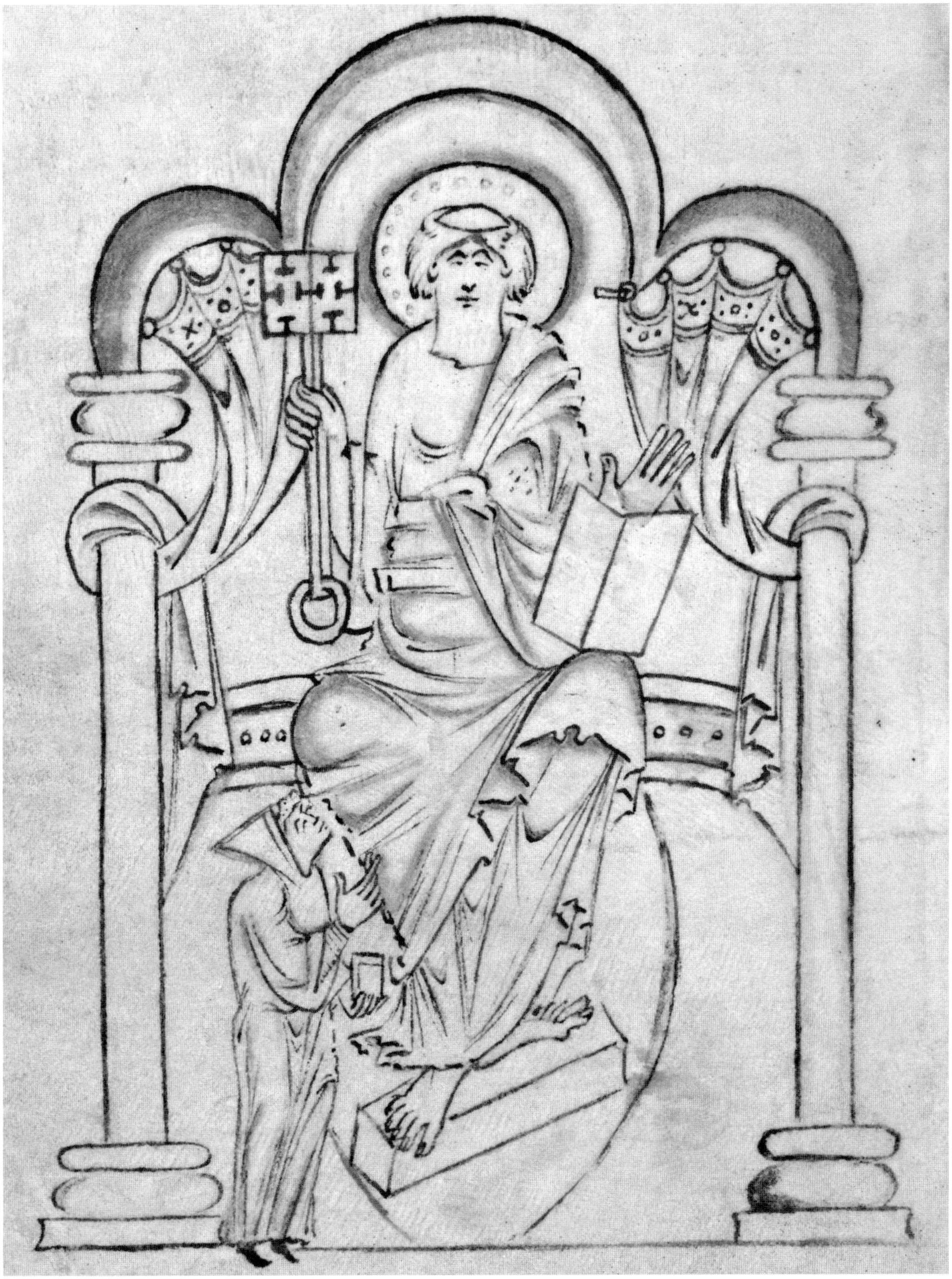

Fig. 6.4: St Peter and Ælfwine from Ælfwine's prayerbook (London, British Library, Cotton Titus D.xxvi, f. 19v). Reproduced with permission from the British Library.

indicates. By the thirteenth century, angelology had become part of the formal curriculum at the Universities (Keck 1998: 3); before this, ideas about angels were less formal and more fluid. An Anglo-Saxon such as Ælfwine might not have thought of asking how many angels could dance on the head of a pin, but he certainly would have wanted to be reassured that they surrounded him at all times, dancing or otherwise.

References

Andrieu, M. (ed.) 1931–61. *Les "Ordines Romani" de haut moyen âge*. Louvain.

Bartlett, R. 1986. *Trial by fire and water: the medieval judicial ordeal*. Oxford: Clarendon Press.

Bethurum, D. (ed.) 1957. *The Homilies of Wulfstan*. Oxford: Oxford University Press.

Biddle, M. 1969. Excavations at Winchester 1968. Seventh Interim Report. *Antiquaries Journal* 49: 295–329.

Biddle, M. 1970. Excavations at Winchester 1968. Seventh Interim Report. *Antiquaries Journal* 50: 277–326.

Bredehoft, T. A. 2006. Filling the margins of CCC 41: textual space and a developing archive. *Review of English Studies* 57, 232: 721–32.

Brown, E. A. R. 1981. Death and the Human Body in the Later Middle Ages: the Legislation of Boniface VIII on the Division of the Corpse. *Viator* 12: 221–70.

Buckberry, J. L., and D. M. Hadley. 2007. An Anglo-Saxon Execution Cemetery at Walkington Wold, Yorkshire. *Oxford Journal of Archaeology* 26, 3: 309–29.

Bynum, C. W. 1995. *The Resurrection of the body in Western Christianity, 200–1336. Lectures on the history of religions*. New York: Columbia University Press.

Cameron, M. L. 1993. *Anglo-Saxon medicine*. Cambridge: Cambridge University Press.

Clemoes, P. (ed.) 1997. *Ælfric's Catholic Homilies: the First Series. Catholic Homilies I*. Oxford: Oxford University Press for the Early English Text Society.

Cockayne, O. 1864. *Leechdoms, wortcunning and starcraft of early England; being a collection of documents, for the most part never before printed, illustrating the history of science in this country before the Norman Conquest. 3 vols*.

Colgrave, B. 1940. *Two lives of Saint Cuthbert: a life by an anonymous monk of Lindisfarne and Bede's prose life*. Cambridge: Cambridge University Press.

Crawford, S. 2009. *Daily life in Anglo-Saxon England*. Oxford: Greenwood World.

Cross, J. E., and A. Hamer. (ed.) 1999. *Wulfstan's canon law collection*. Vol. 1. Cambridge: D.S. Brewer.

Daniélou, J. 1953. *Les anges et leur mission: d'après les Pères de l'Eglise. Collection Irenikon, n.s. 5*. Paris: Chevotogne.

Dümmler, E. L. (ed.) 1895. *Epistolae Karolini Aevi*. Berlin: Weidmann.

Fadda, A. M. L. (ed.) 1977. *Nuove omelie anglosassoni della rinascenza benedettina*. Firenze: F. Le Monnier.

Fehr, B. (ed.) 1914. *Die Hirtenbriefe Ælfrics in altenglischer und lateinischer Fassung, Bibliothek der Angelsaechsischen Prosa IX. Pastoral Letters*. Hamburg: Verlag von Henri Grand.

Foot, S. 1992. "'By water in the spirit': the administration of baptism in early Anglo-Saxon England," in J. Blair and R. Sharpe (eds) *Pastoral care before the parish*. 171–92. Leicester: Leicester University Press.

Foxhall Forbes, H. 2009. The development of the notions of penance, purgatory and the afterlife in Anglo-Saxon England, PhD disertation, University of Cambridge.

Foxhall Forbes, H. forthcoming. *Heaven and Earth in Anglo-Saxon England: theology and society in an age of faith. Studies in Early Medieval Britain*. Farnham: Ashgate.

Frantzen, A. J. 1983. *The Literature of Penance in Anglo-Saxon England*. New Brunswick, N.J.: Rutgers University Press.

Frantzen, A. J. (ed.) 2008. *The Anglo-Saxon Penitentials: a Cultural Database*.

Gittos, H. 2002. "Creating the Sacred: Anglo-Saxon Rites for Consecrating Cemeteries," in S. Lucy and A. Reynolds (eds) *Burial in Early Medieval England and Wales*. 195–208. London: Society for Medieval Archaeology.

Graham, T. 1993. The Old English Liturgical Directions in Corpus Christi College, Cambridge, MS 422. *Anglia* 111: 439–46.

Grant, R. J. S. 1978. *Cambridge, Corpus Christi College 41: The loricas and the missal. Costerus. New series 17*. Amsterdam: Rodopi.

Günzel, B. 1993. *Ælfwine's prayerbook (London, British Library, Cotton Titus D.xxvi + xxvii)*. London: Henry Bradshaw Society.

Hadley, D. M., and J. Buckberry. 2005. "Caring for the dead in late Anglo-Saxon England," in F. Tinti (ed.) *Pastoral care in late Anglo-Saxon England, Anglo-Saxon studies 6*. 121–147. Woodbridge: Boydell and Brewer.

Hall, A. 2007. *Elves in Anglo-Saxon England: matters of belief, health, gender and identity*. Woodbridge: Boydell Press.

Hecht, H. 1900. *Bischofs Waerferth von Worcester Übersetzung der Dialoge Gregors des Grossen über das Leben und die Wunderthaten italienischer Väter und über die Unsterblichkeit der Seelen. Bibliothek der angelsächsischen Prosa, 5. Bd.* Leipzig: G. H. Wigland.

Hill, T. D. 1985. When God Blew Satan out of Heaven: The Motif of Exsufflation in Vercelli Homily XIX and Later English Literature. *Leeds Studies in English*: 132–41.

Hurst, D. 1955. *Bedae venerabilis opera, pars III: opera homiletica*. CCSL 112. Turnhout: Brepols.

Hurst, D. 1960. *Opera exegetica 3: In Lucam evangelio expositio. In Marcam evangelio expositio*. Corpus Christianorum Series Latina. Turnhout: Brepols.

Johnson, D. F. 2006. "The Crux usualis as apotropaic weapon in Anglo-Saxon England," in E. K. Catherine, K. Sarah Larratt, and J. Karen Louise (eds) *The Place of the Cross in Anglo-Saxon England*. 80. Woodbridge: The Boydell Press.

Johnson, R. F. 1998. Archangel in the Margins: St. Michael in the Homilies of Cambridge, Corpus Christi College 41. *Traditio* 53: 63–91.

Jolly, K. L. 1996. *Popular religion in late Saxon England: elf charms in context*. Chapel Hill (NC): University of North Carolina Press.

Jolly, K. L. 1998. "Elves in the psalms? The experience of evil from a cosmic perspective," in A. Ferreiro (ed.) *The devil, heresy and witchraft in the middle ages: Essays in honor of Jeffrey B. Russell*. 19–44. Leiden: Brill.

Jolly, K. L. 2005. "Cross-Referencing Anglo-Saxon Liturgy and Remedies: the Sign of the Cross as Ritual Protection," in H. Gittos and M. B. Bedingfield (eds) *The liturgy of the late Anglo-Saxon church*. 213–43. Woodbridge: Boydell Press.

Jolly, K. L. 2006. "Tapping the power of the cross: Who and for whom?," in K. L. Jolly, C. E. Karkov, and S. L. Keefer (eds) *The place of the cross in Anglo-Saxon England*. 58–79. Woodbridge: Boydell Press.

Jolly, K. L., C. E. Karkov, and S. L. Keefer (eds) 2008. *Cross and culture in Anglo-Saxon England: Studies in honor of George Hardin Brown*. Morgantown, WV: West Virginia University Press.

Jones, C. A. 1998. *Ælfric's Letter to the Monks of Eynsham. Cambridge Series in Anglo-Saxon England*. Cambridge: Cambridge University Press.

Karkov, C. E. 2003a. "Exiles from the kingdom: The naked and the damned in Anglo-Saxon art," in J. Wilcox and B. C. Withers (eds) *Naked before God: Uncovering the body in Anglo-Saxon England*. 181–220. Morgantown, WV: West Virginia University Press.

Karkov, C. E. 2003b. "Judgement and salvation in the New Minster Liber Vitae," in K. Powell and D. G. Scragg (eds) *Apocryphal texts and traditions in Anglo-Saxon England*. 151–63. Woodbridge: D.S. Brewer.

Karkov, C. E. 2006. "Text as Image in Ælfwine's Prayerbook," in H. Magennis and J. Wilcox (eds) *The power of words: Anglo-Saxon studies presented to Donald G. Scragg on his seventieth birthday*. 95–114. Morgantown, WV: West Virginia University Press.

Karkov, C. E. 2010. "Abbot Ælfwine and the Sign of the Cross," in S. L. Keefer, K. L. Jolly, and C. E. Karkov (eds) *Cross and Cruciform in the Anglo-Saxon World: Studies to Honor the Memory of Timothy Reuter* 103–34. Morgantown: West Virginia University Press.

Karkov, C. E., S. L. Keefer, and K. L. Jolly (eds) 2006. *The place of the cross in Anglo-Saxon England*. Woodbridge: Boydell Press.

Keck, D. 1998. *Angels & angelology in the Middle Ages*. Oxford: Oxford University Press.

Keefer, S. L. 1995. "Manuals," in R. W. Pfaff (ed.) *The Liturgical Books of Anglo-Saxon England, Old English newsletter*. 99–109. Kalamazoo, MI: The Medieval Institute Western Michigan University.

Keefer, S. L. 1998. "*Ut in omnibus honorificetur Deus:* The *corsnaed* Ordeal in Anglo-Saxon England," in J. Hill and M. Swan (eds) *The community, the family and the saint: patterns of power in early medieval Europe: selected proceedings of the International Medieval Congress, University of Leeds, 4–7 July 1994, 10–13 July 1995*. 237–64. Turnhout: Brepols.

Keefer, S. L., K. L. Jolly, and C. E. Karkov. 2010. *Cross and Cruciform in the Anglo-Saxon World: Studies to Honor the Memory of Timothy Reuter*. Morgantown: West Virginia University Press.

Kelly, H. A. 1985. *The Devil at Baptism: Ritual, Theology, and Drama*. Ithaca: Cornell University Press.

Keynes, S. 1996. *The Liber vitae of the New Minster and Hyde Abbey, Winchester: British Library Stowe 944: together with leaves from British Library Cotton Vespasian A. VIII and British Library Cotton Titus D. XXVII. Early English manuscripts in facsimile 26*. Copenhagen: Rosenkilde and Bagger.

Kjølbye-Biddle, B. 1992. "Dispersal or concentration: the disposal of the Winchester dead over 2000 years," in S. R. Bassett (ed.) *Death in towns: urban responses to the dying and the dead, 100–1600*. 210–47. Leicester: Leicester University Press.

Knowles, D., and C. N. L. Brooke. (ed.) 2002. *The Monastic Constitutions of Lanfranc*, Rev. by Christopher N. L. Brooke. edition. *Constitutiones*. Oxford: Oxford University Press.

Krapp, G. P. 1932. *The Vercelli book*. New York: Columbia University Press.

Krapp, G. P., and E. V. K. Dobbie. 1936. *The Exeter Book*. New York: Columbia University Press.

Lapidge, M., P. Monat, and P. Robin. (ed.) 2005. *Histoire ecclésiastique du peuple anglais = Historia ecclesiastica gentis Anglorum*. Paris: Editions du Cerf.

Liebermann, F. (ed.) 1903. *Die Gesetze der Angelsachsen. Bd.1, Text und Übersetzung*. Halle: Niemeyer.

Mango, C. A. 1980. "The Invisible World of Good and Evil," in C. A. Mango (ed.) *Byzantium: the empire of New Rome*, Pbk ed edition, *History of civilisation*. xiii, 334 p., [24] p. of plates. London: Weidenfeld and Nicolson.

McLaughlin, M. 1994. *Consorting with saints: prayer for the dead in early medieval France*. Ithaca, NY: Cornell Univ Press.

Meaney, A. L. 1992. "The Anglo-Saxon view of the causes of illness," in S. Campbell, B. Hall, and D. Klausner (eds) *Health, Disease and Healing in Medieval Culture*. 12–33. Basingstoke: Macmillan.

Moffat, D. 1990. *The Old English Soul and Body*. Woodbridge: Brewer.

Morris, R. (ed.) 1874. *The Blickling homilies of the 10th century: from the Marquis of Lothian's unique MS. A.D. 971*. Vol. 58. *Blickling Homilies*. London: Trübner for E.E.T.S.

Napier, A. S. (ed.) 1883. *Wulfstan: Sammlung der ihm zugeschriebenen Homilien nebst Untersuchungen über ihre Echtheit*. Berlin: Weidmann.

Neville, J. 1999. *Representations of the natural world in Old English Poetry. Cambridge Studies in Anglo-Saxon England*, 27. Cambridge: Cambridge University Press.

O'Brien O'Keefe, K. 1998. Body and Law in Late Anglo-Saxon England. *Anglo-Saxon England* 27: 209–32.

Page, R. I. 1978. Old English Liturgical Rubrics in Corpus Christi College, Cambridge, MS 422. *Anglia* 96: 149–58.

Paravicini Bagliani, A. 2001. "The corpse in the Middle Ages: the problem of the division of the body," in P. Linehan and J. L. Nelson (eds) *The Medieval World*. 327–341. London: Routledge.

Paxton, F. S. 1990. *Christianizing Death: the Creation of a Ritual Process in Early Medieval Europe*. Ithaca, N.Y.: Cornell University Press.

Paxton, F. S. 1995. Curing bodies – Curing souls: Hrabanus Maurus, medical education, and the clergy in ninth-century Francia. *Journal of the History of Medicine and Allied Sciences* 50, 2: 230–52.

Pfaff, R. W. 1995. "Massbooks: Sacramentaries and Missals," in R.W. Pfaff (ed.) *The liturgical books of Anglo-Saxon England, Old English newsletter*, 7–34. Kalamazoo, MI: The Medieval Institute Western Michigan University.

Raw, B. 2004. "Pictures: books of the unlearned?," in C. Paul (ed.) *The Christian Tradition in Anglo-Saxon England*. 103–119. Woodbridge, Suffolk: Boydell & Brewer.

Raw, B. C. 1997. *Trinity and incarnation in Anglo-Saxon art and thought. Cambridge studies in Anglo-Saxon England*. Cambridge: Cambridge University Press.

Reynolds, A. J. 2002. "Burials, boundaries and charters," in S. Lucy and A. J. Reynolds (eds) *Burial in early medieval England and Wales*. 71–94. London: Society for Medieval Archaeology.

Reynolds, A. J. 2009. *Anglo-Saxon deviant burial customs*. Oxford: Oxford University Press.

Richards, M. P. 2003. "The body as text in early Anglo-Saxon law," in J. Wilcox and B. C. Withers (eds) *Naked before God: uncovering the body in Anglo-Saxon England*. 97–115. Morgantown, WV: West Virginia University Press.

Schmid, J. 2002. *Et pro remedio animae et pro memoria: bürgerliche repraesentatio in der Cappella Tornabuoni in S. Maria Novella. I Mandorli Bd. 2*. München: Deutscher Kunstverlag.

Skeat, W. W. (ed.) 1966. *Aelfric's lives of saints: being a set of sermons on saints' days formerly observed by the English church*. London: For the Early English Text Society by Oxford University Press.

Symons, T. (ed.) 1953. *Regularis Concordia: the Monastic Agreement of the Monks and Nuns of the English Nation. Regularis Concordia.* London: Thomas Nelson and Sons Ltd.

Tangl, M. (ed.) 1916. *Die Briefe des Heiligen Bonifatius und Lullus.* Berlin: Weidmannsche.

Thompson, V. 2002. "Constructing Salvation: a Homiletic and Penitential Context for Late Anglo-Saxon Burial Practice," in S. Lucy and A. Reynolds (eds) *Burial in Early Medieval England and Wales.* 229–40. London: Society for Medieval Archaeology.

Thompson, V. 2004. *Death and Dying in Later Anglo-Saxon England.* Woodbridge.

Turner, C. H. (ed.) 1916. The Churches at Winchester in the early Eleventh Century. *Journal of Theological Studies* 17: 65–8.

Verheijen, L. (ed.) 1981. *Sancti Augustini Confessionum Libri XIII.* Vol. 27. *Confessiones.* Turnhout: Brepols.

Zycha, J. (ed.) 1900. *Sancti Aureli Augustini De fide et symbolo; De fide et operibus; De agone christiano; De continentia; De bono coniugali; De Sancta virginitate; De bono viduitatis; De adulterinis coniugiis lib. II; De mendacio; Contra mendacium; De opere monachorum; De divinatione daemonum; De cura pro mortuis gerenda; De patientia.* Vol. 41. *De cura mortuis pro gerenda.* Vienna: F. Tempsky.

7. The role of healing in the Jesuit mission to China, 1582–1610

Mary Laven

> A Christian woman was nine months pregnant when her baby died within her belly. Her female neighbours applied various remedies in response to the problem, but began to despair of the health of the woman, since they knew full well from experience that when a baby dies in such a manner, the mother never survives. The husband, distressed and afflicted on account of his double loss, made recourse to Our Saviour, whose image he had in his house, saying: "You, Lord, know that with all my power I try to observe your law ... and, equally, I know that, if you want, you can deliver my wife from death; I pray you to grant this to me, so that my daughters are not abandoned, and if you do this, I promise you I shall not be ungrateful". Following the prayer of the good man, the dead baby was immediately born, leaving the mother alive and healthy; a notable marvel to everyone who heard or saw it (Ricci 2001: 437).

This miraculous tale emanated from the southern Chinese city of Nanchang. It was recounted by the Italian Jesuit, Matteo Ricci, in a letter sent from Beijing to Claudio Acquaviva, the General of the Society of Jesus in Rome, in October 1607: a round-up of the last two years' events in the four Jesuit mission stations of Shaozhou, Nanchang, Nanjing, and Beijing. The story of how God intervened to save a Chinese Christian woman from certain death by effecting the expulsion of her dead baby is one of 12 miracles narrated by Ricci in his letter. These include three cases of sick children being saved by baptism; the story of a blind man who regained his sight after the removal of idols from his home and his subsequent baptism; three cases of the possessed being liberated, thanks to the instruction of Christians (demons fled when idols were replaced by Christian images and the victims of possession were enlightened with Christian truth); two cases of divine punishment (a blasphemer was blinded and an idolator stricken with illness); one case of a Christian man falsely accused of homicide, who prayed before an image of Christ, and was found innocent; and, finally, a showpiece of divine mercy, in which God answered the appeal of Chinese Christians and cured a sinful man of tuberculosis (Ricci 2001: 431–454).

Ricci speaks of these events as 'marvels' and 'miracles', 'meraviglie' and 'miracoli' in Italian, and he informs his superiors back in Rome of the value of these tales to the

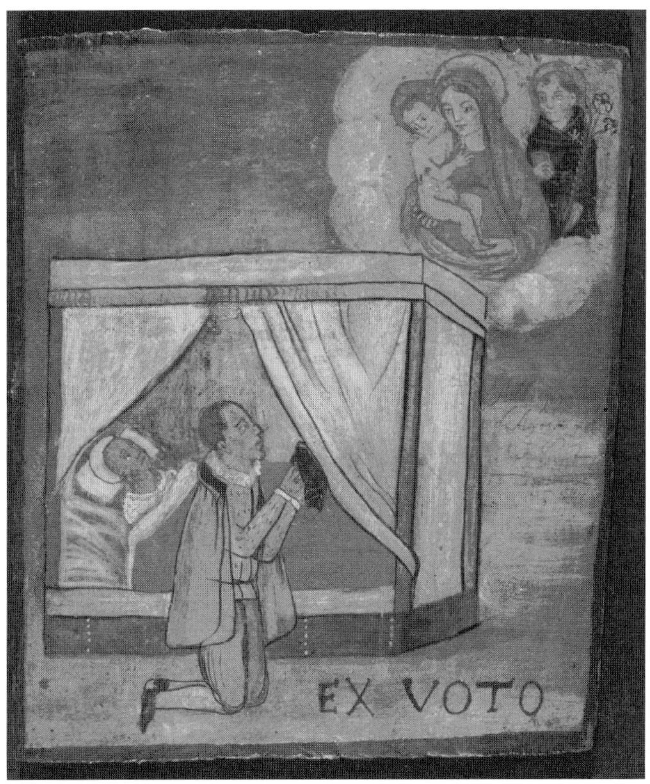

Fig. 7.1: A man on his knees, by the side of his wife's bed, prays to the Madonna and Child and to San Nicola. 1582. Reproduced with permission from the Museo di San Nicola a Tolentino.

reputation of the Jesuits in China, mentioning on more than one occasion that 'this was publicised throughout the city with no small increase in credit to the law of God' (Ricci 2001: 439, 441). Upon the Jesuits' arrival in Shaozhou in 1589, Ricci commented on the fact that their reputation for working miracles preceded them (ibid. 139). When Ricci's colleague Michele Ruggieri travelled to Shaoxing in January 1586, he also remarked that the Jesuits' reputation for curing the sick brought them 'great credit' (Ruggieri: 46v–7r). Each of the tales clearly carried a strong didactic message, which was inscribed upon the bodies of the protagonists. While the bodies of Christians were seen to enjoy the protection of God, the bodies of idolaters were wracked and tormented. Baptism was shown to heal as well as to save. Put simply, conversion was not only the prerequisite of salvation; it was also the route to a healthier, happier life. As such, one might consider Ricci's reference to miracles of healing to be an unsurprising aspect of his mission strategy. And yet it is no accident or anachronism that healing has been relegated to a back seat in the historiography of the Jesuit mission in China. Matteo Ricci and his colleagues wished to be remembered not as miracle-men, but as practitioners of science and reason.

In this essay, I hope to shed light on the place of miraculous healing in the Jesuit mission, and to situate the events described within their European and Chinese contexts. In so doing, I shall point to tensions in the Jesuit historical narrative. In one sense, Jesuits are a historian's dream, because they were and are the most avid recorders and publishers of their own history. As a highly mobile community, dispersed across Europe and across the globe in pursuit of their missionary goals, they were forced to create institutional bonds through letters, reports and travelogues, which later became the trophies of a heroic past, to be further broadcast through the medium of print. For the same reason, Jesuits are also a historian's liability; because of their tireless self-documentation, their monopoly on their own institutional record, it can be extremely hard to escape their version of events. By focussing on healing, a central but under-acknowledged feature of the Jesuit mission to China, it may be possible to dodge the exclusive control of the Jesuit record. Healing responded to profound human needs, and transcended cultures. It takes us to the heart of the Jesuit encounter with the Chinese people.

The Italian Jesuit Matteo Ricci arrived in August 1582 in Macao – the Portuguese-run peninsula at the southern tip of China – having spent the previous four years in Goa and Cochin on the west coast of India. He was summoned to China by his old novice-master, Alessandro Valignano – now Visitor to the Jesuit missions in the East Indies – and he went there to assist another Italian, Michele Ruggieri, who – notwithstanding the ban on foreigners throughout the Chinese empire – had already made his first tentative sorties onto the mainland. In 1583, Ricci and Ruggieri founded the first Jesuit residence in China at Zhaoqing, a small walled town some 100 km to the west of Canton. In 1588, Ruggieri returned to Rome to gather support for the Chinese mission. Ricci continued on his journey, attended by a succession of Jesuit companions, most of whom he outlived; heading northwards to the Imperial City, he established residences in Shaozhou, Nanchang, Nanjing and Beijing. From the start, Ruggieri and Ricci were avid students of the Chinese language, and they adhered to Valignano's policy of 'accommodation'; that is to say, wherever possible they tried to accommodate or assimilate aspects of the host culture, rather than insisting on European customs as a precondition of conversion. At the beginning of the mission, in the early 1580s, they took the advice of a local official and adopted the dress of Buddhist priests. In 1595, having realised that Buddhist priests were not universally respected, they rethought their sartorial tactics, and assumed instead the robes of Confucian scholars. Now they were casting their net at the elite caste of literati – scholar-officials – and they made their way slowly but determinedly to the imperial court, where a residence was established in 1601. There Ricci died in May 1610. In his final surviving letter to Claudio Acquaviva (General of the Society of Jesus) written in August 1608, he reported that there were now over two thousand Christians throughout China. The total population of the 'mighty kingdom' at this time may have exceeded two hundred million.[1]

The Jesuits who hit China in the 1580s were a cerebral bunch. They had enjoyed the benefits of a top-quality education in the Jesuit schools and colleges of Europe. The pyramidal curriculum consisted of a strong grounding in the 'lower disciplines' of grammar, classical languages, poetry, history, and rhetoric, followed by 'philosophy' (a term that encompassed logic, metaphysics, ethics, mathematics, and physics), and finally – at the apex of the Jesuits' training – theology. Lucid expression, persuasion and communication were highly prized, as was the ability to learn languages, both classical and modern. And so it was natural enough that these able and articulate young men, upon their arrival in China, should try to reach out to the intelligentsia – the 'literati,' steeped in Confucian learning, who were products of an equally rigorous education, and who were appointed to administrative office on account of their success in the gruelling civil service exams.

In a long letter written by Ricci in 1609, a year before his death, and addressed to the Vice-Provincial in East Asia, Francesco Pasio, he laid down in some detail his vision of how to convert the Chinese (Ricci 2001: 516–517). Drawing on his own extensive experience in order to advise on the future of the mission, Ricci first emphasised the hard-won status the Jesuits enjoyed as virtuous men and as literati, 'the two things that here are most highly esteemed.' He then explained how

> ...in this kingdom ... sciences and opinions founded on reason are greatly prized ... And consequently it seems that it will be easy to persuade the principal men of the kingdom of the things of our holy faith, confirmed with so much evidence of reason, and when the most learned men agree with us, it will be easy to convert the rest (Ricci 2001: 517).

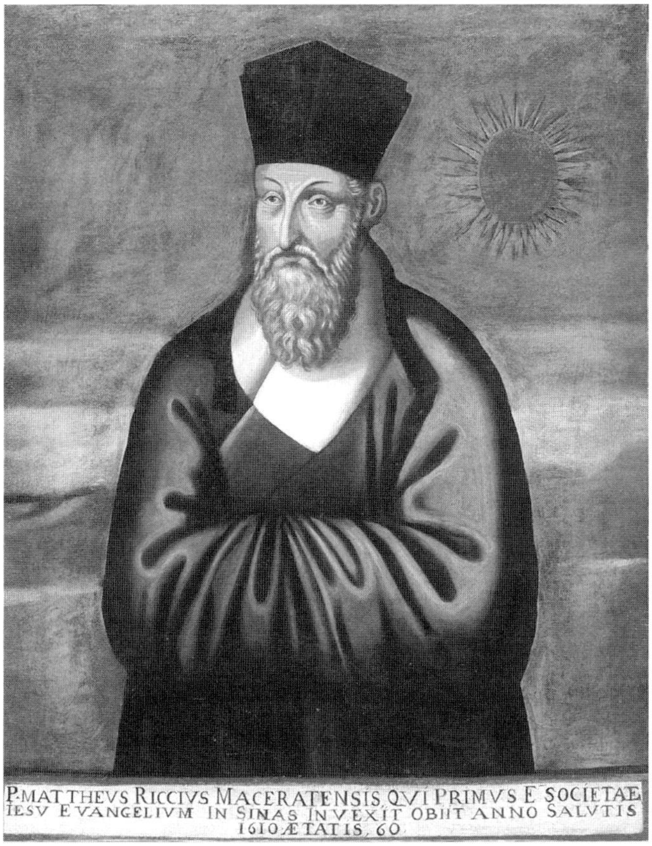

Fig. 7.2: Emmanuele Yu Wen-Hui (known as Pereira), Portrait of Matteo Ricci, 1610. Church of the Gesù, Rome. Reproduced with permission from the Direzione Centrale per l'Amministrazione del Fondo Edifici di Culto.

For Ricci, the optimal medium of evangelisation was print, which he celebrated for its ability to convey a precise and lasting message: 'Books spread more easily, they speak to more people and they continuously voice their truths in a more considered fashion and with more precision than that which is spoken with the mouth.' Ricci turned next to the importance of Western science in appealing to the Chinese:

> The Chinese have a wonderful intelligence, natural and acute ... From which, if we could teach our sciences, not only would they have great success among these eminent men, but it would also be a means of introducing them easily to our holy law and they would never forget such a great benefit (Ricci 2001: 517).

The document shows Ricci's mission strategy at its most well-formulated and coherent. This was a thoroughly elitist plan that targeted the imperial court, the ruling class of literati, and the highly-educated. Ricci's programme was a cogent one, powered by reason, books, and science. Educated men would speak to educated men. Through reason, the Jesuits would convert the Chinese.

As has already been suggested, given Ricci's skills as a rhetorician and publicist, it is no surprise that his own version of events has dictated his posthumous image. The most famous portrait of Ricci (Fig. 7.2), executed in Beijing in 1614, and based on sketches of his corpse, made immediately after his death, shows the Jesuit priest posing as a Confucian scholar – a pose he maintained for the last fifteen years of his life. It is an image that celebrates Ricci's accommodation to Chinese life, his mastery of the language, his acceptance by the scholarly elite of China. Ricci's intellectual achievement was immense. His published oeuvre, the majority of which was composed in Chinese, included a collection of adages on friendship, a treatise on memory, selections from Epictetus, and another stoic-influenced piece entitled *Ten Discourses by a Paradoxical Man*. Ricci's ability to integrate the canons of Chinese and European literature was demonstrated by the catechism that he published in 1603. Grounded in the traditional Western theology of Thomas Aquinas, Ricci skilfully reconfigured the truths of Christian doctrine to appeal to a Chinese readership versed in the Confucian classics.

Besides his expertise in philosophy and theology, Ricci also boasted a sophisticated understanding of nature. He was proud of having studied under the Jesuit astronomer Christopher Clavius at the Roman College during the years 1572–75. Clavius, with whom Ricci kept in touch throughout the China years, was a central figure in Roman astronomy, and a prolific author, who was invited to serve on the committee charged with reform of the Gregorian Calendar in 1582. One of the last serious defenders of the Ptolemaic tradition, he nevertheless engaged in serious and open dialogues with Galileo in the 1580s, and was one of the earliest to endorse his telescopic discoveries. It was as 'scientist' that Ricci was depicted in an image dating from 1749 (Fig. 7.3), printed shortly after the Jesuits' position in China had become untenable as a result of the Chinese Rites Controversy.[2] Here, Ricci is shown in a trio with his successors Adam Schall (1591–1666) and Ferdinand Verbiest (1623–1688), surrounded by astronomical instruments. The picture celebrates the Jesuits' entry into the imperial court, bearing the passport of Western scientific knowledge. From the outset, Ricci was determined to woo the Chinese with his superior knowledge of the universe, and his scientific works included Chinese versions of Clavius' writings on spheres and astrolabes, as well as tracts on arithmetic and geometry, and a commentary on Euclid. He always travelled with a collection of maps, prisms, astrolabes, and sundials, the material bearers of the knowledge that Ricci hoped would work to convert the Chinese literati. As he commented contemptuously in 1597, 'the Chinese have no science' (Ricci 2001: 348). The perceived technological and spiritual superiority of the West were to work together for the salvation of the Chinese.

On the plaque marking Ricci's birthplace in Macerata, erected to commemorate the four hundredth anniversary of his entry into China, he is described as 'missionary, scientist, humanist.' A great deal of the scholarship on Ricci is devoted to linking those three identities; to showing how his work as a scientist and a humanist contributed to his vocation as a

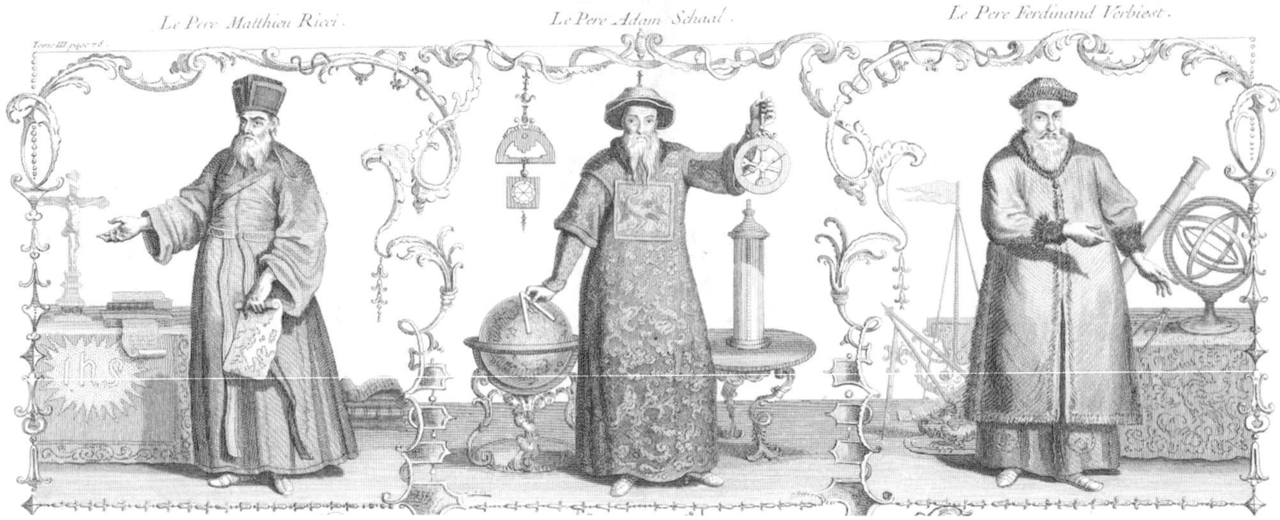

Fig. 7.3: From Jean-Baptiste Du Halde, Description géographique (Paris, 1735). Reproduced with permission from the syndics of the Cambridge University Library.

missionary. Historians have explored the ways in which Ricci boosted his credibility and status in China by means of his science and scholarship; they have demonstrated the ways in which Ricci's knowledge opened doors for him, and gained him the all-important dinner invitations that created the discursive space within which he could introduce Christian teachings (Laven 2011: 109). They have examined Ricci's participation in Chinese print culture, and his masterly manipulation of intellectual patronage, on which the accreditation of Christian ideas rested. And they have attempted to unpick the Confucian, Christian and European classical strands that are woven together so cleverly in his Chinese publications. For Ricci's science – in the sense of 'scientia,' 'learning,' 'knowledge' – has been seen to rest at the heart of his mission. Of course, these images have their basis in fact. Ricci's befriending of and by Chinese literati is well-attested. His successes at the imperial court were indeed predicated on his flaunting of Western science. Many of Ricci's Chinese language publications have survived, demonstrating his extraordinary intellectual agility and formidable literary productivity. But choices are inevitably made in the way that individuals (and institutions) represent themselves, and come to be represented by others. Ricci could have been portrayed, like a medieval saint, surrounded by comic-strip images of the healing miracles over which he presided; he was not.

How might that alternative picture have looked? The canvas could perhaps have been divided into five sections: four lunettes, surrounding the central portrait of Ricci. In the top left hand corner, the artist could have begun with fertility miracles. As early as 1584, Ricci testified to the first of these: the long-awaited birth of a baby, a daughter, to the wife of the Governor of Zhaoqing. In Ricci's mind, the unexpected birth was a gift of God, in recognition of the favours shown to the missionaries by the Governor. 'God wanted to pay him in part for the love that he had shown towards us' (Ricci 2001: 92). By 1586, the Governor's wife had produced a male heir, provoking a rush of local women to go to the Jesuits in search of their remedy for infertility. It was said that the European priests had given the Governor an image of the Madonna, and that she had intervened to bring about the providential pregnancies.

The top right-hand corner could then have been dedicated to the topic of healing sick children. There were several examples to choose from, including the crop of children who had fallen victim to the measles epidemic in Nanchang in 1607, and who miraculously recovered following their baptism. But the most telling example to depict would perhaps have been that of the Chinese woman who had watched six of her new-born children die, and who was now nursing her two surviving infants, both of whom were struggling for their lives. A Christian woman, who had assumed the name of Maria, advised her to baptise the children, since 'God, the author of life', would allow them to live long and prosper. The mother proceeded cautiously, initially allowing only the older of her two children (a three year old) to be baptised; within a few days, he was such a picture of health that the second child was baptised along with the rest of the family. At the same time, the new converts removed all the idols from their house (Ricci 2001: 437–438).

In the bottom-right hand corner, the artist – depending on his beliefs and iconographic impulses – might or might not have chosen to conflate the healing of the mentally ill with the expulsion of demons from the possessed. The Jesuits in China were called upon repeatedly to assist with cases of demonic possession: to exorcise demons both from individuals and from properties. But they were also rated for their abilities to assist those suffering from psychiatric complaints. In 1595, Ricci told his colleague, the Portuguese Jesuit Duarte de Sande, who was based in Macao, of how a mandarin passing through Shaozhou en route to the capital had learnt of the priests' reputation for healing, and had conceived a hope that the Jesuits 'as men of proved virtue and great knowledge', possessed of astronomical expertise, might be able to cure his son of madness. As Ricci explained to de Sande, his reaction was opportunistic: 'seeing the possibility that had arisen to enter into the grace of such a powerful mandarin, I flew at the opportunity'; 'for, besides the desire that I could have to serve him, by curing his son I would have satisfied that other desire that I had, that is to pull back the curtains and go to the court of Beijing, (Ricci 2001: 199–200).

In the final lunette, the artist (or his patron) might have combined the themes of healing and conversion. Particularly compelling was an image of the thirteen-year-old, who – having been struck by lightning – was in a coma for three days, during which time God appeared to him, and brought him back to consciousness. After this episode, the boy of course converted, and was known by the Christian community as 'Michael of the lightening bolt' (Ricci 2001: 390).

A possible objection to this imagined representation would be that Ricci, unlike my putative medieval saint, was not in fact the instrument of the miracles described. These were, as Ricci himself commented, 'miracles granted by God to help the work' of the China mission – a more fitting title for my image than San Matteo; that, after all, he never was. Another objection would be that not all of the 'marvellous events' described are true miracles – a woman getting pregnant after many years of trying, for instance, might indeed be interpreted as providential, but not necessarily miraculous. And as for the Jesuits being called upon to help a mentally sick boy, there is no evidence in this case of a cure, let alone a miracle. What does seem clear, however, was that the Jesuits in China were associated with healing disease and establishing control over the body, and that their perceived successes in this domain brought them admiration and fame. Moreover, we know that, for all Ricci's emphasis on 'reason, books, and science,' and for his insistence that the future of Chinese Christianity lay with the literati, only a tiny handful of the two thousand or so neophytes counted by the end of Ricci's life fitted his ideal profile (on the small percentage of literati converts, see Standaert 2008: 175).

In 1586, writing from Zhaoqing, Ricci complained that although there had been forty baptisms, 'we have not yet converted to Christianity a single person of serious consequence' (with the possible exception of the elderly father of a local magistrate); this despite the fact that Ricci had made numerous friends and contacts among the local literati (Ricci 2001: 122). Moreover, instead of the learned Confucians sought out by Ricci (whose moral premises could, he believed, be readily assimilated in a Christian context), Ricci's Chinese flocks were made up in large part of converts from Buddhism. They were 'men of penitence', who turned to Christianity because it seemed to chime with their ascetic lifestyle. This of course irked Ricci, who always stressed that – of the three strands of Chinese religion – Buddhism, Daoism, and Confucianism – the last was the one which the Jesuits should aim to accommodate. Furthermore, his Buddhist converts had an unfortunate commitment to vegetarianism, something Ricci could never see the point of. One such vegetarian was a convert known as 'Gioseppe', a prosperous merchant in Shaozhou, presiding over a household of thirty to forty people, who was baptised in 1591. Unfortunately, reported Ricci, 'he was persuaded that becoming a Christian meant leaving the world and adopting the hermit's life, relinquishing the things of this world, as he has done, leaving his wife and all the traffic of his house, and only attending to matters of his [spiritual] health' (Ricci 2001: 173).

The situation would improve with time. The most famous and prestigious among Ricci's converts were Xu Guangqi, also known by his Christian name as 'dottor Paolo', a scholar-official who was baptised in 1603, and who worked continuously with Ricci over the next few years on the translation and composition of scientific works in Chinese; and Li Zhizao, a.k.a. 'Leone', who was also a high-flying civil servant, and who worked with Ricci to render Clavius' works into Chinese; he was baptised by Ricci in 1610 (on these high profile converts, see Jami, Engelfriet, and Blue 2001). But these model converts were, as I have suggested, a tiny minority of the two thousand Christians who were living in China at the close of Ricci's life. Besides a smattering of nobles, merchants, and doctors, we learn of carpenters and boatmen becoming Christian, and of a remarkable number of female converts. 'Remarkable' because women are so glaringly absent from Ricci's narrative (let alone his pronouncements on mission policy) until the closing years of his life, when suddenly they burst onto the scene: a Christian woman named Elena shuns the devil; Maria, wife of prominent Nanchang Christian Pietro, insists on seeking Christian instruction and baptism, despite the disapproval of her neighbours; a mother has her sickly children christened; a female convert drives out demons from a neighbour's house; another – baptised without the permission of her husband – endures cruelty and beatings, but remains faithful to Christianity and a dutiful wife (Ricci 2001: 431–454).

It is my contention that the majority of Chinese men and women who actually became Christian were more impressed with the Jesuits' pastoral work – their ministry to the sick and needy – than they were by their intellectual achievements. One old man became a Christian following the conversion of his son, so amazed was he by the 'marvellous' transformation in the younger man's behaviour, from ruinous slob to dutiful child. He admitted to a friend: 'although I have not heard the [Christian] law preached, I hold it to be true and holy having seen my son'(Ricci 2001: 442). 'The word' counted for little, in comparison with the visibly ameliorative effects of religion

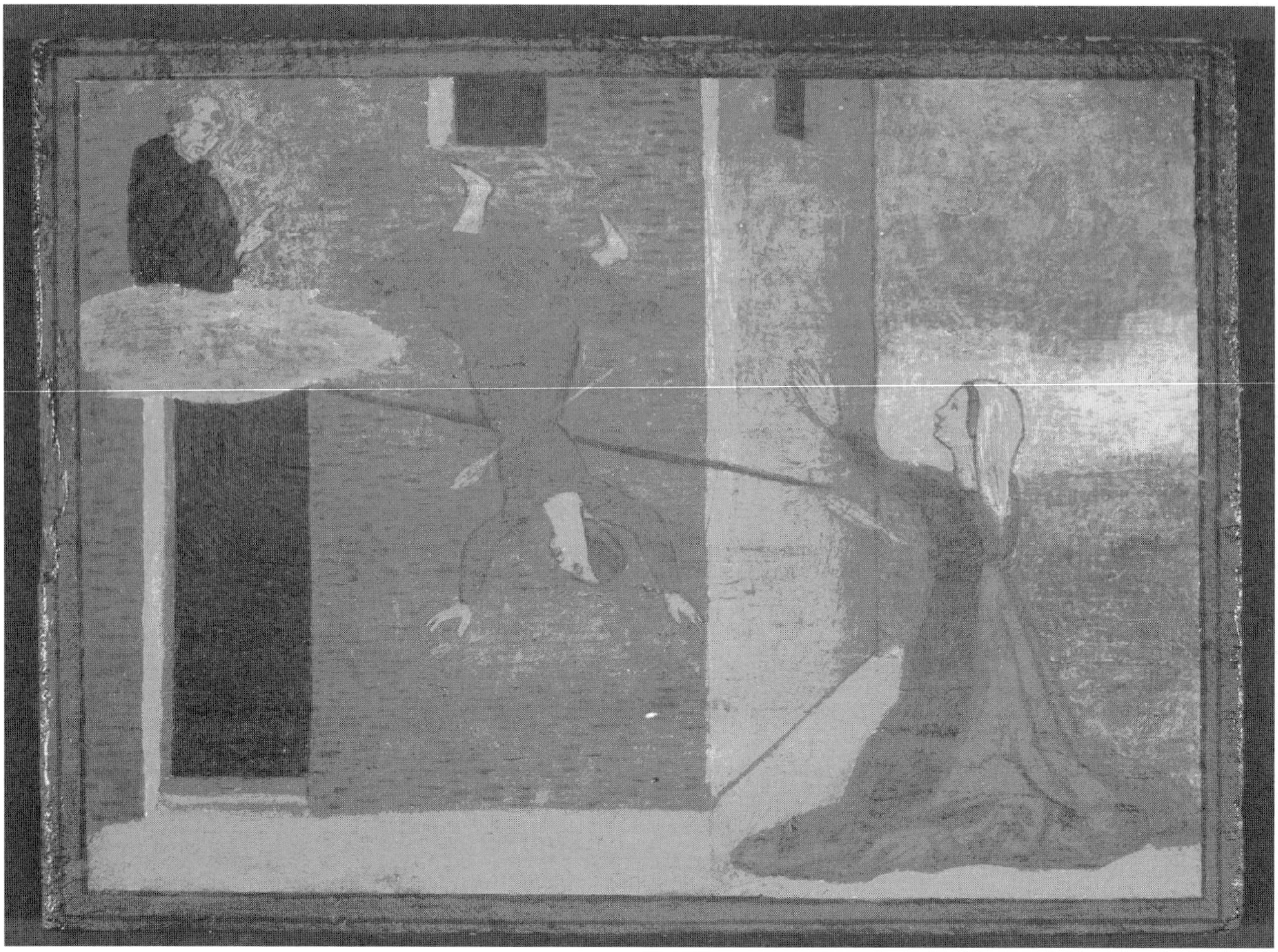

Figure 7.4: 'A woman who has fallen from a window is saved by San Nicola.' Sixteenth-century. Reproduced with permission from the Museo di San Nicola a Tolentino.

upon a close relative. Contrary to the Jesuits' insistence on dispassionate, disembodied understanding, their real successes were founded on an emotional, 'embodied' experience of Christianity. (For a powerful call 'to admit the psychic and the corporeal' into the realms of history and cultural anthropology, see Roper 1994: 27).

These miraculous tales were rooted in Chinese soil, but were typical of Catholic belief in early modern Europe – in the distant Western lands that Matteo Ricci and his Jesuit collegues had left behind. There, when a man feared for the safety of his wife, when the wise women and medics had failed, he could turn to Christ, the Virgin, and other saints. Or rather, as historians of 'medical pluralism' have taught us, he would probably pull out all the stops at once, combining the medical and the spiritual, often in a variety of forms (Gentilcore 1998). He looked to God and his intermediaries for a result, and when he got it, he gave thanks. This was a religion of credits and debits, of favours exchanged. And the expectations of the Catholic faithful that their requests might have automatic efficacy, or that they could do a deal with God, pervaded the religious culture that so profoundly offended Protestant reformers.

Spurred on by Protestant disapproval, early modern Catholics vaunted their miracles in countless representations, both verbal and visual. 'Ex votos,' pieces of art work commissioned in gratitude for miracles wrought, could take many forms: sailors whose boats had been spared during a storm might donate a model ship to hang in their church's nave, while wax models of body parts were used to commemorate the divine forestalling of corporeal afflictions. A particularly rich collection of 'ex votos' is preserved in the museum of the Basilica at Tolentino, in central Italy (Fig. 7.1, 7.4 and 7.5). Although this includes some three-dimensional artefacts, the majority of the representations are paintings on wooden board: small, cheap, and engagingly simple. Many depict women, kneeling by a crib, clutching their rosaries or crucifixes, praying to their local saint, St Nicholas, to heal their ailing infants. But there are fathers as well as mothers, their faces wracked with

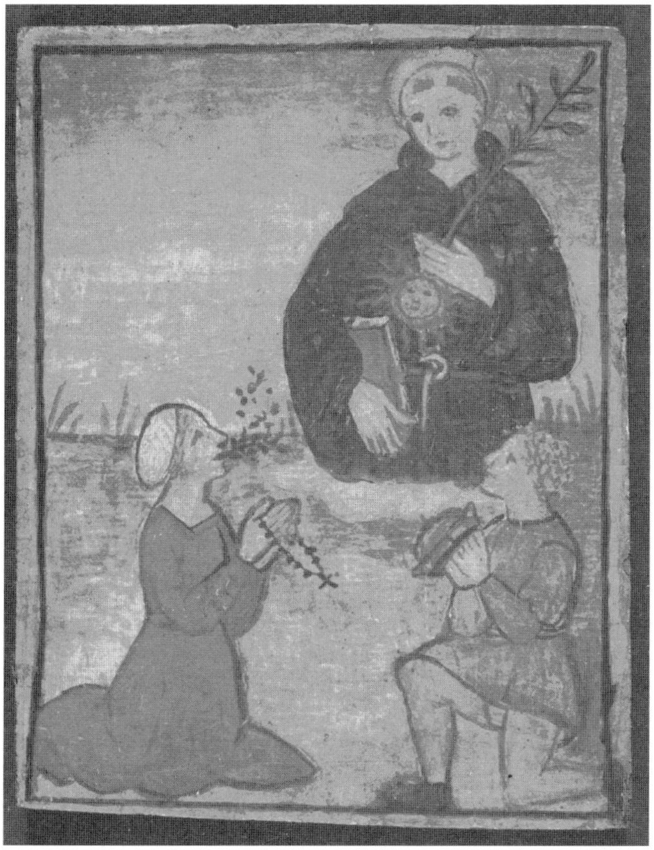

Figure 7.5: A bride and groom, on their knees, give thanks to San Nicola. Late fifteenth- or early sixteenth-century. Reproduced with permission from the Museo di San Nicola a Tolentino.

anxiety, keeping watch by the sick-bed, waiting desperately for the intervention of their favourite saint. More dramatic scenes are also conveyed: a young man caught up in a wool-winder, a woman falling out of a first-floor window, a bride possessed by the devil, a girl witnessing her father being attacked by armed brigands, fires, ship-wrecks, and accidents in the work-place involving boiling water or dropped tools.

Such images reminded Catholics of the steps they must take in order to set their disordered world aright. Theirs was a religion that provided numerous rituals and props that could be employed in times of crisis. It was a type of devotion that Matteo Ricci – the Jesuit who grew up in the Italian Marches, but spent the greatest part of his career attempting to convert the Chinese – would have been familiar with. His home town, Macerata, in the papal states, where Ricci was born in 1552, stood roughly midway between Tolentino and the far more famous pilgrimage destination of Loreto, resting-place of the Virgin's house, which had flown under the direction of Christ and a team of angels from Nazareth to central Italy in the thirteenth century, in order to save it from destruction by Muslim invaders. The Holy House is still there today, standing in the centre of a vast and beautiful fifteenth- to sixteenth-century complex. From the Piazza della Madonna, designed by the architect Bramante, one enters the basilica, in the centre of which stands the House itself. Modest and brick-built inside, the Virgin's apartments are enclosed by an exquisite sculpted husk. This miraculous building continues to draw the faithful in search of miracles. Chartered trains bring the sick on three-day missions of hope. The site teems with wheel-chairs.

The Jesuits were already big in Loreto by the mid-sixteenth century. This was one of the factors that prompted them to set up a college in nearby Macerata in 1561. Here, in the same year, the young Matteo Ricci would commence his schooling; later Alessandro Valignano would spend a year as Rector of the college, before he was appointed Visitor to the Far East in 1573. Ricci was not the only Macerata boy to end up in East Asia. Among his compatriots, Oliviero Toscanelli accompanied Valignano to the East in 1573, and was appointed master of the elementary school in Macao in 1592; Giulio Piani, who served as Valignano's secretary, played a key role in the baptism of 12,000 people in Japan in 1577, while Giulio Mancinelli became Rector of the College of Macao in 1589. Years later, Ricci would remember his holy homeland with pride. Writing in 1599 to Girolamo Costa (another Maceratese who was by now Procurator General of the order), Ricci told him: 'Sometimes I boast to these barbarians that I come from the land where Christ Our Lord transported the house that his mother had in this world, many miles away. And they are stunned when I tell them these and other marvels that God has made happen in these western lands' (Ricci 2001: 361).

That the Jesuits, erudite and rational as we like to think of them as being, were not above the appeal of 'marvels', has been amply demonstrated by historians of Catholic culture in early modern Europe. Trevor Johnson, in his ground-breaking work on eighteenth-century Bavaria, describes how Jesuits distributed holy water infused with the relics of Saints Ignatius Loyola and Francis Xavier – water that was then dabbed on to ailing parts of the body or sprinkled on caterpillar-infested fields. Rather than seeing a gulf between orthodox and popular religion, Johnson insists that the Jesuits were participating in a 'shared culture of the miraculous' (Johnson 1996: 185). David Gentilcore, in his work on southern Italy, has also noted the involvement of Jesuits, alongside other new orders, in handing out religious medals, cheap prints of saints, and rosaries; the appeal of these objects and images was in their perceived power to solve the problems of everyday life – a power more forceful than that of sermons (Gentilcore 1992). The time has surely come to apply these revisionist insights to the work of the Jesuits in China.

Miracles belonged to the Europe that Ricci and his colleagues had left behind, but they also belonged to China. The similarities between European and Chinese concepts of the miraculous struck Michele Ruggieri and his Portuguese colleague Antonio d'Almeida forcefully during their travels in southern China in 1586. Visiting a temple near Nanjing,

Ruggieri observed: 'inside this church there hung many tablets on which miracles were recorded; there were also eyes and feet modelled in wax, silver and gold, as one sees in the famous houses of the Madonna in our Europe' (Ruggieri: fo. 44v). He might well have been thinking specifically of Loreto.

Of course, one might wonder if Ruggieri was misreading the evidence before him and, in the spirit of European travellers throughout the world, desperately familiarising the exotic. (After all, according to Jesuit commentators, Canton was just like Florence, Suzhou indistinguishable from Venice). However, Susan Naquin's study of Beijing temples provides some reassurance. She comments on the presence in late Ming temples of 'votive tablets' – 'inscribed plaques given by grateful believers' – and also has evidence that 'figures or objects' (usually made of plaster or paper) 'were left as evidence of prayers made and prayers answered: baby shoes, silk painted with eyes, figures of babies or dogs (before a god that cured dog illnesses).' According to Naquin, 'The active presence of a God was indicated not merely by prayers answered but also by miraculous events.' Some temples and their images had survived disastrous fires or floods. One temple had pillars in the Buddha hall that repainted themselves! The bitter water in the well of a courtyard of another temple had suddenly become wonderfully drinkable. There were temples with healing objects, for example, basins in which you could bathe your eyes and be cured of eye disease (Naquin 2000: 40–45).

We can add to Naquin's evidence from city temples that from Chinese pilgrimage sites, such as the Buddhist monastery at Nanhua, visited by Ricci in 1589. Such shrines were the locales for numerous apparitions and miracles, chronicled and publicised in gazetteers (Naquin and Chün-fang Yü 1992). The eighteenth-century chronicler of Wudang Shan, the mountainous site of a Daoist complex, erected in the early fifteenth century, recorded that two years after the publication of his 'mountain monograph' – a guide to the pilgrimage site – 'I was blessed with the birth of a boy'; he further affirmed that 'every generation has transmitted tales of divine miracles (Lagerwey 1992: 308).' Daoists did not only believe in miracles; they also believed in a type of alchemy: the transformation of mercury sulphates into silver. When Ricci and Almeida arrived in Shaozhou in 1589, they were deluged by visitors 'on account of the fame of the miracles that the people had heard that we performed in Zhaoqing' and of 'the ability' that Ricci was thought to have in 'making silver (Ricci 2001: 139).' So there were miraculous strands in both Buddhist and Daoist beliefs, and given the essentially syncretic nature of Chinese religion it is plausible that people were willing to accommodate marvels wrought by the Christian God within their repertoire of miraculous solutions. Chinese beliefs in divination and in demons, in alchemy and in supernatural healing, and their votive practices at shrines and temples meant that European-style miracles were easily assimilated. They transcended language barriers and they built bridges between cultures. They appealed not so much to the successful in life – to the literati whom Ricci courted – as to the vulnerable.

The Jesuits brought knowledge of Christianity to the Chinese in books and treatises. They also inscribed that knowledge upon the bodies of the people they encountered, and – in particular – upon the bodies of those they cured. Although I have argued in this essay for the transcendental power of healing, I must conclude by stressing that the body – just like any other site of meaning – is culturally constructed. Just as the doctrines recorded in Ricci's catechism could be contested, so too could knowledge inscribed upon the body. Christianity was, of course, a peculiarly embodied religion. Its doctrine focussed on the 'Passion' – the physical suffering and death – of Christ. Its central practice was the Eucharist, at which bread and wine were transformed into the body and blood of Christ before being ritually consumed by laity and priesthood. In Catholicism at least, the themes of bodily suffering ran deeper. Saints suffered martyrdom; the pious revered their corporeal remains. The frail were blessed; so too were those who cared for them. Hospitals played a prominent role in Catholic programmes of charity. And if suffering did not occur naturally, it could always be imposed artificially through acts of self-mortification and denial: in Europe, the Jesuits were renowned for their acts of flagellation (Johnson 1996: 187). Pain and healing were both vocations of the holy. (Camporesi 1988, Walker Bynum 1987).

The Jesuit missionaries operating in Asia during the sixteenth and early seventeenth centuries were not detached from this corporeal conception of religion. In particular, they enjoyed a strong sense of their calling to martyrdom, and they were quick to interpret even minor illnesses and indispositions – an arthritic knee or an injured foot – in this light (Ruggieri: 82v). In China, the Jesuits' sense of foreboding was increased by the news – from India – of the death of their colleague, Rodolfo Acquaviva (nephew of Claudio), who was famously martyred at Salsette in 1583. Michele Ruggieri looked back to the time that he had shared a room in Lisbon with Acquaviva while waiting to board a ship to Goa, recalling how every night Acquaviva had suffered from unstoppable nose-bleeds. According to Ruggieri, he himself had told his room-mate that these bleedings were surely a premonition of 'the blood that he would spill in India for the love of Christ (Ruggieri: 81v).' In 1592, when Ricci was suffering from 'a twisted foot,' which he had got when he had jumped out of a window following an attack on the Jesuit house in Shaozhou, the Jesuit reflected: 'If it please His Divine Majesty to conclude all this with a happy death, such as that of Father Rodolfo ... what happy news that would be' (Ricci 2001: 162).

The Jesuits were also, as we have seen, committed to healing. Matteo Ricci, the son of an apothecary, took a particular interest in the availability of rhubarb, prized by Europeans for its medical properties, and in the herbal remedies used locally (Ricci 2001: 166, 178). In common with other members of religious orders, Jesuits were sometimes called upon to nurse one another. In Shaozhou, in 1589, Ricci wrote that he and his three companions were recovering from a long period of sickness. Ricci's Portuguese colleague António de Almeida was the worst afflicted, and on the seventh day of his illness – recorded as the Feast of San Luca (the patron saint of physicians and surgeons) – he had reached the brink of death. Although there appeared to be no hope for him, Ricci entrusted his case to God, and stayed up all night attending to him, feeding him meat sauce and strong broth. And so 'it pleased God that the following day he was out of danger' (Ricci 2001: 147). Given that both sickness and healing were integral to the Jesuits' brand of piety, it is perhaps small wonder that they should apply this embodied experience of religion to their missionary work. Thus when Matteo Ricci attempted to explain the role of Christ's apostles, a role that he expressly saw himself as continuing, he told the Chinese that 'The Lord of Heaven gave them the special gift of understanding the language of all nations, of saving souls, of cancelling sins and performing miracles (Ricci, 'The Catechitic Conversations' (c. 1585), quoted in Malek 2002: 746). It was by means of further miracles that the Jesuits would drive forward the healing apostolate.

On the other hand, there were ways in which the corporeal nature of Christianity was damaging rather than advantageous to the Jesuit mission. It is notable that Ricci's catechism, *The Lord of Heaven*, first published in 1603, includes no reference to Christ's suffering and death on the cross. Consistent with Ricci's decision to remain silent was the position taken by Alessandro Valignano in 1588, in a document sent back to Rome that itemised the gifts that should accompany an embassy from the pope to the Chinese emperor. While Valignano on the whole saw this as an opportunity to present the emperor with the finest works of European craftsmanship, he was most particular in his advice on the images that would be appropriate. Whether producing stained glass or oil paintings, he insisted that the artists who received the commission for these gifts must be instructed to represent 'neither wars and martyrdoms, nor the mysteries of the passion' and should depict only 'happy things' (Archivum Romanum Societatis Iesi, Fondo Gesuitico, 722/2, 5r, 7r). Finally, in the winter of 1600, when the eunuch Ma Tang discovered a crucifix in Ricci's luggage, there was trouble. Part of the problem was that – in keeping with the style of Baroque Catholicism and its taste for drama and gore – the wooden figurine was made to look life-like. Ricci described it as 'a beautiful crucifix, carved out of wood, and painted with blood that looked real.' At first, Ma Tang imagined that the Jesuits were engaged in an act of voodoo, and that they were torturing a wooden doll with the intention of making the Chinese emperor suffer. Ricci knew not what to say in response to this, since he considered the eunuch and his colleagues too ignorant to launch into the truth of the crucifixion. He settled therefore on a rather vague explanation: that 'this was a great saint of our land, who had wanted to suffer the pains of the cross for our sake' and, for this reason, it was common to keep painted or sculpted images of him, 'to have him always in front of our eyes,' and to give thanks for the great benefit that he had bestowed upon us. But the eunuchs responded contemptuously to this account (Ricci 1942–49: II 115–117). By 1616–17, the time of the Nanjing lawsuit against heterodoxy, the true nature of Christ's demise was all too clear to critics of Christianity: 'They also say that he died nailed by evil administrators to a structure in the form of a character that denotes ten [i.e. the cross]. So he is just a barbarian convict condemned to death. How could an executed barbarian convict be called the Master of Heaven?' (Gernet 1985: 56–56). So the blood and the suffering and the vulnerability of Christ that characterised Baroque Catholicism had little to recommend to a Chinese audience.

Ricci and his colleagues were on safer ground with healing, where the disjunctions between Chinese and Christian belief and practice were less profound than one might expect. For the sick and weak, the Christian embrace of the vulnerable appealed. The story of Christ's passion represented an altogether different proposition. In Ricci's catechism, the Christian God was therefore presented to the Chinese not in his human frailty, but in his superhuman power. Like the Chinese emperor, the 'Lord of Heaven' was omnipotent and disembodied (Ricci 1985: 57).

Acknowledgements

I am grateful to seminar audiences in Cambridge for their responses to earlier versions of this paper; to Jason Scott-Warren for reading and commenting on this essay in draft; to Stefania Santini and Barbara Placido for their help in acquiring images; and to the Lyne family for introducing me to the devotional culture of the Italian Marches.

Notes

1 Brook (1999: 162) cites the 'standard estimate' for the Chinese population in 1600 as being 150 million. Mote (1999: 745) proposes a far higher estimate of 231 million people by the end of the seventeenth century.

2 The papal bull of 1715, issued by Clement XI that prohibited the Jesuits' tolerant attitude towards Chinese ancestor worship was followed by the 1721 decree of the Kangxi Emperor that banned Christian missionaries from China.

References

Brook, T. 1999. *The Confusions of Pleasure: Commerce and Culture in Ming China*. Berkeley: University of California Press.

Camporesi, P. 1988. *The Incorruptible Flesh: Bodily Mutilation and Mortification in Religion and Folklore*. Cambridge: Polity Press.

Du Halde, J.-B. 1735. *Description géographique*. Paris.

Gentilcore, D. 1992. *From Bishop to Witch: The System of the Sacred in Early Modern Terra d'Otranto*. Manchester: Manchester University Press.

Gentilcore, D. 1998. *Healers and Healing*. New York: Manchester University Press.

Gernet, J. 1985. *China and the Christian Impact: A Conflict of Cultures*. Cambridge: Cambridge University Press.

Jami, C., P. M. Engelfriet, and G. Blue (eds) 2001. *Statecraft and Intellectual Renewal in Late Ming China: The Cross-Cultural Synthesis of Xu Guangqi*. Leiden: Brill.

Johnson, T. 1996. "Blood, Tears and Xavier-Water: Jesuit Missionaries and Popular Religion in the Eighteenth-Century Upper Palatinate," in B. Scribner and T. Johnson (eds) *Popular Religion in Germany and Central Europe, 1400–1800*. 183–202. London: Macmillan.

Lagerwey, J. 1992. "The Pilgrimage to Wu-tang Shan," in S. Naquin and Chün-fang Yü (eds) *Pilgrims and Sacred Sites in China*. 293–332. Berkeley: University of California Press.

Laven, M. 2011. *Mission to China: Matteo Ricci and the Jesuit Encounter with the East*. Mary Laven. London: Faber and Faber.

Malek, R. (ed.) 2002. *The Chinese Face of Christ*. Sankt Agustin: Institut Monumenta Serica and China-Zentrum.

Mote, F. W. 1999. *Imperial China: 900–1800*. Cambridge, Mass.: Harvard University Press.

Naquin, S. 2000. *Peking Temples and City Life, 1400–1900*. Berkeley: University of California Press.

Naquin, S., and Chün-fang Yü (eds) 1992. *Pilgrims and Sacred Sites in China*. Berkeley: University of California Press.

Ricci, M. 1942–49. *Fonti Ricciane*, 3 vols. Edited by P. M. D'Elia. Rome: La libreria dello stato.

Ricci, M. 1985. *The True Meaning of the Lord of Heaven*. St Louis: Institute of Jesuit Sources.

Ricci, M. 2001. "*Lettere 1580–1609*." Edited by F. D'Arelli. Macerata: Quodlibet.

Roper, L. 1994. *Oedipus and the Devil: Witchcraft, sexuality and religion in early modern Europe*. London: Routledge.

Ruggieri, M. Relaciones. *Archivum Romanum Societatis Iesu, Jap. Sin. 101*.

Standaert, N. 2008. "Jesuits in China," in T. Worcester (ed.) *The Jesuits*. 169–185. Cambridge: Cambridge University Press.

Walker Bynum, C. 1987. *Holy Feast and Holy Fast: The Religious Significance of Food to Medieval Women*. Berkeley: University of California Press.

8. Protest re-embodied: shifting technologies of moral suasion in India

Jacob Copeman

This chapter explores embodied protest as a shifting form of knowledge in modern India and elsewhere. Insofar as the shifting nature of this knowledge is made manifest in a transition between different and yet at the same time complexly entwined types of embodied protest, we may speak of a process of *re*-embodiment. The *locus classicus* of embodied protest in the Indian subcontinent is fasting (Hopkins 1901, Laidlaw 2005, cf. Gandhi 1949). The most famous practitioner of political fasting in India is, of course, Mahatma Gandhi, who fasted on numerous occasions as part of a whole host of political campaigns including, but not limited to, that aimed at removing the British colonial government. What I identify and explore here, however, are recent usages of blood donation as a means of political protest and the relationships between these two embodied protest forms: the one classic (fasting), the other emergent (blood donation). I shall argue that the form blood donation takes as a mode of protest is as the social reform of protest, i.e. that blood donation as embodied protest represents a significant shift in knowledge about how protest should be enacted. But the re-embodiment of protest is by no means a process of unilinear transformation. The continuities I identify between the types of embodied protest just mentioned demonstrate how embodied knowledge may be transformed even while it is maintained (Toren 2002: 112). Indeed, Jenkins (2010) has recently argued that transformation can act as the very condition of continuity. As he puts it, 'long-term forms [may be] prolonged through the adoption of new ways' (ibid. 2). This chapter, then, provides evidence that blood donation is increasingly coming to both conceptually connect with and in some cases challenge the status of the fast in Indian public life as *the* virtuous technique of moral suasion. I also consider the ways in which both fasting and blood donation may be considered Foucauldian 'technologies of the self'. They are technologies in the sense of 'tools' – modes of political asceticism through which subjects may modify or 'work upon' their own bodies and conduct in order to transform themselves and thereby attain political power. Yet in important ways they are also contrasting technologies of the self. I shall draw on recent important work by Street (2009) in order to argue that protest blood donation, in a manner that implicitly critiques the political fast, is an explicitly *relational* technology of the self.

As a high-profile medical practice fraught with controversy, blood donation possesses a close association with protest and contestation worldwide. In particular, exclusionary practices in blood donation policy often give rise to keen protest. Strong (2009) provides a vivid account of protests by gay men in the United States and in Hong Kong against regulations excluding them from giving blood, and Seeman (1999) has described the violent protests of Ethiopian Jews on learning that their donated blood had over a number of years been secretly destroyed by the Israeli blood service. In both cases, fears about higher incidence of infection among the populations in question were the grounds for exclusion.[1] In addition, national scandals over the contamination of blood supplies, whether through technical negligence or bureaucratic indifference, have occasioned impassioned protest (Feldman and Bayer 1999). Such protests are protests *about* blood donation, or about its effects. This chapter differs in being concerned with blood donation itself as a form of protest. The occasion for protest in the instances explored below is highly variable; what unites them is the manner in which protest is enacted in each case, i.e. through the giving of blood for transfusion. The multiple and diverse expressive possibilities of blood donation as a medical technique with strong political connotations have been documented in impressive works by Cohen (2001), Weston (2001) and Simpson (2004). These contributions have drawn particular attention to the politics of race and nation invoked by blood donation in different regional contexts. My aim here is to contribute to this literature by drawing attention to a hitherto unexplored way in which blood donation is being mobilised politically as a technique of moral suasion.

As I have noted, the emergence of blood donation as a protest form in some ways represents the re-embodiment of protest in the subcontinent. Blood donation both diverges from and is continuous with fasting as an embodied form of protest. I hope to show below that it is from this dual relationship with fasting that blood donation gains part of its power as a contestatory method. The two forms share the same 'corporeal ethical register' (Davies 2006: 261) insofar as both are acts of political asceticism in which ascetic self-denial is mobilised as a moral weapon. That is to say, protest is overtly embodied, and embodied in comparable ways in each case. This is one aspect of the continuity I have mentioned. But the increasing prevalence of protest blood donation also signals a transition in knowledge about how to engage the body in protest that pertains to social reformist trends and their valorisation of utility.

Sant blood donation

Reflecting and responding to the internationalisation of the public policy orthodoxy that voluntary blood donation provides the safest blood for transfusion, India's Supreme Court banned paid donation from 1 January 1998 and directed the government to begin actively encouraging voluntary, non-remunerated blood donation. The government's subsequent National Blood Policy (2002) additionally required the phasing out of family-replacement donation, where family members are expected to donate to replace the blood withdrawn from the blood bank to treat their ailing relative, within five years.[2] The banning of paid donation and the phasing out of replacement donation has required innovative strategies on the part of blood banks, supported by government institutions, to radically increase voluntary blood donation. Despite the assortment of campaigns, however, replacement donation still accounts for more than 50 percent of all donated blood in India. The low voluntary figures are partly accounted for by the apprehension, widespread in the subcontinent, that blood donation results in the draining away of one's strength and a permanent volumetric deficit. This was frequently expressed to me during fieldwork (2003–2005) in the formulation: 'If I donate blood, then I will need a transfusion, so why should I donate?' From within this bleak collection landscape, however, there has emerged one critical success story: religious movements, in particular devotional orders in the *sant* tradition, have become essential providers of voluntarily donated blood. The Dera Sacha Sauda devotional order, which currently holds the Guinness world record for most units of blood donated in a single day, and which forms my principle case study in this essay, is a particularly high profile contributor.

Indeed, one of the most striking features of blood donation practices in contemporary India is the embracing of voluntary blood donation as a key focus of organised spiritual service by major devotional orders associated with the north Indian *sant* tradition. These orders have emerged over the last 15 to 20 years as some of India's highest-profile proselytisers of blood donation, which they have come to view as a critical act of service to humanity, world and nation. As with many other movements in the *sant* tradition, the Dera Sacha Sauda is presided over by a living saint, commonly known as '*satguru*' (literally 'true master'), who is worshipped as god. Like the comparable Radhasoami movement studied by Juergensmeyer (1991) and Babb (1986), the Dera Sacha Sauda emerged out of the *sant* heritage, 'the creed of the saints, a tradition associated with such figures as Kabir and Nanak' (Babb 1986: 17).[3]

Common to most *sant* movements is *guru-bhakti* – devotion to a spiritual master, devotee constituencies made up of both Hindus and Sikhs, an emphasis on the recitation of sacred words, a conception of transcendence as being open to all regardless of caste or gender, a social reformist agenda and a set of teachings genealogically derived from a family of non-sectarian *sants*, or saints, which began to emerge in the medieval period.[4] Many *sants* were themselves from low-status and generally lower caste backgrounds, teaching that all human births are rare and valuable (Lorenzen 1995: 18, 24). This kind of social criticism persists in present-day *sant* movements which uniformly criticise elaborate ritual, idol-worship and virtuoso displays of asceticism. Devotion, not ritual prowess or caste, so the *sants* proclaimed, is the determinant of one's worth and ability to attain liberation. Though the roots of this social criticism lie in teachings promulgated by medieval *sants*, it connects the agenda of contemporary *sant* movements with that of present-day reformist thinkers whose aims are to purify and rationalise religiosity: 'The characteristic model for these reformers was and still is a religion of doctrinal coherence whose prevailing mood or disposition was one of reason, sobriety, and selfless service' (Bayly 2004: 117).

The point about non-sectarianism is important. The *sant* tradition is not exclusively Hindu or Sikh but venerates the teachings of *sants* who have been important and influential in each religion. In this devotional context, distinctions between Hindus and non-Hindus and indeed distinctions of caste and other internal differentiations of 'community' are downplayed in favour of shared devotional attachment to a spiritual master.

Worship among these movements consists largely of devotional singing (*bhajan*), regular gatherings (*satsang*) at which devotees listen to the guru's discourses, and service of the guru (*guru-seva*).[5] The notion of *guru-seva* is particularly important. Devotees' donation of their blood is undertaken as an aspect of this devotional prerequisite. Though gurus say blood donation is *manav-seva* (service of humanity), devotees view it just as much as *guru-seva* since it is their gurus who ask them to do it and who, in effect, their donation activities serve to glorify. Juergensmeyer (1991: 127–146) has demonstrated the central role of *seva* activities in the Radhasoami movement, and this is equally true of the Dera Sacha Sauda and Nirankari Mission. Analogous to the making of offerings and sacrifices in a Hindu temple, *seva* is offered as worship and praise and is a critical means of obtaining the guru's blessings (Juergensmeyer 1991: 142). Warrier (2003), Mayer (1981), Babb (1986) and McKean (1996: 9, 140) have additionally highlighted the importance of *seva* activities in more obviously 'Hindu' guru-led institutions outside the *sant* tradition. As this would suggest, these *sant* movements form part of a much wider landscape of gurus, and they define themselves as much in relationship to them as to each other.

Not all *sant* movements are based around revered living gurus, but the Dera Sacha Sauda most certainly is, and this has created tensions with mainstream Sikhs for whom the guru is an abstract, transhistorical entity, meant to reside within and not outside the devotee. The final living Sikh guru, Gobind Singh, proclaimed in 1708 that the spiritual energies of the guru had been transferred into the Sikh sacred book, the *Adi Granth*, thus effectively proscribing the emergence of future human gurus. As I illustrate elsewhere (Copeman 2008), violent clashes between one *sant* movement – the Sant Nirankari Mission – and mainstream Sikhs over this and other matters played a significant role in the institutionalisation of blood donation as an important feature of Nirankari religious life. Indeed, the fraught relationship between orthodox Sikhs and *sant* movements with living masters has been central to the recent history of blood donation in India. Not only did the assassination of a Nirankari guru by a militant Sikh group in 1980 lead to the inauguration of frequent blood donation events at which devotees both commemorate and re-enact the guru's violent bloodshed, it was orthodox Sikh attacks on the Dera Sacha Sauda in May 2007 that precipitated the giving of blood by its devotees as a novel protest form.

Devotion and reform

Founded in 1948 and based on the outskirts of the city of Sirsa in Haryana state, the Dera Sacha Sauda has twice broken the world record for most units of blood collected in a single day, first in December 2003 on the death anniversary of a former guru (attracting 15,432 donors), and then again in October 2004 as part of the mortuary ceremonies of the present guru's father (17,921 donors). As with the Nirankaris and Radhasoamis, the guru is of pivotal importance as devotional cynosure of the movement, and the institution of *guru seva* is strikingly prominent. Recalling south Indian guru Sathya Sai Baba's Sarvadharma symbol, made up of symbols from the principal world religions, the spiritual master's name, Guru Maharaj Gurmeet Ram Rahim Singh Ji, combines linguistic inflections from several religious traditions. When addressing *satsang* (worship) gatherings, the guru sits beneath co-joined Hindu, Sikh, Muslim and Christian symbols which, through a contrivance of directing lines, appear to emanate from the figure of the guru, thus depicted as these religions' incarnate union.

'Dera' – the extended residential site of an influential figure – has similar connotations to 'ashram', while 'Sacha Sauda' means literally true deal or dealings, and is an allusion to the fact that monetary gifts from devotees are not accepted by the organisation which claims to be fully autarkic, owning extensive lands in Haryana and in other states on which its resident '*sadhus*' (renouncer-monks) work '18 hour days'. From the produce of the land, the movement runs a supermarket; it also manages a petrol pump, a biscuit factory, and one of its ashrams houses a revolving restaurant and boating lake, said to be for the benefit of the hundreds of thousands of pilgrims who annually converge on the site. Strong disapproval of monetary donations is expressed in all the literature it produces and in *satsang* gatherings by the guru himself: 'Man...wants to bribe God. Man does not understand that God, who made this world, and made this entire universe, can he not make your few small rupees? Just think about it! But nobody thinks!'[6] *Satsang* orations consist of many such exposés of supposedly exploitative religious practices – the offering of money in temples, but also the donation of funds to high profile, media savvy gurus such as Sathya Sai Baba. Demanding that devotees be reflexive and critical ('Use your mind!'), the guru constructs himself as their honest protector.

Dera Sacha Sauda devotees are predominantly working and middle-class Hindus hailing from towns and villages in Haryana and Rajasthan, though there are also many Sikh adherents, and the gurus have all, thus far, been of Sikh origin. The organisation claims to possess more than a million devotees and 38 ashrams throughout India, though most of these are in the north of the country. Like those of the Radhasoami tradition (Juergensmeyer 1991), Dera Sacha Sauda teachings emphasise the importance of reciting sacred words (*ram nam*) for the achievement of transcendence. The guru exhorts devotees to abstain from alcohol and meat, to be faithful in marriage, never to lie and, as was noted above, never to make religious offerings of money. The group also expostulates a strong anti-ritual message: 'Dera Sacha Sauda does not believe in any kind of false practices, false pretensions, misguidance or any kind of show off which has nothing to do with spiritualism and those ritual practices which take you away from your real goal'.[7] In my discussions with them, devotees repeatedly drew attention to the absence of distracting 'rituals' (*rasmen*), something they saw as one of the movement's key defining features. Instead, they argued, it is through the more direct method of *guru-bhakti* that spiritual progress results. The movement's professed aversion to ritual and 'show off' (*tamasha*) situates it in a reformist tradition that has been determined to undermine 'superstitious ritual'.[8]

The guru extols offerings of medicine and blood because they encode usefulness and non-convertibility. This encoding of donated blood results from the assumption that only those who need them receive transfusions.[9] Money, however, is proscribed because it can be put to any kind of nefarious use and therefore encodes not usefulness, but dangerous ambiguity: 'O! You give a donation of money and say it is given to God. Does God sign on that?' The guru valorises the cow not because it is a physical abode and amalgamation of the gods as espoused in classic Hindu texts, but because of its utility: 'Animals are ahead of humans in many ways: the body of the cow is used after death [i.e. for leather], and while they are living their dung is also used [for fuel and manure].[10] Their bones are also used. But out of these, nothing is used of a human being'.[11] Blood donation, however, makes a human being useful and is thus 'really a great donation and God must bless you for this'.[12] The point I wish to underline here is that the guru's exaltation of blood donation on account of its usefulness is a means by which devotees can do what the movement supposedly bans in the name of its austere brand of purified, 'reformed' religiosity: i.e., give gifts and conduct spectacular ritual. 'Virtuous utility' both sanctions and sanctifies the taboo.

As I have suggested, blood donation is centrally placed within the movement's social reformist agenda. This is especially so in regard to efforts to reform the nature of giving. I offer an example. Articles published in regional newspapers by journalists covering a blood donation event staged on the death of the present guru's father in October 2004 recount the guru's address at the inauguration, in which he reportedly declared that the decision to hold the camp was taken at the *pagri* ceremony of his father, Sadar Magghar Singh.[13] It was staged in order to 'contribute towards the service of mankind rather than having a public feast'.[14] In another article on this 'great *yagna*[15] of blood donation', the guru is quoted as claiming that *manav seva* (service of humanity) such as blood donation is 'of greater significance than organizing any type of feast'. The reflective replacing of customary feasts and rites with blood donation in 'the service of mankind' I term 'substitutive ennoblement'. In addition to making similar statements to those cited above from the local media, devotees in Sirsa told me explicitly that they thought of themselves as performing blood donation in replacement of *pind-dan*: the giving of food to the deceased. As one male devotee told me: 'Guru Ji said our aim is to serve society (*samaj*). So [when his father died], instead of *pind-dan*, which is actually just a waste (*bekar*) of money, we did *rakt-dan* [blood donation] which is useful (*upyogi*) and a great donation (*rakt-dan maha-dan hai*)'.

There is, though, ambivalence at the heart of these processes of reform. Given the movement's conversion of the 'useless' gift of food to the deceased into the 'useful' gift of blood for humanity, there are seemingly grounds for defining the Dera Sacha Sauda as a 'religion of utility' (Copeman 2009a: 4, 65–7). But the order's commitment to utility is enigmatic – that is, both within the rules of utility and against them (Bhabha 1984) – for its excessive blood collection leads to disutility and wastage. Firstly, more blood is collected at its events than is possible to transfuse, and secondly, the rush to surpass world records results in the abundant collection of 'quality not sufficient' (i.e. under-filled) blood units. Blood donation thus permits a double movement of reform and reversion, allowing what claims to be a restrained and austere devotional order, opposed to *tamasha* (spectacle) and 'show off', to enact spectacles of donor excess, with the utility of blood donation helping sanction that which would otherwise be proscribed.

The example just presented does not negate the status of blood donation as an important instrument of social reform; in fact, it attests to the elevated status of blood donation within the reformist repertoire, for the power of 'blood donation' is demonstrated as a circulating store of virtue in its being employed to sanctify or at least facilitate forces oppositional to its own reformist message. I do not wish to suggest here a usage of blood donation as mere disguise or concealment of the movement's 'true' aim of spectacular enactments of conspicuous extraction. Rather, what we are presented with here are coterminous objectives concerning a burgeoning spectacular religiosity and the reformist valorisation of utility which blood donation helps mediate.

I suggest now that if gift-giving needed to be 'reformed' by the Dera Sacha Sauda – that is, instilled with virtuous utility – in order to be made legitimate, the movement has sought to do something similar in the arena of *protest* styles. The critical events to which I now turn showcase the extent to which blood donation has developed into a multilayered instrumental and expressive act both within the Dera Sacha Sauda and beyond.

Donating blood in protest

In May 2007, the Dera Sacha Sauda was headline news around the world. A global presence has been the aim of the movement for a number of years ('the day will come soon when every child in this world will have the name of Sacha Sauda on his tongue', says the movement's website, http://www.sachasauda.com), but its ambition was achieved in a manner it would neither have sought nor ever envisaged. 'Cult leader sparks Sikh riots with "guru" stunt' was the headline in the British *Independent* newspaper on 21 May 2007. The allegation was that the Dera Sacha Sauda guru had, in an advertisement in a local newspaper, impersonated the tenth and – according to orthodox Sikhs – final living Sikh guru, Guru Gobind Singh. The Sacha Sauda guru was reported to have been shown administering nectar (*amrit*) to devotees in the manner (and dress) of Gobind Singh. As was noted above, relations between *sant* movements headed by living gurus and orthodox Sikhs are tense. The alleged ritual and sartorial imitation of Gobind Singh by the Dera Sacha Sauda guru precipitated violent clashes in Punjab in May 2007 which left at least one person dead, and many more injured. News reports proclaimed the situation to be 'perhaps the most serious law and order crisis since the militancy days', a period of threat and instability that began in the late 1970s with the Sikh-Nirankari clashes and which continued into the 1980s with agitations for a separate Sikh homeland.[16] Indeed, the parallels with the Nirankari controversy of several decades before, which, as has been noted, culminated in the killing of the Nirankari guru, are striking, with the Dera Sacha Sauda now taking the place of the Nirankaris as the orthodox Sikh object of opprobrium.

News reports commented on the hard line taken against the devotional order by the Akal Takht, the primary seat of Sikh religious authority, based at the Golden Temple in Amritsar, and by various factions of the Sikh nationalist political party, the Akali Dal. Some Sikh organisations called for suicide squads to 'carry on the task of uprooting all Sacha Sauda *deras* in Punjab', while the Akal Takht somewhat more measuredly demanded a *bandh* (in many ways like a strike, though frequently forcible rather than voluntary) and protest marches in addition to the closure of all Dera Sacha Sauda premises in the state.[17]

The Dera Sacha Sauda guru, under intense pressure, eventually made a not very fulsome apology for the offence caused, which was rejected by the Akal Takht *jethedar* (spiritual leader), who 'quoted the "Gurbani" (religious scriptures), saying that even the Guru, who was bountiful in pardoning, would not accept anything from a person whose spirit did not reflect a mindset suitable for a religious way of life'.[18] The marches and boycotts continued for a time, but after several weeks an uneasy calm descended which has largely persisted until the present time of writing.

What was not reported in the international news coverage was that the Dera Sacha Sauda staged a blood donation camp in order to protest against the Akal Takht's decree (*hukamnama*) that its ashrams in the Punjab be closed. One Indian news article quoted the Sacha Sauda's senior vice chairman: '"The question of vacating *deras* does not arise at all"...Prithviraj told a gathering of followers immediately after Sikh priests gave a fresh ultimatum to the Punjab government to shut down all *deras* in the state. He said the group will protest against the *hukamnama* in a peaceful manner with its followers donating blood and organs after death'.[19] A television news report documenting a camp in Sirsa on May 21 at the height of the tensions quoted a devotee: 'We believe in non-violence, following that we have organized this blood donation camp. We hope the central government and the judiciary will definitely do justice'.[20]

This usage of blood donation as a mode of protest alerted me to what is in fact a wider emergent trend in India, that of giving blood as a form of contestatory behaviour. Blood donation has, in fact, been incorporated into the repertoire of protest techniques throughout the country.

It is clear that the Dera Sacha Sauda's episodic enactments of conspicuous extraction enable the movement to achieve a number of different things, from the spectacles of excess its austere theology officially repudiates, to renown and pre-eminence among the different devotional movements which likewise engage in high-profile social service activities. As was noted above, on the death of the present guru's father the movement conducted blood donation as the substitutive ennoblement of *pind-dan* and 'wasteful feasting' in what might be termed the contrastive achievement of virtue ('useful' vs. 'useless' giving, this-worldly vs. other-worldly giving, and so on). This provides a helpful starting point for beginning to understand how blood donation has come to possess purchase for the Sacha Sauda and others as a mode of contestatory behaviour. The Dera Sacha Sauda's giving of blood was unlike the styles of protest called for by Sikh organisations in two important respects: First, in contrast to the Akal Takht and Akali Dal-inspired protests which took the form of *bandh*s, marches and even calls for the deployment of suicide squads, the Sacha Sauda's protest behaviour was demonstrably 'socially productive'; and second, in contrast with orthodox Sikh threats of violent bloodshed, Sacha Sauda devotees also shed their blood, but *non-violently*.

The Indian repertoire of protest techniques has been and continues to be the subject of much criticism among social reformers in modern India. Nehru, for instance, criticised hunger strikes and *bandh*s in the 1950s as 'obsolete' political methods: 'I feel that the time is gone when we could solve our problems in this way in India or anywhere else' (quoted in Chakrabarty 2007: 37). The *bandh* attracts criticism as being a particularly unproductive style of contestatory behaviour. Indeed, *bandh*s have been officially proscribed since 1998 (though they still occur with great frequency).[21] A recent newspaper editorial disparaged *bandh*s as 'fully paid holidays' which cause misery and destruction.[22] Another declared that they 'demean the very cause they are supposed to promote' and spoke of the unfortunate vortex effect of calling a *bandh*: 'another *bandh* to protest the first *bandh*, followed by a strike by the people against whom the first *bandh* had been called'. The writer proposes alternative, 'more effective' modes of protests such as the wearing of black armbands and, if the protest is in response to an atrocity, a fund to help the victims. Significantly, blood donation is also suggested as a more constructive way than a *bandh* to express discontent.[23]

The 'positive action' of the Sacha Sauda's blood donation protest thus contrasted with the *bandh*s called by Sikh organisations, the latter being a mode of protest frequently scorned as being benighted and self-indulgent. Protest blood donation may in consequence be considered a style of protest congruent with a social reformist agenda that promotes virtuous utility and as being the substitutive ennoblement of key existing styles in the protest repertoire such as *bandh*s, marches and fasting. Other examples support this view. For instance, members of the United Forum Against Consumption of Liquor in Hyderabad are reported to have 'constructively' donated their blood in protest against the preponderance of 'anti-social' liquor shops and bars.[24] Of particular interest is the Oasis NGO, based in Vadodara, Gujarat, which organises workshops for 'character building' and 'healthy living', and which in 1999 conducted what it termed a 'constructive *dharna*'[25] in protest at a 'smear campaign' against it: 'Managing trustee Sheeba Nair said a blood donation camp was organized...as Oasis wanted to keep away from rallies, slogan shouting and *bandh*s, the usual modes of protest'.[26] The example corresponds with that presented above concerning *pind-dan*: whereas in that example it was forms of gift that were subject to reform via blood donation, here it is the *dharna* that is renovated via exactly the same means.

What the Oasis example underlines so well is that blood donation, in contrast with *bandh*s, disruptive rallies, and so on, is, as a protest form, the antithesis of civil disobedience. After all, blood donation is something the government persistently (if somewhat ineffectively) enjoins Indians to do. Whereas *bandh*s potentially disturb state functioning (witness Nehru's comments above), blood donation is a state-loving mode of protest. What has been termed 'avoidance protest' consists of 'dissatisfied groups seek[ing] to attenuate their hardships and express their discontent through flight, sectarian withdrawal, or other activities that minimise challenges to or clashes with those whom they view as their oppressors' (Adas 1981: 217). On one level, the Dera Sacha Sauda's protest blood donation did indeed reproduce the logic of avoidance protest insofar as it probably did help minimise clashes; and yet, unlike strikes (*hartals*), non-payment of taxes and election boycotts, protest blood donation constitutes an extreme *positive engagement*. I do not suggest these other protest styles are not politically engaged (the manner of their engagement takes the form of disruptive avoidance); however, blood donation engages not through avoidance but through intensified involvement, i.e. through the active pursuit of civic virtues. Unlike 'abstemious acts' (Haynes 1991: 276) such as fasting and strikes, where the structural logic is one of closure and physical subtraction, blood donation subtracts (from veins) only in order to make an addition (to 'society'), and hence is structurally open and engaged. Blood donation extends the body into the world, while fasting withdraws the body from it.

The Sacha Sauda blood donation protest was not overtly directed towards the state, though it was probably designed to impress several different audiences, including state governments. What is the logic of a blood donation protest when it *is* directed at the state, given that blood offerings, far from disrupting state functioning, actually contribute to it? An example would be the blood donation protests, discussed below, of medical trainees agitating against state reservations (i.e. positive discrimination in favour of 'backward castes') in medical training institutions. Blood donation as contestatory behaviour in disputes with the state, it might be suggested, draws the state into a close embrace by doing exactly what the state wants, with control being exerted through overbearing compliance and suffocating propinquity rather than disobedience. This model of protest bears a resemblance to what Appadurai (1990: 97) has termed 'coercive subordination', the attempt being to neutralise or control the intentions of the state through exalting it with gifts of blood. Gifts of blood in protest may thus be seen as a kind of stockpiling of a physical form of moral asset – the attempt is to store virtue and/or to force on others a moral debt redeemable through redress to the grievance in question. Nandy (1970: 72) rightly points to a distinction between 'formal means of demand articulation and protest' such as recourse to the law and 'extra-systematic pressures' such as fasting and *bandh*s. The interesting thing about blood donation protests is that they appear to hang midway between the two means of demand articulation. Of course, blood donation protest is not a 'formal' means of redressing a grievance, but neither is it at all 'disobedient' or antithetical to state systems; quite the reverse. It bridges, in this respect, systematic and extra-systematic forms of demand articulation.

The second contrast I identified above concerns modes of bloodshed. At the height of the tensions in May 2007 a pro-Sacha Sauda blogger ('samiinsaan') wrote: 'The followers of DSS are expressing their dissatisfaction by donating blood but bad tempered people are flowing the blood of innocent people'.[27] The comment invites a comparison with the Sant Nirankari devotional order. I alluded earlier to the Nirankari Mission's response to the taking of the life of its former guru (the present guru's father): in an example of the reflective mobilisation of blood donation as the non-violent response to an act of extreme violence, Nirankari devotees reacted to the life-taking bloodshed which claimed their guru's life by donating their own blood to help save lives. The Nirankari Mission thus sought to convert the martyrdom of its guru from an experience of victimhood into one of self-initiated ennobling virtue, attributing to the successor guru himself the aphorism that is now used as an exhortatory slogan about the transformation of violent bloodshed into spiritually meaningful donation: 'Blood should flow into veins (*nari*), not drains (*nali*)'. The slogan serves rhetorically to elevate the Mission above petty everyday violent disturbances, the contrast between veins and drains drawing attention to distinctions between the peace-loving Nirankaris and other religious communities, such as extremist Sikhs and Hindu and Muslim 'fundamentalists', who they characterise as being wedded to violent outrages.

In a similar way, the Sacha Sauda's positively reconstituted blood shedding enabled the movement to achieve virtue contrastively. To quote 'samiinsaan' once more: 'They [Sikh protesters] are forgetting...that Respected Sri Guru Gobind Singh Ji has given the swords for saving [the] helpless, not to make the blood shed out of helpless people.[28] These terrorists have taken the lives of many innocent people in the last decade when there was [the] demand of [a separate Sikh homeland]... Again same scene is in Punjab, whereas on the other hand Dera Sacha Sauda devotees are proving themselves to be *real Sikhs* by donating blood, by giving their kidneys, bone marrow and eyes after death. Even some devotees are donating their whole body for the sake of humanity' (my emphasis). 'Samiinsaan' thus provocatively suggests that the Dera Sacha Sauda's protest in the form of 'saving the helpless' makes it *really* Sikh, whereas the violent nature of the orthodox Sikh response represents a perversion of the true Sikh faith. The claim is also implicitly made that Dera Sacha Sauda devotees' donation of body parts (blood, kidneys, bone marrow, eyes and even the totality of their bodies) reanimates a consecrated template laid down by Guru Gobind Singh in a way that orthodox Sikhs fail to do. As is well known, Gobind Singh is said in 1699 at the foundation of the *khalsa* (orthodox Sikh community) to have demanded that five devotees offer him their heads (Mahmood 1997: 43). But now, says 'samiinsaan', it is only the Dera Sacha Sauda which fulfils Gobind Singh's teachings in this respect, with its bounteous corporeal offerings 'for the sake of humanity'. The contrast thus becomes one between 'Sikhs' and *real Sikhs* – as defined (or revealed) by those who are willing to excavate their bodies most deeply.

Gandhian protest styles

The donation of blood as contestatory behaviour thus acquires reformist virtue by means of contrast with other forms of protest and bloodshed. I argue now, however, that what really gives blood donation force as a mode of protest derives from the ways in which it both reconnects with and surpasses Gandhian styles of protest. (I do not wish in the following to suggest that what I am calling Gandhian protest styles are necessarily exclusive to the Gandhian nationalist tradition. But while Gandhi undoubtedly learnt and drew deeply from numerous existing colonial and pre-colonial archetypes (see Spodek 1971), he also adapted them in all sorts of significant ways – not least in wielding them on a far broader canvas than had hitherto been the case).

The Gandhian leaders of the merchant protests against British rule in early twentieth–century Surat 'emphasized the virtue of acts of abstinence and self-control. Involvement in protest was an act of renunciation (*tyag*), of self-suffering, and of penance (*tapascharya*)... Rhetoric stressing the selflessness of resistance essentially confirmed the merchants' sense of the moral rightness of their actions, and contributed to a growing feeling of ethical ascendancy vis-à-vis the British'. Central to Gandhian resistance was 'the control of personal emotions' (Haynes 1991: 276). As Gandhi (1961: 6) explained, *satyagraha* – literally 'truth firmness', Gandhi's method of seeking justice through passive resistance as a 'weapon of the weak'[29] – does not admit 'of violence being inflicted on one's opponent, [rather] he must be weaned from error by sympathy and patience'. This designation finds a reflection in the account of the pro-Sacha Sauda writer cited earlier, his suggestion being that the blood donation protest was characterised by similar principles of calmness and restraint: 'They [Sikh protesters] started breaking *namcharcha ghar*s [prayer meeting houses], setting fire to the homes of *sadh sangat* [devotees; lit. fellowship of the seekers after truth], and misbehaving with the effigies of guru ji and this way hurting the emotions of crores of people who are the believers of Dera Sacha Sauda...On the other hand, the followers of Dera Sacha Sauda are so calm and silent that to protest they organize blood donation camps, filling forms of body parts donation after death, eye donation after death, and donating kidney, liver and bone marrow alive'.[30] It was not only 'samiinsaan' who connected the blood donation protest with Gandhian styles.[31] So the first association between protest blood donation and 'Gandhianism' pertains to the exhibition of restraint and non-violence.

The second relates to asceticism. Like fasting-as-protest, blood donation-as-protest is enacted both by and on the physical person of the protestor. Moreover, *satyagraha* – 'truth-firmness' – was conceived by Gandhi (1961: 6) as 'vindication of truth not by infliction of suffering on the opponent but on one's self'. There are various ways in which Nirankari and Sacha Sauda devotees have translated blood donation into ascetic practice, with blood donation coming to be considered by them an act of ascetic self-denial. During fieldwork I witnessed or heard about attempts being made by devotees to donate blood despite being physically unfit to do so, or to give twice at one donation event, or three times on consecutive days. Indeed, as I explain elsewhere (Copeman 2009a: 91), it would appear that some devotee-donors welcome the thought that their physical frailness may make blood donation physically taxing or dangerous for them. They are thus inclined to treat blood donation as an austerity like those practiced in the form of fasts and other meritorious acts of ascetic discipline. Asceticism may be viewed as a kind of 'enlightened self-interest', with the practice of seemingly self-denying austerities in the present ensuring the self's future spiritual purity and advantage (Laidlaw 2002: 321). So too, seemingly, with some devotee blood donors. Needless to say, attempts made by physically frail devotees to donate contravene all official guidelines. However, if one's priority is to achieve spiritual benefits through meritorious acts of bodily austerity, such a situation may appear as a welcome opportunity. Therefore, in instances where blood donation protests are performed by populations who understand the action as inflicting suffering more on themselves than on their opponents, the connection to a Gandhian idiom is established all the more forcefully.

The third connection to Gandhian styles concerns the directional orientation of protest. For Gandhi, it was essential that fasting and other modes of protest transcend self-interest – he was unequivocal in stating that fasting should be 'for the good of others' (Alter 2000: 41). As was noted above, the protest actions of Surat merchants were similarly pervaded by a rhetoric of selflessness which lent their acts of civil disobedience a sense of 'moral rightness' and 'ethical ascendancy'. The important point here is that blood donation is continuous with this rhetoric whilst, at the same time, subtly surpassing it, since its 'being for the good of others' is built into or made explicit in its enactment in a less ambiguous way than, say, fasting. As Alter describes so well, Gandhi continually downplayed the political utility of the fast; the means (self-purification) were the primary end of its performance: 'Our concern is the act itself, not the result of the action' (Alter 2000: 52, 49). And yet, the fast in present-day agitational strategies *is* used as a means to an end: the principles of 'self-abnegation' and 'impulse-control' are critical tools for the achievement of political potency (Nandy 1970: 72).[32] For all the talk of protest 'for the good of others', techniques that foreground self-abnegation remain open to the charge that they simply force the will of the protester upon those with whom he or she disagrees.

Blood donation possesses efficacy here as a protest technique which 'self-evidently' produces effects which are 'for the good of others', being possessed of a transparent medical utility function which the protester seeks to harness in order to avoid charges of 'coldly calculated blackmail' (Alter 2000: 29). Both the fast and blood donation can plainly be used as techniques of moral suasion for the furtherance of *any* kind of cause. But blood donation differs from the fast in one vital way: counteracting its indeterminacy vis-à-vis protest usages is its predetermined status with respect to its medical usage. Ambiguity on one level is met with clarity of encoding on another. As was noted above, blood donation is proclaimed by the Dera Sacha Sauda guru to be a far greater offering than that of dangerously ambiguous money because of its predetermined usage, and it is this usage which also counts in its favour as a protest technique. Thus, in fulfilling more transparently than fasting the injunction that protest be 'for the good of others', blood donation not only connects with Gandhian styles but improves upon them. Bleeding under the spotlight at media-friendly donation camps, the protesting blood donor exhibits his or her beneficent helpfulness with a clarity that the protesting faster would find it difficult to achieve. Even if an observer were to question the validity of the grievance in question, they would likely concur that the means of redressing it were 'socially useful and out-reaching rather than inward-turning' (Vanita 2002: 155). This perhaps begins to explain why there are signs that blood donation is the emergent successor to the Gandhian fast as moral suasion technique of preference.

Several examples seem to bear out this argument. When, for instance, in November 2007 trainee medics in Tamil Nadu protested against a year-long extension of their Bachelor of Medicine/Surgery (MBBS) courses because of a new compulsory rural service scheme, blood donation was central to their agitation, for it put on view their legitimating commitment to utility. The protest as a whole consisted of a virulent conjunction of signifiers, with medics not only striking and undertaking a 'fast unto death' but also sporting 'saffron dress and carr[ying] books on their heads to drive home the point that they have to forego everything if the Centre went ahead with its decision and extended the MBBS course by a year'. The critical supplement to these protest styles, however, was the students' donation of blood, given in order 'to make it public they were not against rural service but only opposed compulsory service'.[33] Thus did they attempt – by means of giving their blood – to preserve their selfless virtue even as they fought for their self-interest. In a similar case in 2006, trainee medics in Bangalore fasted, conducted numerous boycotts, formed a silent human chain and donated blood in protest against proposals to reserve 27 percent of places in elite medical institutions for OBCs (Other Backward Classes). This campaign, wrote one medic, 'provided blood to the needy, irrespective of him being SC/ST/OBC [Scheduled Castes/Schedules Tribes/Other Backward Classes] and negated the propaganda of the Government saying that the strike, protests are against public and against patients'.[34] Again, blood donation was enacted by medical trainees as a means of demonstrating that they cared about the good of all as a counter to the widespread perception that a dominant group was protesting in order to further its domination.

There are, then, several connective threads that suggest congruence between protest blood donation and protest fasting. Like the archetypal fast, blood donation is suggestive of non-violence and ascetic discipline. Where it surpasses Gandhian protest is in its making explicit (for purposes of exhibition) the external directional orientation meant to be attached to acts of Gandhian protest. Contestatory blood donation thus combines Gandhianism with demonstrable utility in a new and powerful protest mode. Its constructiveness contrasts with the forms of protest that it replaces: *hartals*, *bandhs*, non-payment of taxes, election boycotts, and so on. Its inner logic is not one of avoidance but of intensified involvement in a new combination of systematic and extra-systematic forms of demand articulation. Unlike the threat of dying to redress a grievance (Hopkins 1901), protest blood donation operates through imposing a moral debt to be repaid through redress to the grievance in question. The significance of protest blood donation lies in its being situated in an existing milieu of moralised and embodied protest techniques, while at the same time elaborating on this milieu by making utilitarian, other-oriented claims. We can frame this productively in terms of Foucault's delineation of 'technologies of the self', defined by the French philosopher and historian as 'permit[ting] individuals to effect, by their own means, a certain number of operations on – their own bodies, their own souls, their own thoughts, their own conduct – and this in a manner so as to – transform themselves, modify themselves, and to attain a certain state – of perfection, happiness, purity, supernatural power' (Foucault 2000: 177). The Gandhian fast is a classic technology of the self insofar as the self-transforming operation of curbing food intake is undertaken in order, in part at least, to obtain a measure of political power. Protest blood donation may be defined similarly as a political technology of the self, but with one major difference to the Gandhian fast. Here

I borrow Street's (2009) concept of 'relational technology': protest blood donation is a *relational technology of the self* because it symbolically draws in (is performed *for*) the 'other' that fasting is considered by some to occlude, thereby both mirroring and inverting the self-transforming operation of the fast.

As a still embryonic protest form, it is not possible to say whether contestatory blood donation will ever succeed the Gandhian fast as *the* virtuous technique of moral suasion. What can be said about protest blood donation, however, is that it vividly attests to the wide range of cultural forms open to reformist renovation by means of the incorporation of blood donation. The Dera Sacha Sauda has been at the forefront of such renovations, as evidenced both by its performance of protest blood donation and by its substitutive ennoblement, via blood donation, of existing giving forms.

Conclusion

This chapter has concerned the shifting nature of knowledge about how to engage the body in protest. Protest blood donation achieves reformist virtue by means of contrast with those other forms of protest and bloodshed in comparison to which its modernist constructiveness is thrown into sharp relief. But I have also depicted the emergence of protest blood donation as a process connoting the *re-embodiment* of protest. This is because of the simultaneous intimacy and distance that characterises the relationship between 'novel' blood donation and 'archetypal' fasting. To cite again Jenkins (2010: 161), we may say that protest re-embodied in blood donation 'consists neither in the transmission of an essence, nor in the invention of the new, but has a third form of existence'. The two embodied protest forms possess common features: they share a corporeal ethical register which centres on ascetic self-control, and both are enacted by and on the body of the protesting self. Blood donation, however, extends the corporeal ethical register of protest into an engagement with utility to an extent that fasting would seem unable to achieve. Perhaps in this respect, then, the re-embodiment of protest amounts to the de-traditionalising of protest. A further insight, though, is needed in order to complete the picture: the recognition that 'de-traditionalising' can indeed be 're-traditionalising in another guise' (Strathern 1994: 18).

Acknowledgements

My thanks to Kriti Kapila, Norbert Peabody, Nathaniel Roberts, Alice Street, Rupa Viswanath and the editors of this volume for helpful comments on versions of this chapter.

Notes

1 These works by Strong and Seeman on populations excluded from blood donation thus provide examples of what elsewhere (Copeman 2009b) I term bioineligibility. The term is meant to complement Cohen's (2004: 169) concept of 'bioavailability', a contingency 'organized variously around the loving or charitable gift, the commoditized sale, or the authoritarian or piratical forced extraction or seizure'.

2 Most medical opinion is now strongly opposed both to paid and to replacement donation. Paying donors is said to provide an incentive to conceal disqualifying factors such as HIV/AIDS (Brooks 2004: 282), while replacement donation is thought to pressurise patients' relatives unduly, pushing many to seek paid donors to donate in their stead, thus threatening those who can not arrange for this kind of donation with denial of life-saving treatment.

3 On the *sant* tradition and *bhakti* movements in north India, see Lele (1981), Schomer and McLeod (1987), Lorenzen (1995).

4 Though the words *sant* and saint are extremely close in meaning, there is apparently no etymological connection between them (Juergensmeyer 1991: 22).

5 *Satsang* means literally 'association of the good'. *Sant* and *bhakti* movements are 'congregational religion[s]' (Lorenzen 1995: 24).

6 Retrieved 28 June 2009, from http://www.sachasauda.com/satsangs/English2.htm.

7 Retrieved 28 June 2009, from http://www.derasachasauda.org.

8 Since its inception in 1875, the Hindu reformist movement Arya Samaj has been particularly active in pursuing campaigns against idolatry, caste and popular ritual traditions (Hansen 1999: 71).

9 The assumption is logical but problematic in the Indian context in which many doctors are said to administer transfusions as a 'tonic' rather than on the basis of need. An everyday example of the encoding of utility in gifts is when passers-by give food instead of money to beggars in order to forestall their spending of the money given to them on drugs or alcohol.

10 Specifically, 'the cow is seen as an embodiment of Laxsmi. She is also a maternal figure associated with nourishment and nurture. Eating her flesh is therefore suggestive of matricide' (Babb 2004: 233n).

11 *Dhan Dhan Satguru Tera Hi Aasra, The Truth*. DSS Video CD Vol. 1 (Dera Sacha Sauda n.d.).

12 Retrieved 28 June 2009, from http://www.sachasauda.com/satsangs/English2.htm.

13 *Pagri* ceremonies are held for the passing of the deceased male's turban to the new head of the family, ideally the deceased's eldest son.

14 *Rupana Times*, 11 Oct. 2004.

15 A *yagna* is a popular fire offering ritual.

16 Retrieved 28 June 2009, from http://www.sify.com/news/fullstory.php?id=14454113.

17 *Times of India*, 21 May 2007. Responses to the controversy were highly politicised, with the Congress – which has been the beneficiary of Sacha Sauda electoral support – defending the order, and the BJP, which is in an alliance with the Akali Dal, heavily criticising it.

18 *The Hindu*, 30 May 2007.
19 *Times of India*, 21 May 2007.
20 Retrieved 28 June 2009, from http://rtv.rtrlondon.co.uk/2007-05-21/37340eb3.html.
21 Most frequently in West Bengal, according to the *Economic Times* (18 Dec. 2006), where the average number of *bandh*s per year is 40-50.
22 *The Hindu*, 31 Jan. 2005.
23 Retrieved 28 June 2011, from http://www.rediff.com/news/2003/jul/30diary.htm.
24 *The Hindu*, 13 June 2007.
25 *Dharna* usually refers to a hunger protest, classically performed by a creditor at the door of a debtor in order to exert moral pressure for repayment. Though officially proscribed during British rule, protestors still frequently characterise their protests as *dharna*s.
26 *Indian Express*, 26 Oct. 1999. See also http://www.oasiswebsite.com.
27 Retrieved 28 June 2009, from http://www.ndtvblogs.com/views/viewblogs.asp?gl_guid=&blogname=sami&q_userid=12887
28 The carrying of a sword (*kirpan*) is a recognisable marker of the orthodox Sikh, along with unshorn hair, a steel bracelet (*karha*), and a comb (*kanga*).
29 See Gandhi (1949: 266) for full definition.
30 Retrieved 28 June 2009, from http://www.ndtvblogs.com/views/viewblogs.asp?gl_guid=&blogname=sami&q_userid=12887. A crore is 10 million.
31 Devotees' 'calm' and 'silent' giving of their blood as a method of protest led to newspaper declarations that the Dera Sacha Sauda action was an attempt at '*gandhigiri*' (e.g. see http://sify.com/news/fullstory.php?id=14454113, http://www.ibnlive.com/news/dera-takes-to-gandhigiri-as-sikh-clergy-talks-tough/40955-3.html ('Dera takes to Gandhigiri')). The term *gandhigiri* (loosely, action/actions characteristic of Gandhi) was popularised by the 2006 bollywood film *Lage Raho Munna Bhai* (Rajkumar Hirani) which has been credited with reviving interest in Gandhi throughout the country. The main protagonist in the film protests through the delivery of bouquets of flowers to the 'oppressor' he seeks to challenge, and there has apparently been a spate of 'flower protests' in the wake of the film. As Ghosh and Babu (2006) put it, 'Gandhi, the man, was once the message. In the India of the post-liberalization brand, *gandhigiri* is the message'.
32 See, in this respect, Cohen (2004) on Vajpayee and *hijra*s.
33 *The Hindu*, 30 Nov. 2007.
34 Retrieved 28 June 2009, from http://www.savebrandindia.org.

References

Adas, M. 1981. From Avoidance to Confrontation: Peasant Protest in Precolonial and Colonial Southeast Asia. *Comparative Studies in Society and History* 23, 2: 217–247.

Alter, J. 2000. *Gandhi's Body: Sex, Diet and the Politics of Nationalism*. Philadelphia: University of Pennsylvania Press.

Appadurai, A. 1990. "Topographies of the Self: Praise and Emotion in India," in C. Lutz and L. Abu-Lughod (eds) *Language and the Politics of Emotion*. 92–112. Cambridge: Cambridge University Press.

Babb, L. A. 1986. *Redemptive Encounters: Three Modern Styles in the Hindu Tradition*. Berkeley: University of California Press.

Babb, L. A. 2004. *Alchemies of Violence: Myths of Identity and the Life of Trade in Western India*. Delhi: Sage.

Bayly, S. 2004. "Conceptualizing from Within: Divergent Religious Modes from Asian Modernist Perspectives," in J. Laidlaw and H. Whitehouse (eds) *Ritual and Memory: Toward a Comparative Anthropology of Religion*. 111–134. Walnut Creek: Altamira Press.

Bhabha, H. K. 1984. Of Mimicry and Man. *October* 28: 125–133.

Brooks, J. P. 2004. The Rights of Blood Recipients Should Supersede Any Asserted Rights of Blood Donors. *Vox Sanguinis* 87: 280–286.

Chakrabarty, D. 2007. 'In the Name of Politics'. Democracy and the Power of the Multitude in India. *Public Culture* 19, 1: 35–58.

Cohen, L. 2001. The Other Kidney: Biopolitics Beyond Recognition. *Body and Society* 7, 2–3: 9–29.

Cohen, L. 2004. "Operability: Surgery at the Margin of the State," in V. Das and D. Poole (eds) *Anthropology in the Margins of the State*. 165–190. Santa Fe: School of American Research Press.

Copeman, J. 2008. Violence, Non-violence, and Blood Donation in India. *Journal of the Royal Anthropological Institute* 14: 277–295.

Copeman, J. 2009a. *Veins of Devotion: Blood Donation and Religious Experience in North India*. New Brunswick, N.J.: Rutgers University Press.

Copeman, J. 2009b. Introduction: Blood Donation, Bioeconomy, Culture, in J. Copeman (ed.) *Body & Society: Special Issue on Blood Donation, Bioeconomy, Culture* 15, 2: 1–28.

Davies, G. 2006. Patterning the Geographies of Organ Transplantation: Corporeality, Generosity, and Justice. *Transactions of the Institute of British Geographers* 31, 3: 257–271.

Feldman, E., and R. Bayer. (eds) 1999. *Blood Feuds: AIDS, Blood, and the Politics of Medical Disaster*. Oxford: Oxford University Press.

Foucault, M. 2000. *Essential works of Michel Foucault 1: Ethics: subjectivity and truth* (ed. P. Rabinow). London: Allen Lane.

Gandhi, M. K. 1949. *An Autobiography. The Story of My Experiments with Truth*. London: Phoenix Press.

Gandhi, M. K. 1961. "What Satyagraha Is: Satyagraha," in B. Kumarappa (ed.) *Non-Violent Resistance (Satyagraha)*. 6. New York: Schocken Books.

Ghosh, A., and T. Babu. 2006. Lage Raho Munna Bhai: Unraveling Brand Gandhigiri. *The Economic and Political Weekly* 41, 51: 5225–5228.

Hansen, T. B. 1999. *The Saffron Wave: Democracy and Hindu Nationalism in Modern India*. Delhi: Oxford University Press.

Haynes, D. 1991. "From Avoidance to Confrontation? A Contestatory History of Merchant-State Relations in Surat, 1600–1924," in D. Haynes and G. Prakash (eds) *Contesting Power: Resistance and Everyday Social Relations in South Asia*. 239–289. Berkeley: University of California Press.

Hopkins, W. 1901. On the Hindu Custom of Dying to Redress a Grievance. *Journal of the American Oriental Society* 21: 146–159.

Jenkins, T. 2010. *The life of property: house, family and inheritance in Béarn, south-west France*. Oxford: Berghahn.

Juergensmeyer, M. 1991. *Radhasoami Reality: The Logic of a Modern Faith*. Princeton: Princeton University Press.

Laidlaw, J. 2002. For an Anthropology of Ethics and Freedom. *Journal of the Royal Anthropological Institute* 8, 2: 311–332.

Laidlaw, J. 2005. A Life Worth Leaving: Fasting to Death as Telos of a Jain Religious Life. *Economy and Society* 34, 2: 178–199.

Lele, J. Editor. 1981. *Tradition and Modernity in Bhakti Movements*. Leiden: E. J. Brill.

Lorenzen, D. N. 1995. "Introduction: The Historical Vicissitudes of Bhakti Religion," in D. N. Lorenzen (ed.) *Bhakti Religion in North India: Community Identity and Political Action*. 1–32. Albany: State University of New York Press.

Mayer, A. C. 1981. "Public Service and Individual Merit in a Town of Central India," in A. C. Mayer (ed.) *Culture and Morality: Essays in Honor of Christoph von Furer-Haimendorf*. 153–173. Delhi: Oxford University Press.

McKean, L. 1996. *Divine Enterprise: Gurus and the Hindu Nationalist Movement*. Chicago and London: University of Chicago Press.

Nandy, A. 1970. The Culture of Indian Politics: A Stock Taking. *Journal of Asian Studies* 30, 1: 57–79.

Sarin, M. L. 2002. "National Blood Policy," in *The Blood Banker's Legal Handbook*. Chandigarh: Sarin Memorial Legal Aid Foundation.

Schomer, K., and W. H. McLeod. 1987. *The Sants: Studies in a Devotional Tradition of India*. Delhi: Motilal Banarsidass.

Seeman, D. 1999. 'One People, One Blood': Public Health, Political Violence, and HIV in an Ethiopian-Israeli Setting. *Culture, Medicine and Psychiatry* 23: 159–195.

Simpson, B. 2004. Impossible Gifts: Bodies, Buddhism and Bioethics in Contemporary Sri Lanka. *Journal of the Royal Anthropological Institute* 10: 839–859.

Spodek, H. 1971. On the Origins of Gandhi's Political Methodology: The Heritage of Kathiawad and Gujarat. *Journal of Asian Studies* 30, 2: 361–372.

Strathern, M. 1994. *New Certainties for Old? The Case of an Enabling Technology*. Lancaster: Lancaster University Centre for the Study of Cultural Values.

Street, A. 2009. "Failed recipients: extracting blood in a Papua New Guinean hospital," in J. Copeman (ed.) Blood Donation, Bioeconomy, Culture. *Body and Society* 15, 2: 193–215.

Strong, T. 2009. "Vital Publics of Pure Blood," in J. Copeman (ed.) Blood Donation, Bioeconomy, Culture. *Body and Society* 15, 2: 169–191.

Toren, C. 2002. "Anthropology as the Whole Science of What It Is to Be Human," in R. Fox and B. King (eds) *Anthropology Beyond Culture*. 105–124. London: Berg.

Vanita, R. 2002. "Dosti and Tamanna: Male-Male Love, Difference, and Normativity in Hindi Cinema," in D. Mines and S. Lamb (eds) *Everyday Life in South Asia*. 146–158. Bloomington: Indiana University Press.

Warrier, M. 2003. "The Seva Ethic and the Spirit of Institution Building in the Mata Amritanandamayi Mission," in A. Copley (ed.) *Hinduism in Public and Private: Reform, Hindutva, Gender and Sampraday*. 254–289. Delhi: Oxford University Press.

Weston, K. 2001. "Kinship, Controversy, and the Sharing of Substance: The Race/Class Politics of Blood Transfusion," in S. Franklin and S. McKinnon (eds) *Relative Values: Reconfiguring Kinship Studies*. 147–174. Durham and London: Duke University Press.

Part II

9. Introduction to Part II: technology as practice

Lise Bender Jørgensen

Technology has traditionally been perceived as fundamentally different from theoretical knowledge and beliefs (Dobres 2000, Ingold 2000: 289ff). The concept of technology has primarily been connected to mechanical aspects, separating intellectual design from its execution, and focussing on the properties of materials and the tools and techniques used to work materials. Technology is thought of as having a material and a practice component. Whereas the former has been extensively investigated through material and technical science-like approaches, the latter has been difficult to deal with in an academic format. In recent years, however, this has increasingly been challenged, and the assumed dichotomies between technology (practice) on one hand and theoretical and abstract knowledge on the other are being conflated. This has rapidly opened new insights and questions. One approach emerging is to focus on knowledge embedded in action and to investigate actual practice to unlock how professionals and craftspeople understand their practice; to them this is not practice separated from reflections, but a fully-fledged form of knowledge where movements, rhythm and dexterity are perceived as integral cognitive features. In Scandinavia, for example, awareness of technology as a knowledge system in its own right is now becoming the acknowledged philosophy behind academic discussions at several institutions.[1]

The distinction between discursive and non-discursive knowledge is deeply ingrained in Western thought. It goes back to Plato and Aristotle (Molander 1993: 65–66), and was further developed by the empiricists and rationalists (Dobres 2000: 54). Questioning its validity began in the second half of the 20th century with the works of people like philosophers Gilbert Ryle (1949) and Michael Polanyi (1958, 1966) and the urban planner, Donald Schön (1983, 1987). Ryle drew attention to the distinction between what he called 'knowing that' and 'knowing how'; Polanyi minted the concept of tacit knowing. Schön opened an inquiry into the epistemology of practice, investigating how professionals think in action and how their skills are transmitted. Further investigations and the development of theories of practice have been carried out by Pierre Bourdieu (1977), Michel Foucault (1970), Anthony Giddens (1984) and Bruno Latour (1987, 1999). The last decade has seen an increasing interest in non-discursive knowledge, and ways to engage with it. The following papers relate to this development by questioning the non-discursive character of technology through a number of specific material practices.

Drawing on examples from areas of practice usually seen as merely technical, and often even mechanical, this introduction to the 'technology' section aims to point out verbalised as well as literate aspects of so-called tacit knowledge, possible links between technological practice and beliefs and superstitions, and the interplay between the body (mind and muscles) and practice. The taken-for-granted assumption that technology is non-discursive knowledge is challenged. Studies of practice show that vocabularies, recipes and other forms of notation systems assist and direct the performance of required movements, and how they are integrated aspects of tacit knowing. It follows that to engage in depth with the studies of technologies and practice we have to develop methods that address these aspects, rather than focussing on the end product or assuming the practice to be irrelevant. Insights into the cognitive aspects of practice open our conceptual landscape and help us to find ways to transcribe embodied knowledge and beliefs into an academic format through which they can be investigated.

Movements, especially when in sequence, assist memory. They also hold memory. This is something we all employ in our everyday life, establishing routines – in laying the breakfast table, packing a suitcase etc. More or less subconsciously, routines serve to stick to an established pattern and safeguard against forgetting, securing the desired outcome. Why is it so? Do movements absorb knowledge during the process of becoming routine? What are the workings behind the interaction between movements and memory? How does this form of knowledge relate to other forms, especially academic, theoretical, and propositional knowledge? Questions like these are increasingly being addressed by a wide range of scholars, professionals, and craftspeople.

A recipe is more than a list of ingredients. Good cooking also requires a clear idea of the intended outcome, on how ingredients are to be treated, mixed and processed, what tools and props to use, and the ability to assess when the consistency, colour and appearance is right. This also applies to other types of performances. At the Norwegian University of Science and Technology, Trondheim, and the Royal Institute of Technology in Stockholm, a number of projects have been carried out investigating product design, improvisation, flow and the workings of creativity (Fyhn 2009, 2011, Göranzon *et al.* 2006, Jørgensen 2004, Rønning 2005).[2] Characteristically, *theatre* is an important ingredient in several of these initiatives. There are good reasons for that. A theatrical production is a complicated product involving a large number of human, technical and artistic components brought together in splendid harmony within the space of a few short weeks (Josephson 2006). In addition, actors are instructed to create and communicate complex ideas such as emotions. The intangible outcome of a theatrical production is based on clear instructions on several levels. Some are embedded in the storyline; others are given by the director and producer. The layout of the stage, scenery and the composition of the cast are further parameters that constitute physical and psychological frameworks. The workings of the theatre and its overlapping spheres of tangible instructions and intangible experiences can therefore be used for developing methodological approaches to how technological practices are based on performed knowledge that result in tangible outcomes.

The theatre helps to underline how part of knowledge may be embodied. Dramaturgy was used in a course on product development for engineering students at the Norwegian University of Science and Technology. Among other things, the students were told to *act* the part of the machines they were trying to construct – first individual parts of them, like screws, springs or pistons, then collectively, for example a machine for chopping wood. The exercise gave much cause for mirth – but also induced a 'physical understanding' of the workings of the machine that led to the students obtaining significantly better marks for their products than had been the case previously (Fyhn 2005). The physical understanding is about the student embodying the woodchopper, and through that process gaining cognitive insights into the movements of the different parts of the machine and their relationship to each other.

One of Sweden's finest actors, Erland Josephson, has described how his craft is handed down (2006). Young actors, he argues, constantly find themselves involved in fruitful dialogues with colleagues – living ones as well as those long dead. As a young man, an old, but still vigorously active actor (like Josephson himself) would have stood on the stage with actors who in *their* youth had played with great actors of a previous generation. They form links in a chain and make it possible that experience is handed down. A sudden outburst, a strange phrasing, some remarkable intonations, or certain evocative movements snatched up by a young actor from a colleague may well go back to one of the giants of the past, making them and their art live on (Josephson 2006). This way of transmitting knowledge is closely related to apprenticeship and has been dubbed 'knowledge carried by action' by Norwegian craftsman Jon Bojer Godal (2007).

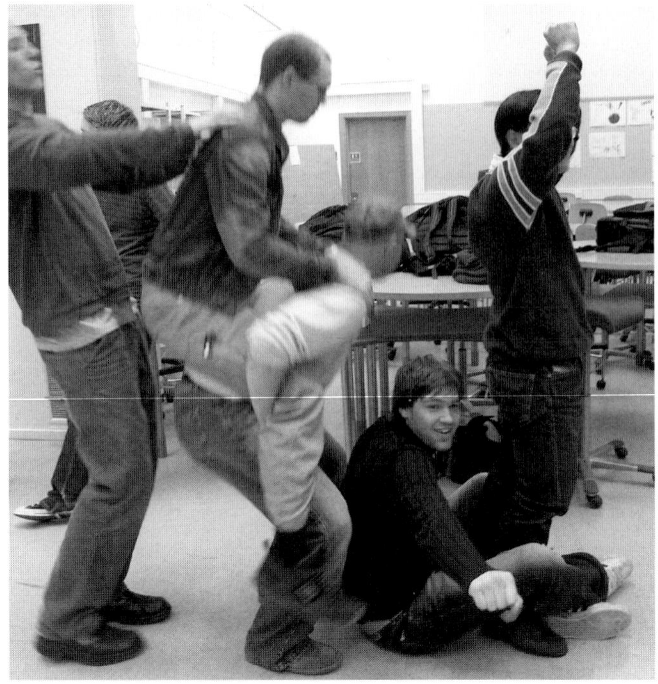

Fig. 9.1: Students acting being a wood-chopper (photo: Håkon Fyhn, 2002).

During his research into professional knowledge, the philosopher Bengt Molander has investigated the role of time and rhythm (Molander 1993: 226ff). Many work routines seem to be designed so that they contain time for reflection. Molander mentions examples from physiotherapy, meteorology and boat builders, and quotes a meteorologist for saying that 'there are short thoughts as well as long thoughts'. The long thoughts are those that draw on the practitioner's experience, recalling all kinds of knowledge, including similar cases in the past, and drawing conclusions from them. This often happens unconsciously, triggered by established routines. The psychologist Pirjo Birgerstam has noted something similar when she investigated how artists and other creative professions, such as architects, work (Birgerstam 2000). She found that the artist constantly alternates between intuitive and rational modes of work, in a kind of hermeneutic spiral. Like Molander's meteorologists and boat builders, the artists also apply routines to facilitate the release of creativity. A break, a contemplative rest, or monotonous physical activity following a period of intense work often does the trick. This phenomenon is also well-known to scholars and scientists – and to take a break is a standard piece of advice to students in the final throes of finishing a dissertation! Birgerstam calls this *incubation*, and describes intuition as resulting from the incubation of an overall idea. This unconscious process of knowledge, also called abduction, or the method of multiple comparisons, is the beginning as well as the climax of creativity (Birgerstam 2000: 58ff). Birgerstam compares intuition and rationality with gin and tonic, perceiving them as complementary rather than opposites (Birgerstam 2000: 96, 2002).

The practice of technology, the performance, takes *time*, short or long, depending on the demands of the craft. There is the time needed for incubation, the processing of thoughts and emotions (assisted by monotonous motion), as well as the time needed to carry out the actions that transform raw materials into a finished object. It also involves elements of rituals, not always very explicit, but in the sense of right or wrong ways of doing things. How were work processes and such rituals created and connected? Were they designed in one go, from scratch, or did they grow slowly out of long experience – like Erlend Josephson's description of how the art of long dead actors live on in an intonation, an outburst, adapted by a younger colleague? And – perhaps the most pertinent question – how can we deal with them, and whatever remains they have left, through our material record? How can we explore and describe such apparently non-discursive, esoteric knowledge and turn it into a format suitable for academic study?

Let us take another look at what defines non-discursive knowledge. According to Molander (2004: 17ff), western philosophy of science perceives theoretical knowledge as something that can be put into books, and taken out of them. He also stresses that knowledge *about* something is qualitatively distinct from the subject; it is a separate element that mirrors or copies reality. Moreover, and of particular importance for our concern, theoretical knowledge requires that it can be formulated in *words*, or in the language of mathematics. Theoretical knowledge is considered as scientific, discursive, in contrast to practical knowledge or skill. It follows that tacit knowledge is non-academic, and it is also often valued differently to academic or abstract knowledge.

A first step towards dealing with non-discursive knowledge is to question the tendency of seeing intuition and rationality as opposite. The second would be to develop methodologies through which we can translate tacit or embodied knowledge into a form through which they become available to textual analysis. Norwegian fishermen have up to 200 different words for waves, designating their shape, strength and behaviour (Eldjarn and Godal 1988: 75ff). This is essential for their ability to assess rough seas, to negotiate them, navigate, and ultimately to survive. Carpenters have a rich vocabulary for timbers *and* the movements and tools through which they work them; oenologists are able to describe the taste of a wine in exquisite detail; perfume-designers do their job mixing composite fragrances and using their sense of smell, and we could go on. Had they no vocabulary to assist them, their jobs would be impossible to do and the transfer of knowledge between craftspeople and over generations would be severely limited. The point is that practitioners already have verbal knowledge, but it often takes a form which is unfamiliar to academic studies. If verbalising tacit knowledge is a step towards turning it into acceptable discursive knowledge, then we should be able to make it literate as well. For this we may again look to the creative professions and in particular the performing arts. The creative process often takes the form of sketching – a search for patterns or for rhythm. It is also a process of abstraction, selecting the elements that hold promise. Birgerstam (2000) perceives sketching as a form of notation, like musical notation. Musical notation is a form of writing designating sequences of sounds, including their key, tempo and force. Choreography charts the movements of a dance. Scriptwriting and storyboards are further ways of writing action. None of these may be considered the final way of describing sound or movement; modern music demands sounds, rhythms or punctuations that cannot be transmitted through traditional music notation. How about instructions like: 'irregular, optional sounds', 'optional notes, as quick as possible', 'knock music rest with bow or chair with frog', or 'blow into the bottle'? All of these appear as signs in modern music scores (Nordström 1989: 52–53). As regards dance notation, a whole range of systems may be listed – Benesh, Choreology, Labannotation, Sutton Movement Writing etc, all to be found on the Internet and employed to write movements.[3] They are used for many other purposes besides dance, so why should we not use them for studies of technologies as well, turning all kinds of so-called tacit, embodied knowledge into a scientific format? This has already been successfully attempted in a study of medieval carpentry (Høgseth 2007). Further help may be found in the digital world that opens a wide range of new possibilities for referencing movement, time and haptics. Going through the motions and emotions, looking for ways of writing them down, may well be an important step towards accepting the 'non-discursive' as an acknowledged, academic form of knowledge.

Technology is never just an automatic performance. Embodied technology is about knowledge, motions, beliefs and timing. The papers following will show some of these aspects through the case studies of particular crafts in the past, as well as more recently.

Notes

1 The *Dialogue Seminar* at the Royal Institute of Technology in Stockholm, Sweden, is one example, exploring professional skills with an emphasis on modern industry (Göranzon *et al.* 2006). Another is the *Craft Laboratory* at Göteborg University, Sweden, that started as a vocational college and is now part of the Department of Conservation (Almevik *et al.* 2011). In Norway, the Sør-Trøndelag University College in Trondheim has established a BA in technical restoration of historical buildings aimed at students with a background in crafts. All three institutions have awarded grants to PhD students with a craft background to pursue research based on their craft (Gustafsson 2002, Høgseth 2007, Almevik *et al.* 2011). The *Norwegian Crafts Development* began as an archive of skilled practitioners of traditional crafts and is now engaged in projects to preserve crafts and techniques by awarding craft scholarships to promising young craftspeople (Falk 2007).

2 A doctoral dissertation on musical improvisation was accepted in the form of a concert (Aase 2009).

3 See www.ickl.org/ and www.dancewriting.org for examples (accessed 8 August 2011).

References

Aase, A. 2009: *Improvisasjon i skandinavisk tradisjonsgitar*. Doctoral Dissertation (CD). Trondheim: Norwegian University of Science and Technology.

Almevik, G, Bergström, L. and E. Löfgren 2011: *Hantverkslaboratoriet*. Göteborg: Göteborgs Universitet. Retrieved 24 September 2011, from http://www.craftlab.gu.se/digitalAssets/1328/1328263_antologin-hantverkslaboratorium-2011.pdf.

Birgerstam, P. 2000. *Skapande handling om idéernas födelse*. Lund: Studentlitteratur.

Birgerstam, P. 2002. Intuition – the Way to Meaningful Knowledge. *Studies in Higher Education* 27, 4: 431–444.

Bourdieu, P. 1977. *Outline of a Theory of Practice*. Cambridge: Cambridge University Press.

Dobres, M.-A. 2000. *Technology and Social Agency.* Oxford: Blackwell Publishers.

Eldjarn, G., and J. Godal. 1988. *Nordlandsbåten og Åfjordsbåten I, Båten i bruk: segling, roning, fisking og vedlikehold*. Lesja: A. Kjellands Forlag.

Falk, E. 2007. *Festskrift. Jon Bojer Godal 70 år. Norsk handverksutvikling – NHU 20 år. Maihaugen Årbok 2007.* Lillehammer: Maihaugen

Foucault, M. 1970. *The Order of Things: an archaeology of the human sciences*. London: Tavistock.

Fyhn, H. 2005. "Det skapende mennesket og selvets metafysikk. Teater og produktutvikling ved NTNU," in B. Rønning (ed.) *FLYT. En nøkkel til kreativitet og innovasjon*. 21–47. Trondheim: Tapir Akademisk forlag.

Fyhn, H. (ed.) 2009. *Kreativ tverrfaglighet. Teori og praksis.* Trondheim: Tapir Akademisk forlag.

Fyhn, H. 2011. *Møte med tilstedværelse. Mellom form og tomhet i produktutvikling, aikido og antropologisk erkjennelse*. Doctoral dissertation, the Norwegian University of Science and Technology. Trondheim. Retrieved 24 September 2011, from http://www.hakonfyhn.net/web_documents/m__te_med_tilstedev__relse_aug11_fyhn.pdf.

Giddens, A. 1984. *The Constitution of Society. Outline of a Theory of Structuration*. Cambridge: Polity Press.

Godal, J. B. 2007. "Hjelper omgrepet handlingsboren kunnskap oss til framtid fotr handverket?", in E. Falk (ed) *Festskrift. Jon Bojer Godal 70 år. Norsk handverksutvikling 20 år. Maihaugens Årbok* 2007, 11–22. Lillehammer: Maihaugen.

Göranzon, B., Hammarén, M. And Ennals, R. (eds) 2006. *Dialogue, Skill and Tacit Knowledge*. Chichester: John Wiley and Sons.

Gustafsson, G. A. 2002. *Den inre teatern i lärandet – en studie om kunskapsväxandet inom hantverk*. Doctoral dissertation. Stockholm: Kungliga Tekniska Högskolan. retrieved on 24 September 2011, from http://kth.diva-portal.org/smash/record.jsf?searchId=3&pid=diva2:9167.

Høgseth, H. B. 2007. *"Håndverkerens redskapskasse". En undersøkelse av kunnskapsutøvelse I lys av arkeologisk bygningstømmer fra 1000-tallet.* Doctoral dissertation, the Norwegian University of Science and Technology. Retrieved on 24 September 2011, from http://ntnu.diva-portal.org/smash/record.jsf?pid=diva2:123127.

Ingold, T. 2000. *The Perception of the Environment. Essays in livelihood, dwelling and skill.* London and New York: Routledge.

Josephson, E. 2006. "A dwelling place for past and living voices, passions and characters," in B. Göranzon, M. Hammarén, and R. Ennals (eds) *Dialogue, Skill and Tacit Knowledge*. 69–72. Chichester: John Wiley and Sons.

Jørgensen, S.-H. 2004: *På sporet av improvisasjonens potensiale.* Doctoral dissertation, the Norwegian University of Science and Technology. Trondheim: Tapir Trykk.

Latour, B. 1987. *Science in Action*. Cambridge, MA: Harvard University Press.

Latour, B. 1999. *Pandora's Hope. Essays on the Reality of Science Studies.* Cambridge, MA: Harvard University Press.

Molander, B. 1993. *Kunskap i handling*. Göteborg: Daidalos.

Molander, B. 2004: The meaning of "knowledge" in the history of knowledge. In: A. K. Børesen (ed): *Science, Crafts and Ignorance. Perspectives on the History of Knowledge.* 13–24. Trondheim: Tapir akademisk forlag.

Nordström, S. 1989. *Så blir det musik*. Göteborg: Bra böckers förlag.

Polanyi, M. 1958. *Personal Knowledge. Towards a Post-Critical Philosophy*. London: Routledge and Kegan Paul.

Polanyi, M. 1966. *The Tacit Dimension*. London: Routledge and Kegan Paul.

Ryle, G. 1949. *The Concept of Mind.* London: Hutchinson.

Rønning, B. (ed.) *FLYT. En nøkkel til kreativitet og innovasjon*. 21–47. Trondheim: Tapir Akademisk forlag.

Schön, D. A. 1983. *The Reflective Practitioner. How professionals think in action*. London: Maurice Temple Smith.

Schön, D. A. 1987. *Educating the Reflective Practitioner.* San Francisco: Jossey-Bass Publishers.

10. The language of craftsmanship

Harald Bentz Høgseth

Rather than comparing the structure of a tool mark to fingerprints, in my view, a tool mark should be compared to a human face, because of its distinctive features – wrinkles, furrows etc. The reason for this is that a fingerprint remains the same all through life, whereas our facial features alter by degrees during life; we constantly get new wrinkles and furrows caused by the weather, ageing, and wear and tear. The same description can be applied to an axe. It wears gradually, by degrees, and constantly collects new damage like notches, breakage, bulges, etc. Despite this, we always recognise the trace of a tool mark, with its characteristic features similar to facial attributes. The structure of a particular tool mark, both in terms of its parts and whole, is possible to identify due to the connection between the stop mark of an edge and the side features or signatures. The particular signatures that macroscopic and microscopic damages on an edge create can be characterised as a series of ridges or grooves running in the longitudinal direction of the tool mark, more or less at a right angle after the stop line of the edge. Each pattern of such ridges or grooves corresponds to the irregularity on the edge of the axe. In this way, it is possible to identify damage or production features caused by the edge of a particular tool after a craftsperson has designed the surface of the piece of wood. Together, the series of marks helps us in the process of identifying the tool and the procedure or the rhythm of the tool mark. Analyses of tool marks can likewise be compared to the inspector's search for a murder's weapon – considering motive and the actions that took place during the misdeed. When an axe is used in the preparation of a log, it leaves behind grooves and furrows that are distinctive and can be used to reveal the motives and actions of the craftsperson.

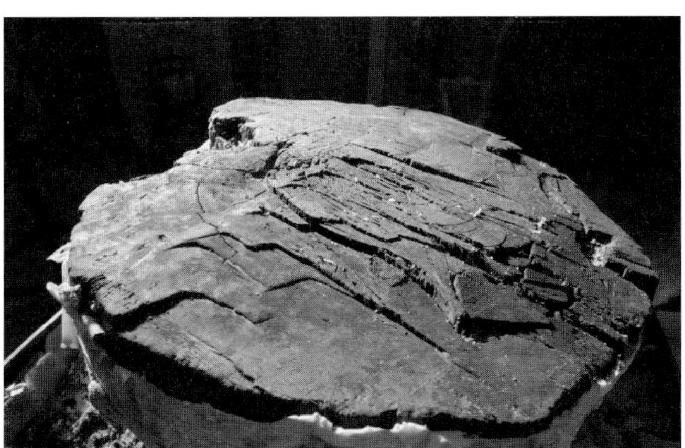

Fig. 10.1: The craftsperson has processed the timber by standing on the left side. The leading edge of the axe has cut its way into the timber to the right.

As a craftsperson and archaeologist, I am particularly interested in the secrets behind tool marks on the surface of wooden building remains of archaeological interest. In this chapter, I will focus on the structure of tool marks and their individual parts. I will aim to develop a better way of documenting tool marks on the surface of wooden building remains, and develop procedures to analyse the working process behind them. I will demonstrate how it is possible to read movements out of their dynamic patterns and try to give a picture of the rhythm of movements and the craftsperson's 'know how'. The framework for my investigation is my perception of tacit knowledge and knowledge in action. As part of this, I aim to build a bridge between tacit knowledge and verbal, academic knowledge, as that is important for preserving and further developing the knowledge of craftspeople. Tool marks represent tangible parts of the craftsperson's knowledge and may give us important information about the connection between the craftsperson, materials, tools and the working process. In this chapter, I will therefore mainly address three approaches. First, I will identify how to document and analyse tool marks. Second, I will analyse the connection between the sequences of tool marks and the dynamics, rhythm and movement patterns behind them. Third, I will show how action behind the movements and the techniques can be translated into an academic verbal language and text.

To identify, document and analyse tool marks

In Norway some archaeological and ethnographical studies of tools from Late Iron Age and Early Medieval periods have been undertaken, among which the works by Oluf Rygh (1999) and Jan Petersen (1951) are best known. Other sources like artefacts, iconography, and physical tool marks are also important for the understanding of the numerous axes that existed during different periods. These sources give us valuable information about the multiplicity of tools, but tool marks supplement this information. They give us both an indirect account of the tools used and inform us of the possible techniques and working processes chosen by the craftspeople.

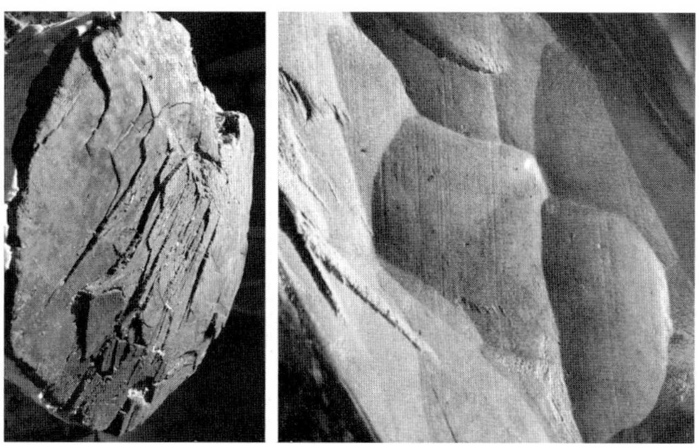

Fig. 10.2: Tool marks from different types of axes. On the left is a tool mark from the late Iron Age (Viking period).

Tool marks from the past, and the movement patterns and procedures behind them, are important evidence. The first step is to identify and analyse the tool marks, then experimental examinations are used in the process of their identification and reconstruction. I work together with craftspeople still holding traditional knowledge. They investigate different types of axes using different traditional techniques. The examinations are based on casts of the original timbers, and their subsequent analysis. We alternate between the cast and the experiment. The aim is to study and compare the marks produced during experiments with those observed on original material.

When we use tools to form timbers they leave signatures on the surface, comprised of particular patterns that characterise the individual tool. If the tool has damages on the edge, they leave a mould in the surface. The timber records the damage, and the damage may then be transcribed. We are able to identify the damages as ridges and grooves in the directionality of the tool mark. Each analysis uncovers patterns corresponding to irregularities in the axe's edge. Damage on the edge of the axe may be either visible or invisible to the naked eye, but whether we can see it or not, the damage always remains, like a signature on the surface of the timber.

Distortion and attrition arise constantly and cause particular grooves and furrows in a tool mark. It is the overall pattern that makes it possible to investigate what kind of axe and what type of movement patterns have caused the particular pattern, and therefore what lies behind the tool marks. Analyses

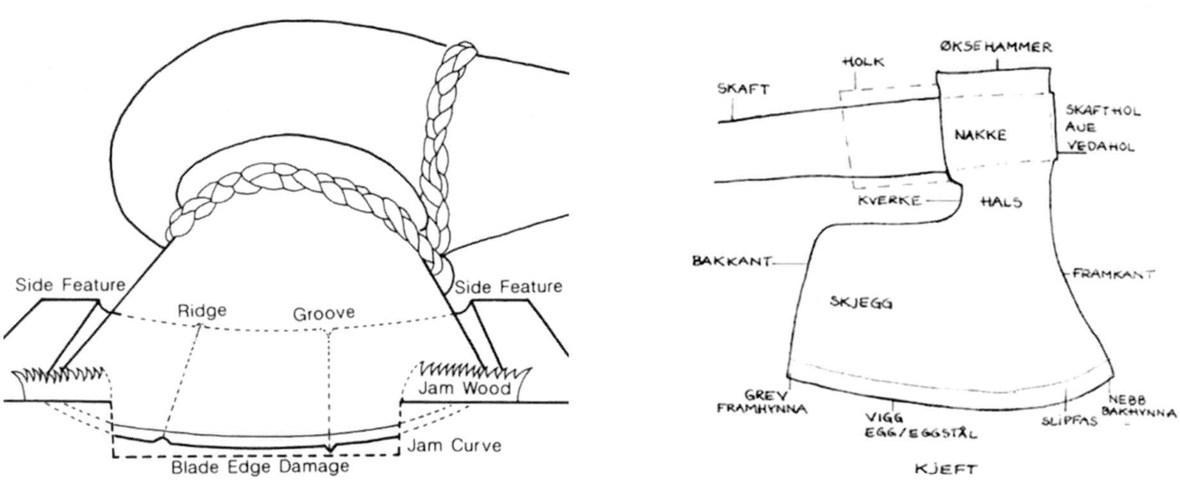

Fig. 10.3: (left) The blade edge always leaves specific patterns, called signatures, in the timbers, like our fingers leave fingerprints, even if there is no clear damage on the edge; (right) an example of traditional definitions of the different parts of an axe (for instance beard, earl, cutting edge, beak, throat, neck, head, throat, eye, handle/shaft, etc.). The definitions are strictly linked to the craftsperson's vocabulary/dialect, which differs from area to area.

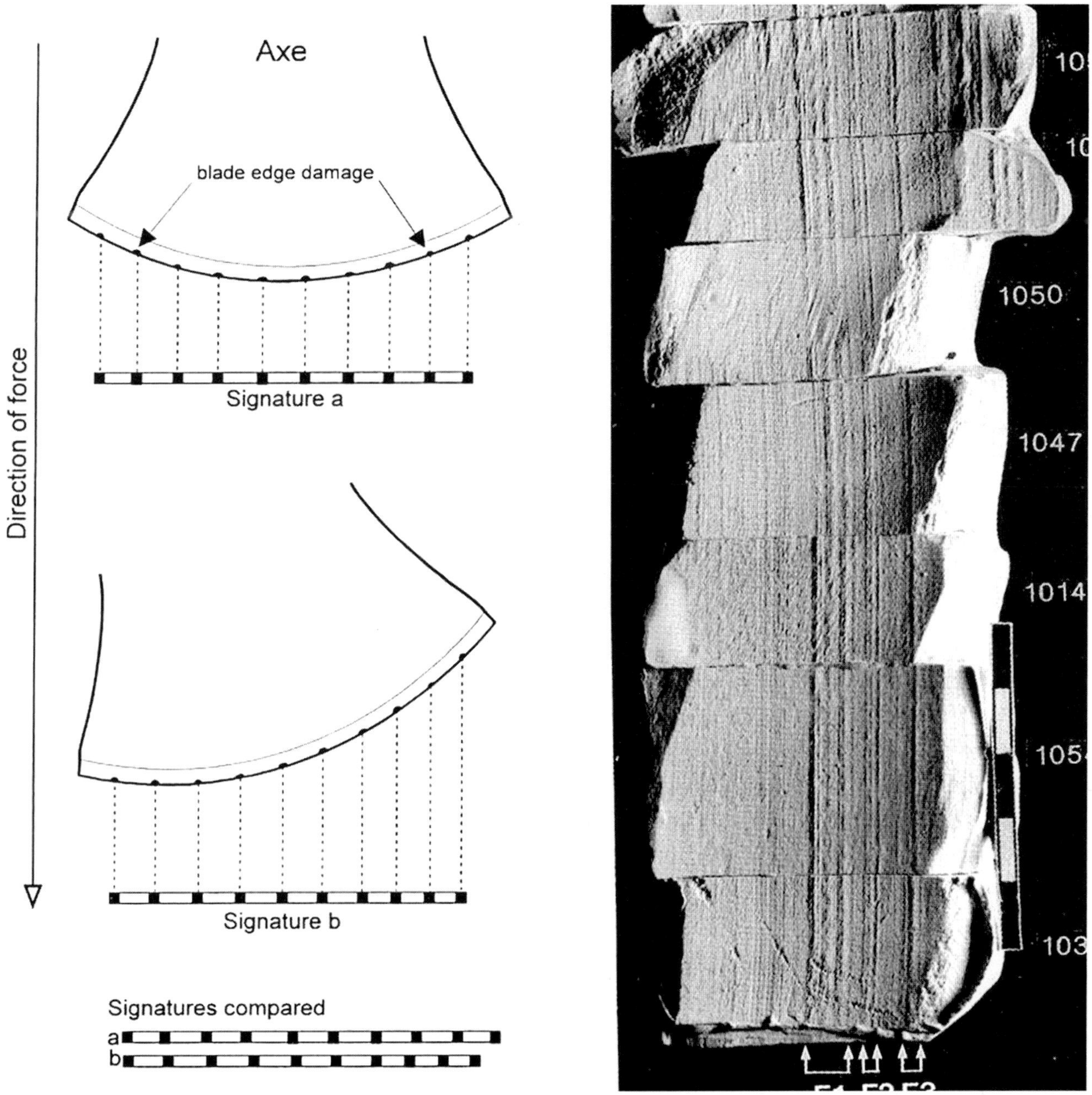

Fig. 10.4: Blade edge damages and signatures on tool marks from different documented castings of the same axe (after Sands 1997).

of the tool marks make it possible to establish relationships between timbers worked with the same tool. The distance between the individual cuts and the sequences of cuts, as well as their depth and angle, give us valuable information about the working procedure. The tool marks and the signatures also reveal whether the cuts were made with the same axe or with different axes. In this way, tool marks connect timbers that are otherwise unrelated and reveal the kind of axe used in the different sequences that were involved with the building of, for instance, a house.

The illustration (Fig. 10.5) indicates the relation between the axe's signature and its stopping point. The stopping point reveals the curve of the edge. The angle between the signature and the stopping point reveals the axe's circular movement. This in turn indicates the craftsperson's movement patterns and working position. When the angle between the stopping point and signature changes, the craftsperson's working position and movements were adjusted.

Before the experimental investigation, I documented the tool marks by casting them in silicon, taking photographs

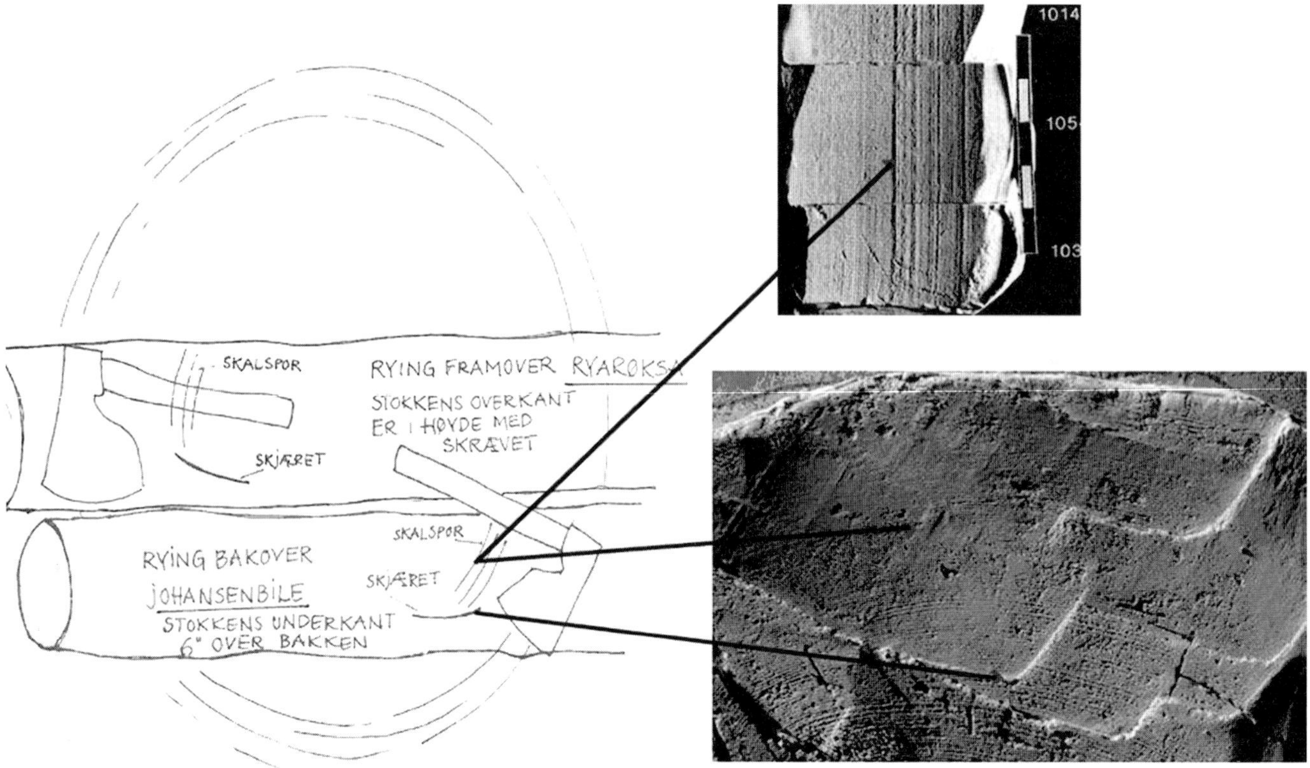

Fig. 10.5: The rotation/movement of the axe and its relationship to the signatures in the tool mark (furrows, grooves), the edge stop line and the axe leading edge.

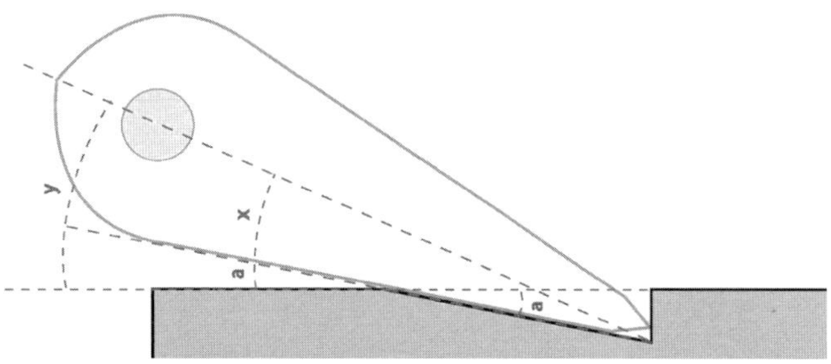

Fig. 10.6: Schematic illustration of the cutting angle.

and using 3-D scanning technology. Three-dimensional photographs were used to identify and map the different tool marks on the surface and analyse which type of tools and actions created them.

The connection between the sequences of tool marks and the dynamics, rhythm and movement pattern behind them

Experimental archaeology is more than just reconstructing artefacts: it is about reconstructing the processes behind artefacts, tool marks, working patterns and actions. In addition, experimental archaeology approaches the craftsperson's choices of materials, tools and actions. It is an interpretative procedure based on original sources through a practical approach. Certain working procedures leave behind particular marks. The process makes us capable of approaching past practices as practices rather than simply results, and it enables us to disprove hypotheses. For example, in comparing the length between the tool marks on an original piece of timber and one made through reconstruction, it is possible to identify changes in the cutting process. The example in Fig. 10.8 shows that the length between the tool marks on the reconstructed and original timber had the same pattern in the beginning of the cutting process, but deeper down, they became different, which means that the length of the movement changed.

Fig. 10.9 demonstrates the tool marks after the first and second cutting process (rough and smooth cutting). We can see the relation between the edge stopping point, the signatures and the angle of the cut, as well as the series of tool marks. The tool marks are a result of different working methods with two different axes. The original timber has tool marks indicating two cutting processes, the rough cutting and the smooth cutting process.

Fig. 10.10 gives us some ideas about the tools, working positions and procedures based on the reconstruction process. The craftsperson started cutting in area A. The timber was

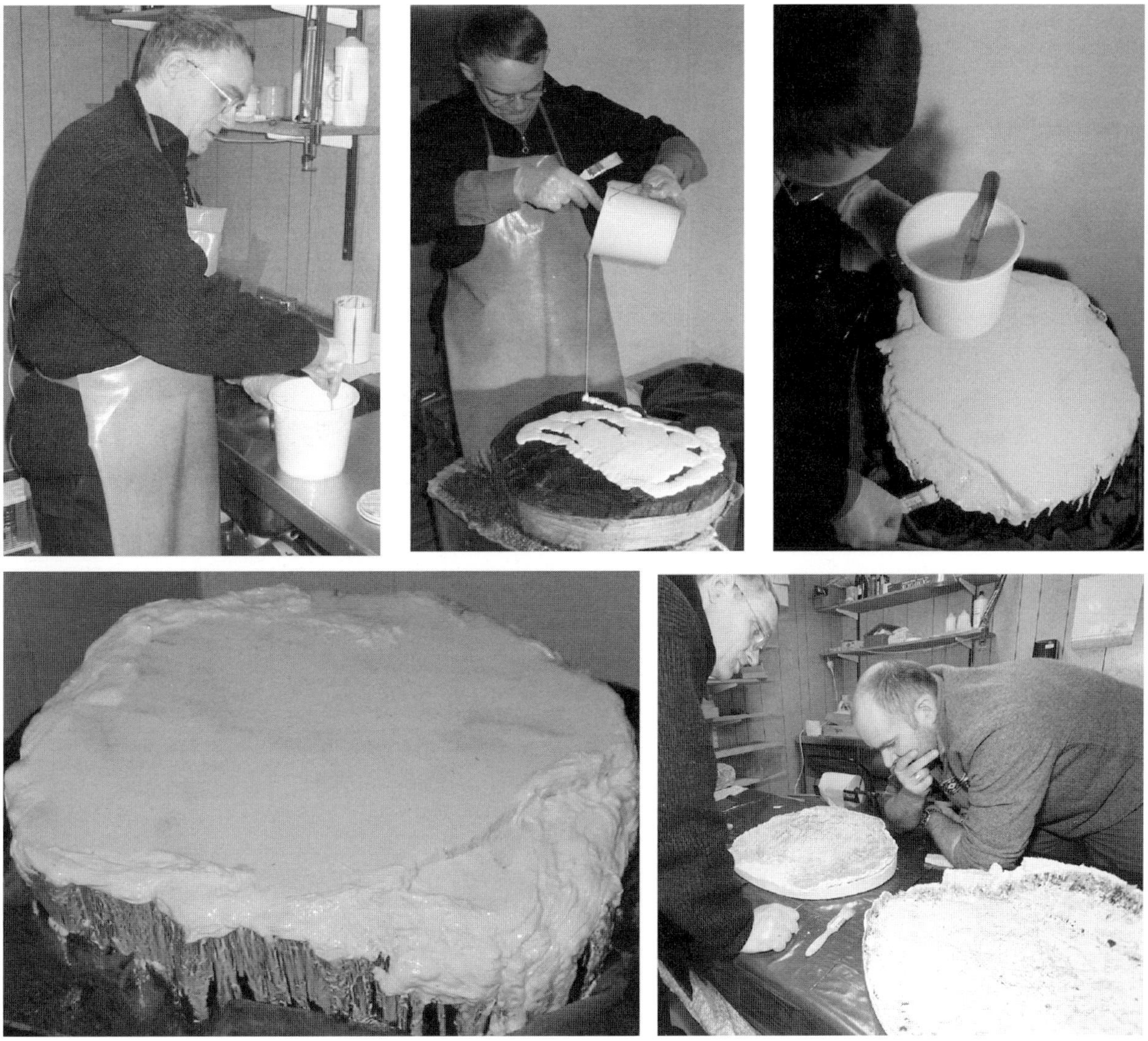

Fig. 10.7: First step of the documentation process: negative castings in silicone and positive castings of the tool marks.

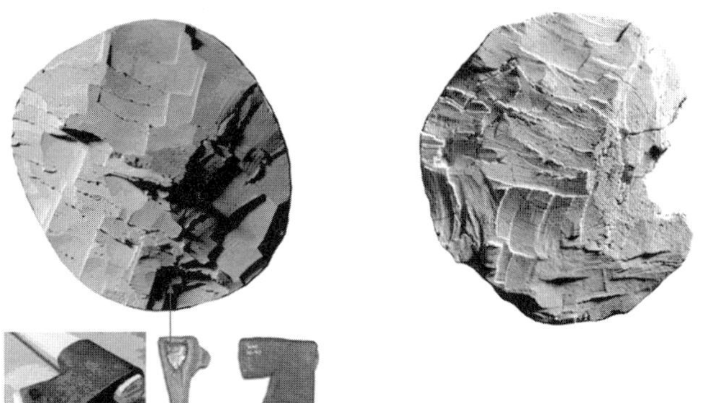

Fig. 10.8: Rough cutting on the original (right) and reconstructed (left) timber. The axe below was used in the cutting process on the timbers to the left.

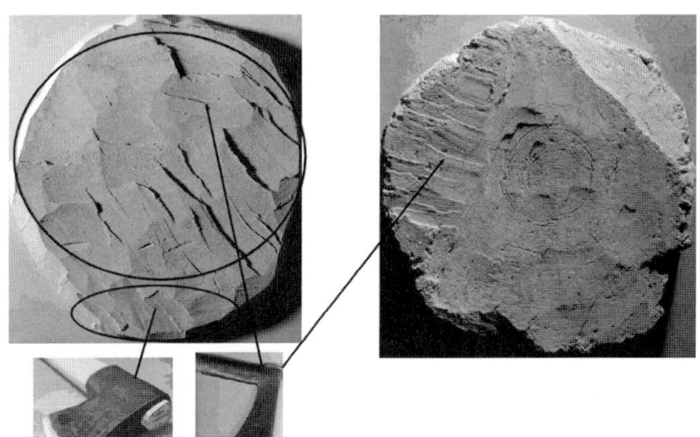

Fig. 10.9: Smooth cutting on original (right) and reconstructed (left) timbers with tracks of tool marks from two axes: the right one for smooth cutting, the left one for rough cutting.

rotated and the procedure continued in area B, then to C and finally to D. The timber was rotated towards the right all the time. The cutting procedure went from the outer edge to the centre. At the end of the process, the timber was lying with area D at the top and the timber was broken in two. The craftsperson ended the process by cleaning the surface working from the outer edge towards the centre with long, controlled and steady movements. Tool marks in area A indicate that the craftsperson was standing on the left side (area B) of the timber. Then he turned the timber 90 degrees to the right. The combination of the axe stopping point, the signatures and the leading edge in area B reveal that the craftsperson's position was changed to the right side (area D) of the timber and he continued cutting in area C. He then turned the timber 180 degrees to the right and finished the procedure in area D. In area D we can see how he finished the job with smooth cutting. In this process, he was working from and standing on the left side. The cutting procedure occurred by turning the timber twice and the craftsperson changed position from the left side of the timber to the right, and then to the left again.

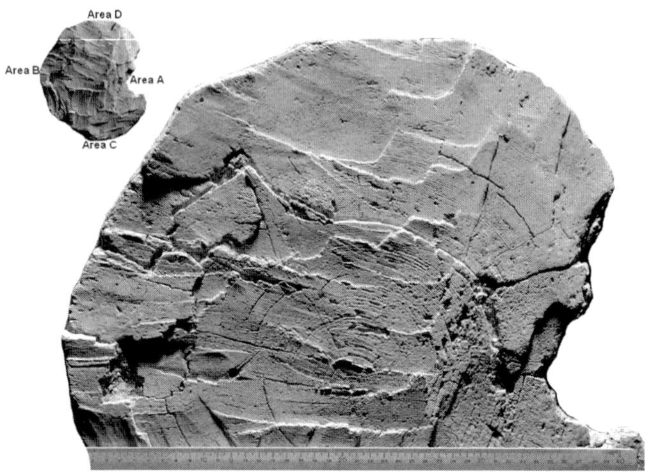

Fig. 10.10: Areas identified in the analysis of working position and procedure.

Fig. 10.11 shows the reconstructed and the original tool marks and is an attempt to reconstruct what tools created the mark. We used a so-called 'beard axe' (skjeggøks) in the end of the process. The tool marks arising from this axe were very similar to the original one, but our axe weighed only 1.2 kg. To cut the timber the axe must have been heavier. The original one was probably over 1.5 kg and with more weight at the front of the edge. The shaft must have been longer than 50 centimetres. The craftsperson had drawn the axe towards himself when cutting and the wood must have been raw.

Translating the action behind the movements and the techniques into an academic format

To succeed in translating movements and techniques into an academic format, it is necessary to invent and develop a notation

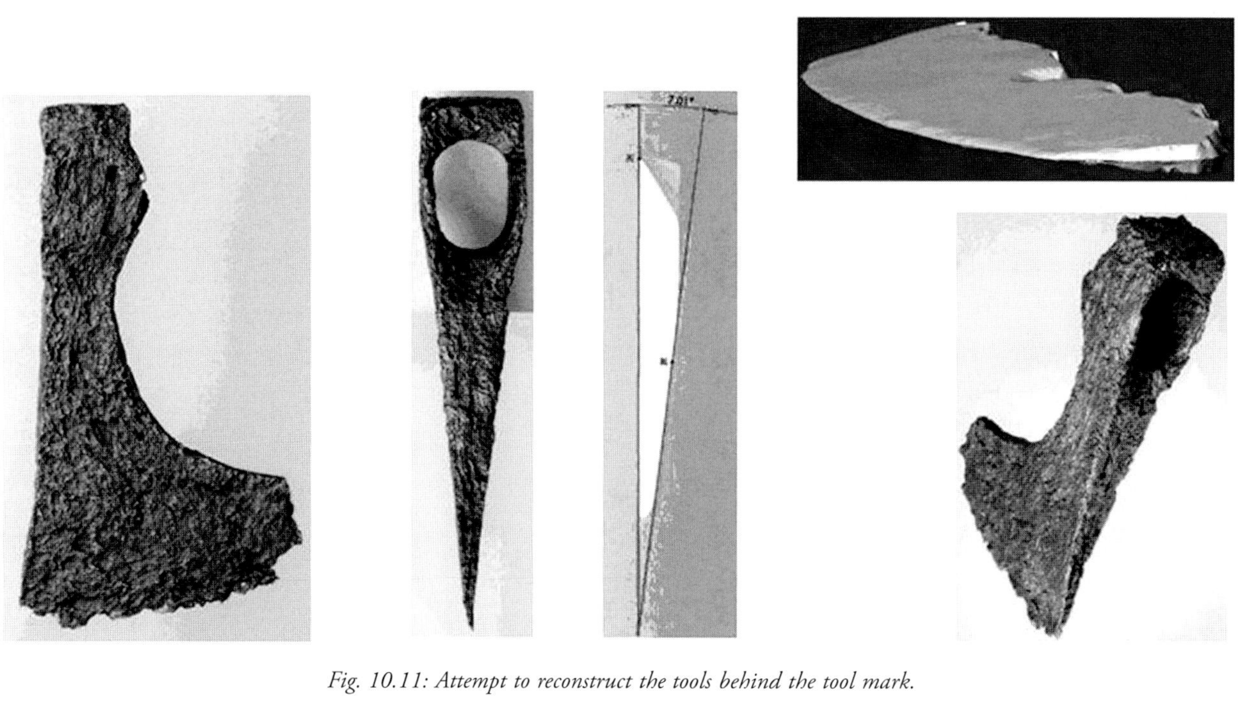

Fig. 10.11: Attempt to reconstruct the tools behind the tool mark.

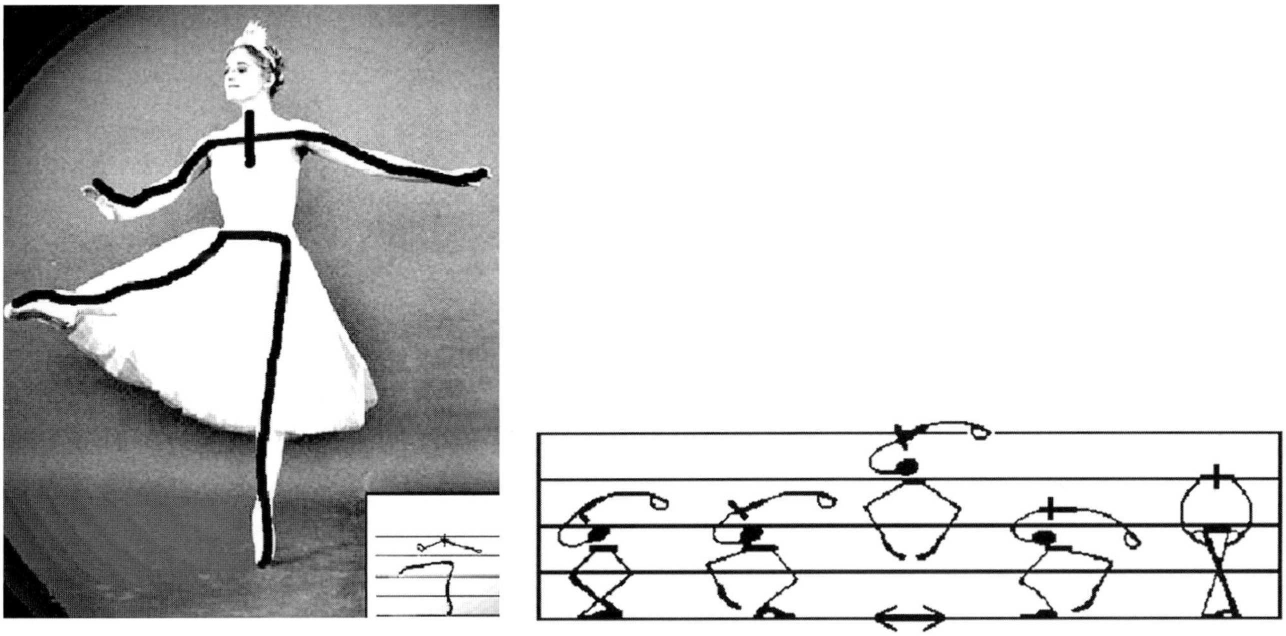

Fig. 10.12: The analysis of movement patterns with the Sutton Movement Writing system.

system that helps analyse details related to the craftsperson, the tool, and the motions. The system developed builds on video documentation of the reconstructing process, where a traditional craftsperson's knowledge was used to analyse the tool marks and to create similar ones with different techniques and tools. In the following I will discuss the resulting method which consists of a notation system that uses signs to focus on and capture the dynamics in the craftsperson's movements and the relations between rhythm, speed, force, technique, etc.; it also points to how the body is tied to materials and the use of tools. The signs do not only describe the dynamics in the craftsperson's movements, they also describe the body's relation to the sliding and curving movements of tools. The notation system makes it possible for us to analyse and transcribe how craftspeople dance with their tools.

The system defines the rhythm and movement patterns that occur within one single cut, as all movements occur within one rhythmic measure. The method is inspired and developed after the Sutton Movement Writing system that is used to note dance choreography (Sutton 2001). The rhythm follows the pattern 1, &, 2, &, 1, &, 2… (see Fig. 10.15). Rhythm (1) represents the first cut making a sound when the axe hits the timber. The symbol over rhythm (1) shows that the craftsperson gives the axe cut a fast jerk when the axe is drawn out. The symbol points out that the action is done with a sense of touch. The next column shows the axe is drawn out and pulled back. The symbol (&) indicates a minor motion before the next movement. The stipple line indicates a soundless motion. The second, third and fourth column shows that the shaft glides in the hand. The craftsperson moves the axe forward and backward in his hand while the axe moves in

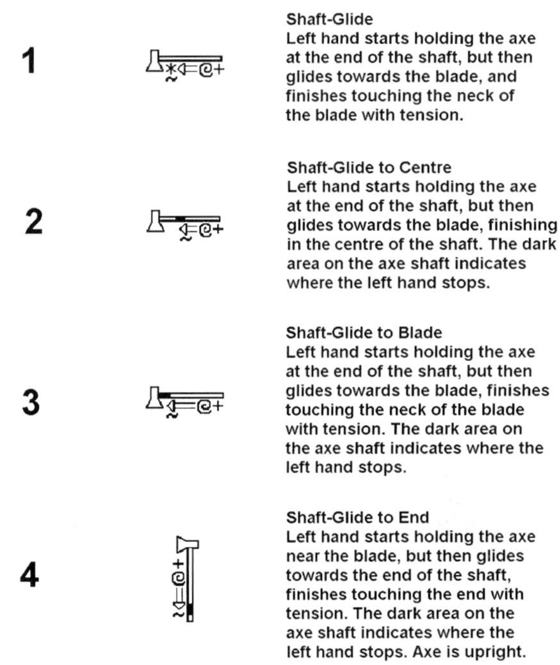

Fig. 10.13: Some examples for symbols used on movement combinations in 'timber writing'.

different angles. Rhythm (2) is also described with a stipple line describing the soundless motion. This sliding movement affects the next step on the rhythm. The left hand slides down on the shaft and stops near the axe neck, while at the same time the axe moves against the craftsperson's upper body. The last column marked with the symbol (&) shows that the axe is

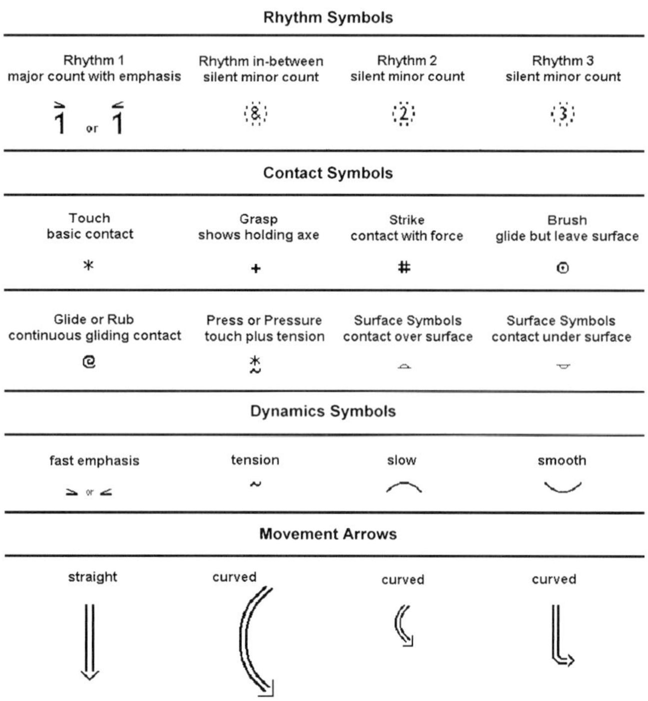

Fig. 10.14: Examples of symbols for rhythm, contact, dynamics and movement.

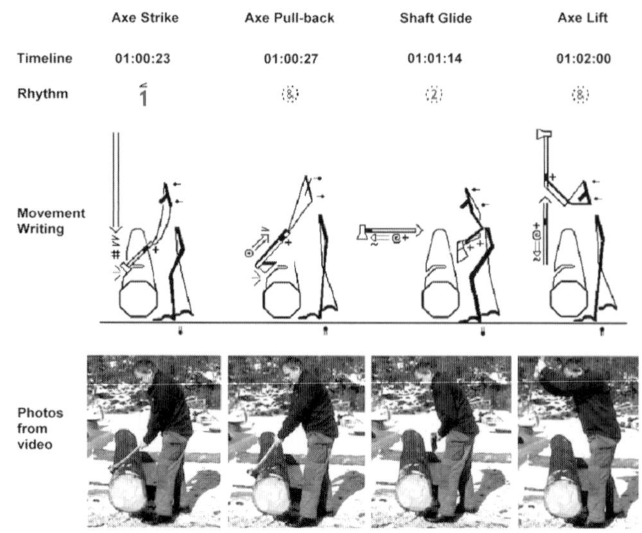

Fig. 10.15: A way to write and analyse movements.

raised while the left hand glides back towards the end of the shaft again. This lifts the axe over the craftsperson's head in a gliding movement. The symbols under the sketch indicate how the body sinks and raises.

I will now give an example from my analysis of the movement pattern, dynamic and rhythm in the different sequences of one single cut. Fig. 10.15 demonstrates the movement's starting point. We can see how the upper body tilts forward. The face looks forward and down against the cut. Both hands hold the end of the axe shaft. The left leg is bent, and placed in front of the right leg near the timber. The right foot is upright. The whole body sinks down while the axe moves fast in a long curved movement against the timber. The axe hits the timber and the wood chips fly. The craftsperson's centre of gravity and balance point is on his left side. The left leg is bent so he can swing his body up and down for each single cut. The axe movement is marked with a vertical arrow showing a fast, powerful and vertical movement, and the plus sign shows how he holds the axe with both hands.

The next two frames of Fig. 10.15 (time 01:00:27 and 01:01:14) demonstrate the movement in between the long and dominant movements that start and finish the rhythm pattern of each single cut. The figure demonstrates how the upper body position changes and tilts backwards. The centre of gravity and balance point change and becomes more balanced, but the position of the legs reveals that the centre of gravity moves up and down, forward and backward. It is a sliding motion. The left leg is no longer bent but straight. The right foot is almost straight in a relaxed way. The body lifts up during this sequence and the craftsperson still holds the axe at the end of the shaft with both hands. The axe hits the cut in an oblique angle. The axe is pulled out of the cut with a sensitive jerk. This movement is short and fast. The oblique angle of the axe is necessary to open the cut enough so the axe does not jam into the timber. If the axe jams into the timber, the craftsperson loses his rhythm. Now the upper body changes position again and tilts forward. The face points towards the cut and the craftsperson focusses. His left hand glides over the shaft towards the blade of the axe and finishes near the neck of the blade with a tense stop. During this glide to the left, the actual axe is moving back towards the craftsperson's waist. Both his legs are bent, the left one a little bit more than the right one. The right leg is behind the left and almost straight. The left foot is nearest to the timber. The balance point is centred and the body sinks down. The fourth figure demonstrates the final sequence in the movement pattern behind the cut. The upper body tilts forward in a gliding movement. The face and glance constantly focus forward and down on the cut. The right leg bends behind the left leg that is straight. The leading movement illustrates the body's sliding motion forward and backward. The body lifts up. The left hand glides down to the end of the axe shaft and stops near the right hand. Both hands now hold the end of the shaft. During this glide to the left, the axe moves up and finishes high.

As time passes, the cut gets deeper. Therefore the craftsperson moves his feet forward, pressing his lower legs and feet against the log, and bends forward and deeper down, over the log, to reach the other side of the log with the axe. This changes the movement patterns considerably. In comparing the movement phases early and late in the cutting process, we can see how the craftsperson's feet are placed under the log, the shin pressing against the log, the upper body bending lower and over the log.

This changes the movement pattern, the body's working position and balance. The change affects the movement of the tool and it is now working at a different angle. This results in the axe hitting the timber with a different curved movement. Such changes can be connected to changes in the tool marks observed on archaeological buildings. The connection between movement patterns and tool marks can be described as following in this case: the deeper down the craftsperson works, the more the angle between the outer edge and the edge stop point opens. The working position becomes more bent (forward). The point of balance and the rhythm change; some of these changes are illustrated in the analysis of the original timber in Fig. 10.17.

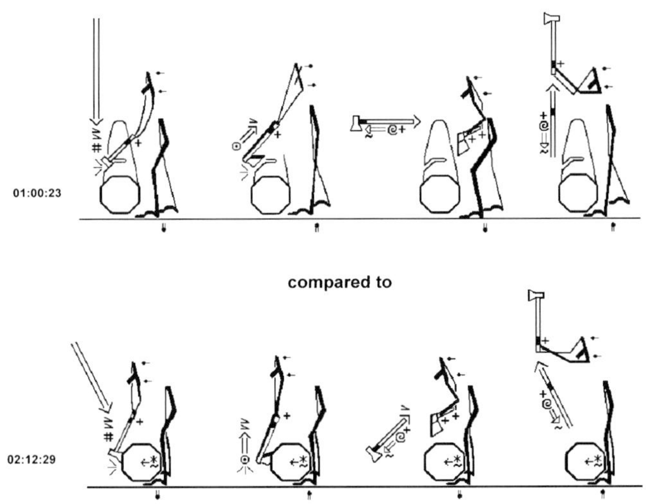

Fig. 10.16: Comparing the movement phases early and late in the cutting process.

The system of annotation which I have demonstrated here has further potential. It is possible to expand it and analyse the movement in more detail. For example, the handgrip could be recorded, noting how it changes in relation to other parts of the body, and how this movement pattern relates to changes in the movements of the tools and whether it is distinct to different people. It is also possible to record how the sounds of cuts change, and how the craftsperson changes his working procedure, position and movement patterns with the sounds of the axe as it digs deeper down into the timber. The notation system shows the movements as we see them. The transcription of the video documentation is from the side, from an observer standing looking at the craftsperson. If the angle had been from another position, the movement symbols would change.

Conclusion

To combine notation systems together with video analyses of the movements may be an important contribution to the verbal and written academic language – and also to the preservation of the living traditional knowledge of the craftsperson. My position is that it is possible to experiment with and reconstruct parts of the knowledge and the dynamics in rhythm and movement patterns behind the working procedures that shaped the tool marks that we now observe in the surface of archaeological timbers. Working with craftsmanship of the past, researchers have to operate with several dimensions when the knowledge that is entailed in the craftsmanship is examined. There are so-called 'unspoken archaeological artefacts' and the problematic relationship between the past and the present (time dimension). The archaeological remains refer to an unspoken knowledge of the past which is in a problematic relation with the present. Should it be accepted that the past is something very distant and different from us?

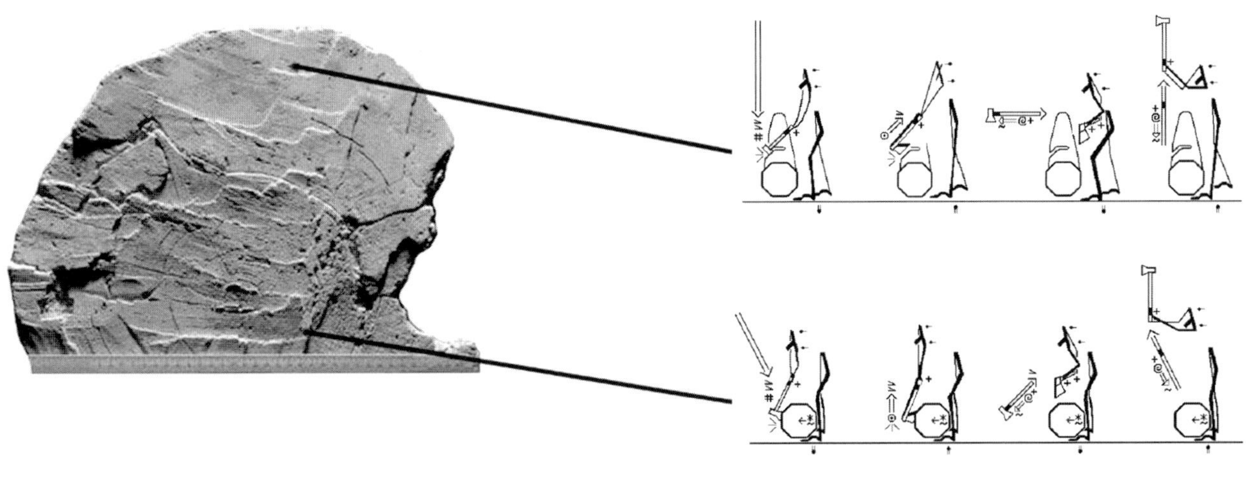

Fig. 10.17: Changes in movements and rhythm and their effects on the tool marks.

Or is it possible to say that the traditional knowledge of the living craftspeople today can contribute to our knowledge about craftsmanship of the past? Are we able to document the complexity of this knowledge by writing it down? Aspects like form, dimensions, the tool marks and quality of materials tell us a lot about craft knowledge. In order to be able to 'read' them, we need, however, to understand the versatile knowledge of the craftspeople, for example, the active movements of a working craftsperson in relationship to the space s/he moves in and his or her intentions. Dance has notation systems, which make it possible to describe movements and sounds through choreography and musical notes. These symbols contain a three-dimensional meaning understood directly by the fellowship that performs the actions and the competent observer. Here it is argued that tool marks and the actions of craftspeople creating them can be noted and examined by the same type of system. In contrast to dance or music, craft is more specific and actual in its physical form. Like dance and music, it is something conducted physically, but is also mental and abstract and the result is concrete and material. The tool marks with their signatures, furrows and grooves express the movement of the craftsperson's tool and trace sequences of action. To describe the depth and the richness of music only in words would be a poor description of the actual performance. The writing gives us just fragments of that particular happening. The music and musicians' actions contain nuances and depth. The writing gives us a good description of the music and the person performing it – but only to a certain extent. In the same way, we can say that notes contain information which writing does not capture. Perhaps we are able to gain an insight into Mozart's inner musical picture by reading his notes, but the notes do not replace Mozart's actual action in terms of his individual characteristic feelings, vulnerabilities and strengths. Nor do the notes cover all aspects of Mozart's performance; together, however, the notes and writings may contribute to a more comprehensive understanding of the musician and the composer. The understanding of the traditional craftsperson's knowledge in today's rational and economic society is about to die. To make the craftspeople and researchers who study this knowledge capable of communicating with each other through a common sign system would be a very valuable step forward in terms of the documentation and preservation of their knowledge, and for analysis of crafts in earlier times. Willeke Wendrich's dissertation 'The World According to Basketry' has attempted to create such a notation systems in order to transcribe craft into an academic format. This article, as part of the wider project of transcribing craft, has attempted to make aspects of the tacit knowledge in craft skills more visible. It does not solve all problems, but offers a possible way towards making tacit knowledge literate, and making non-discursive knowledge discursive.

References

Bourdieu, P. 1998. *Practical Reason. On the Theory of Action.* Cambridge: Polity Press.

Christensen, A. E. 1970. *Båtbyggerverktøy og læreprosess, Sømand, Fisker, Skib og værft.* København

Christensen, A. E. 1985. *The Bryggen papers: main series.* Bergen: Universitetsforlaget.

Christensen, A. E. 1989a. Damascenerstål: litt historikk. *Tidsskrift for Norsk knivforening 2.* Knivbladet.

Christensen, A. E. 1989b. Ljåkniv. *Tidsskrift for Norsk knivforening 3 og 4.* Knivbladet.

Christensen, A. E. 2003. *Middelaldersk båtbyggerutstyr 3.* Oslo: Kysten.

Christensen, A. E. 2005. Osebergfunnet og 1905. *Årbok foreningen til norske ortidsminnemerkers bevaring.*

Cornell, L. 1986. Arbete och arbetformernas utveckling. *Meddelanden från ekonomisk historiska institutionen vid Göteborgs Universitetet 55.* Göteborg.

Crone, P. 1989. *Preindustrial Societies.* Oxford: Blackwell.

Danbolt, G. 1979. *Den estetiske praksis. Gunnar Danbolt, Kjell S. Johannessen, Tore Nordenstam.* Bergen: Universitetsforlaget

Frederiksen, G. 1978. *Redskap for tre. En undersøkelse av redskaper for bearbeiding av tre fra yngre jernalder i Norge.* Magistergradavhandling, Universitetet i Oslo.

Godal, J. B. 1996. *Tre til laft og reis, gamle hus fortell om materialbruk:* Norsk Landbruksforlag.

Goodwin, C. 2003a. "Pointing as Situated Practice". In S. Kita (ed.). Pointing: Where Language, Culture and Cognition Meet, 217–241. Mahwah, NJ: Lawrence Erlbaum.

Goodwin, C. 2003b. "The Semiotic Body in its Environment," in J. Coupland and R. Gwyn (eds) *Discourses of the Body.* 19–42. New York: Palgrave/Macmillan.

Høgseth, H. B. 1998a. *Middelalderske bygningslevninger som kunnskapsformidler: tømmerhus i Nidaros som material- og kulturhistoriske dokument i tiden ca. 1025 til 1475 e.Kr.* Hovedoppgave: Norges teknisk-naturvitenskapelige universitet.

Høgseth, H. B. 1998b. Transmission of knowledge through action: How to interpret timber from archaeological sites. The kaleidoscopic past. Gotarc Serie C, *Arkeologiska Skrifter* 16.

Høgseth, H. B. 2000. Handlingsbasert Kunnskap. Bygningslevninger som kilde til kunnskap i det arkeologiske materialet. *Primitive tider 1.*

Johanessen, K. 1999. *Praxis och tyst kunnande.* Stockholm: Dialoger.

Johanessen, K., and R. Bertil. 1989. *Om tyst kunskap: två artiklar. Didaktisk forskning i Uppsala 7.* Uppsala: Uppsala universitet, Centrum för didaktik.

Lemonier, P. 1993. *Technological choices: transformation in material cultures since the Neolithic.* London: Routledge.

Molander, B. 1996. *Kunskap i handling. 2. omarb. uppl.* Göteborg: Daidalos.

Molander, B. 1997a. *Arbetets kunskapsteori. Sutton Movement Writing System: http://www.movementwriting.org.* Stockholm: Dialoger.

Molander, B. 1997b. "Kunskapsmångfald och olika kunskapstraditioner," in H. Alvsvåg (ed.) *Kunnskap, kropp og kultur. Helsefaglige grunnlagsproblemet, Ad Notam Gyldendal.* 124–146. Oslo

Petersen, J. 1951. *Vikingetidens redskaper. Av Jan Petersen.* Oslo: Dybwad.

Polanyi, M. 1959. *The study of man. The Lindsay Memorial Lectures given at the University College of North Staffordshire.* London: Routledge & Kegan Paul.

Polanyi, M. 1961. *The logic of personal knowledge: essays presented to Michael Polanyi on his 70. birthday 11. March 1961.* Glencoe.

Polanyi, M. 1983. *The tacit dimension.* Glouchester, Mass: Peter Smith.

Polanyi, M. 1998. "Knowing and being: essays by Michael Polanyi." Chicago: University of Chicago Press.

Rolf, B. 1991. *Profession, tradition och tyst kunskap: en studie i Michael Polanyis teori om den professionella kunskapens tysta dimension.* Gyttorp: Nya Doxa.

Rygh, O. 1999. *Norske oldsager.* Trondheim: Tapir.

Ryle, G. 1990. *The concept of mind.* London: Penguin Books.

Sands, R. 1997. *Prehistoric Woodworking. The Analysis and Interpretation of Bronze and Iron Age Toolmarks.* London: The Institute of Archaeology, UCL.

Shetelig, H. 1947. *Arkeologi; historie; kunst; kultur: mindre avhandlinger/ utgitt til syttiårsdagen 25. juni 1947.* Bergen: Griegs forlag.

Shetelig, H. 1950. *Vikingeskipene: deres forgjengere og etterfølgere. Av A. W. Brøgger og Haakon Shetelig.* Oslo: Dreyer.

Sutton, V. 2001. *Sutton's Movement Writing Site.* Retrieved 22 June 2011, from http://www.movementwriting.org/.

Tempte, T. 1982. *Arbetets æra: om hantverk, arbete, några rekonstruerade verktyg och maskiner. Forfatter: Tempte, Thomas Trykt.* Stockholm: Arbetslivscentrum.

Wendrich, W. 1999. *The World According to Basketry.* Leiden: CNWS Leiden Universitet.

11. Conceptual knowledge as technologically materialised: a case study of pottery production, consumption and community practice

Sheila Kohring

How do we come to be knowledgeable social beings? Western ontological traditions situate knowledge as based in individual cognisance and arising from a prioritised, discrete body (see critiques in Knappett 2005, Thomas 2004). Through the body, we internalise and make the world meaningful and in this process, nature and society are co-constructed. However, to make things meaningful, we require engagement with our physical surroundings and the materiality of this physical environment. In order to construct the ontological or conceptual knowledge of the world based on these engagements, it is imperative to acknowledge that engagement is situated in social milieus such as interactions with family, cohorts, colleagues and a broad spectrum of learning communities. It is in these social milieus where, when we first experience the world, someone is there to help make it meaningful and to structure our conceptual knowledge of the world. Without contextual social engagement to structure knowledge acquisition, our individual and embodied habitual movements, inculcated practices and ways of doing things in the world are decontextualised and exist outside a framework of knowledge. If there is no knowledge framework, the reason or ability to conceptualise embodied technical knowledge of 'doing' or 'making' is impossible; both an embodied knowledge and knowledge framework are necessary to 'dwell in the world' (*per* Ingold 2000a: 185)

This paper explores the interlinkage between conceptual social knowledge systems and technological practices. The construction of conceptual knowledge systems, or ontologies, is immediately and inextricably linked to practical engagements. Meaning, knowledge and ontological perspective is shaped by contextualised embodied experiences and the creation of habitual practices and a *habitus* for moving within our social and physical world (Bourdieu 1977, similarly but using a different level of scale, see Ingold 1993 for the concept of taskscape). The interconnectivity between conceptual and practical knowledge weaves together individuals and material objects within a framework of communal, structural meanings. The result of practical engagement is the establishment of knowledge technologies – or ways of structuring our knowledge – that are reinforced and habituated within community of practice frameworks. Through material engagement and a community of practice, conceptual knowledge is linked to technological knowledge as both embodied and materialised practice. In this paper, I will build a case study around Copper Age pottery production in western Iberia. Specifically, I will consider how communities of practice shape technological knowledge and how this reverberates back into the conceptual knowledge framework held by the society to become materialised in future production acts and their material culture results.

Technology and Embodiment Theories

Before exploring potting technologies in Copper Age Iberia, it is fruitful to highlight the theoretical underpinnings for this argument. The last twenty years have provided an excellent framework for thinking through the social aspects of technological implementation, decisions and practices by way of a series of anthropological, archaeological and sociological discussions on the topic (see Dobres 2000, Latour 1991, 1993, 2000, Lechtman 1977, Lemmonier 1993, Pfaffenberger 1992). These approaches have explored the dialectic between individual production events and broader social knowledge systems within societies and the structural principles by which these events are operationalised in meaningful social ways. Increasingly, focus has been placed on the material culture and materiality of the objects as subjects in their own right within the human-material culture relationship; both are inseparable from each other and the relationship can be construed as seamlessly extending between humans and material culture (Latour 1993: 136, 2000: 12). If this is a fundamental aspect of technological engagement and practice, then it has an inescapable effect on our perception of the world (see Ingold 2000b). Technology is a knowledge framework that we use to construct our world, but at the same time, technology shapes us, constructs our world and becomes a principle avenue for communicating social meanings.

Knowledge systems at the societal level are necessarily viewed as broad, abstract and encompassing, however, when focussed on a specific production (or materialisation) act, understanding practical knowledge starts by incorporating the social situatedness of the individual. The social community does not in reality mark out the role of 'producer' as singular and all-encompassing. In fact, the description of 'producer' only scratches the surface of identity and knowledge systems held by a single individual performing a single production act. We must realise that the knowledge a producer brings into production events materialises a whole suite of social as well as technical knowledge informed by the different roles, status positions and genders experienced by the individual. That said, at a community-based scale, these knowledge systems are also shared, manipulated, restricted and reproduced within society, albeit by different people with their own embodied and experienced pasts, and they occur in different and changeable social contexts (see Barth 2002). Turning specifically to the knowledge required when enacting a technical event, embodied and bodily practices affect how this knowledge is individually reproduced. Embodied practices, drawing on Mauss' (1979) concept of *techniques du corps*, emerge as inculcated and habituated bodily movement from learning how to act and move within the social environment. Bodily and conceptual knowledge are inseparable, as both are made meaningful within the social milieu.

Thus, how we view the world and make sense of it is channelled through our body not only from a material engagement aspect as implicated in the approaches above, but also through our 'communities of practice' (*per* Wenger 1998). Groups, to which we belong and are incorporated into, create boundaries for experiential understanding by providing a framework for learning specific techniques and structuring their meanings. These techniques, reinforced through a continued interaction with a community of practice, routinise techniques of doing and in the process also routinise a specific way of understanding the world. Communities of practice enclose us in the comfort of a learnt, shared world-view and, as such, a shared way of making and doing things in this community-constructed world. Communities of practice fluctuate, which adds a dynamic and flexible element to how knowledge changes and techniques are practically enacted. We choose to align ourselves with different communities and groups depending on situation and context, each aspect varies and alters perspective in those different situations and contexts. In essence, our practical engagements within a community, or communities, of practice, structure the conceptual knowledge systems – including social and technological knowledge – that inform our material engagements.

Materialising Technological knowledge as social knowledge

Materialisation theories have also renewed our focus on material culture and the ways social agents/subjects engage with it. Beyond active agents and agency, objects and individuals continually reproduce and transform their social world through technological processes (i.e. Ingold 2000). While this type of materiality and material engagement theory is amorphous enough to incorporate all types of object-subject relationships (see Miller 2005: 4), I want to use the concept of materialisation rather than just materiality or material engagement as it specifically necessitates the enactment of social knowledge within a technical event to produce material culture. Materialisation theory has focussed on the social construction of political institutions and social ideologies through the deployment of material culture, often in a way reinforcing social inequalities within a society (DeMarrais 2004, DeMarrais, Castillo and Earle 1996, Renfrew 2001). I think this focus on social inequalities has been loosened (see DeMarrais 2004) and the idea of materialisation within the technical production process is a useful development. Variation in techniques and styles of production can be seen as the agency of individuals, communities, and objects materialising the history and continuation of social relationships and shared knowledge being reproduced within the milieu of technical processes.

As such, visuality can be a key means of materialising appropriate social meanings, attitudes and relationships within society. Visual differences provide a means for multiple ideologies, institutional facts, and shared conceptions of the world to be materialised within a single engagement event. The application of technical knowledge in *chaîne opératoire* choices draws on a system of socio-technical knowledge that holds the rules structuring convention and variability while at the same time being a reservoir for heterogeneity. This heterogeneity can be used to materialise individual expression and specific contextual meaning, while still materialising broader social meaning through adherence to other visual characteristics (see Gosselain's 2000 discussion of salient production techniques and technical malleability).

Visual attributes, then, are key ways of exploring the materialisation of technological and social knowledge systems within material culture. The variation present in specific visual attributes reflects the inculcated knowledge and practices of individual producers learning within a bounded community of practice. The social choices structuring the standardisation or variability of each specific visual attribute result from both the physical characteristics of the media and the social ways of making and using the finished material objects. Those techniques and features that allow a greater degree of latitude for expression still conform to social knowledge about what makes an appropriate tool or object, but they are also socially acceptable avenues for exploration of technique and practice by producers within that community. At the same time, attributes and techniques held to a greater standardisation reflect deeply seated ideas of appropriateness and aesthetics within the community – change in these areas affect social and technological knowledge. So, as societies innovate, accept and internalise new technologies or big changes in socio-technological knowledge, new arenas for variability in technical practice emerge permitting changes in the finished form of objects as the dialectical influence of social and technological change recreate producer and material culture.

Let us take an example from a more recent technological knowledge system, and consider its innovation and change due to social knowledge and the practical uses for which it is employed by both individuals and society as a whole. Computers often epitomise the concept of science and the 'technological world' within modern society. In actuality, however, computers are enmeshed in the social networks and relationships of our everyday lives; they facilitate and mediate how we communicate, obtain social information and make our world meaningful in a variety of ways. How you engage with computer technologies also often depends on your access to technical knowledge, skill, needs and social history (see Woolgar's 1991 example comparing specialists' and non-specialist trial users' ease of engagement). In essence, computer competency and technical skill are materialised through engagement and practice; computers, just like all material culture, are produced through the intersection of technological and social knowledge systems rather than falling into one single field of discourse.

What conceptual knowledge and meanings are being materialised through the computer and its associated social technology? Do computers materialise a Western prioritising of technology within an ideological framework categorised as 'science'? Do they materialise an ideology necessitating constant communication and information transfer through the quest for increasing speed and accessibility? Do they reflect a belief in social interconnectivity through the diversity of email, blogging and social networking resources available through them? Computers materialise all these conceptual beliefs, often simultaneously, but the material reality and presence of the computer further affect our more immediate and practical daily interactions. As a social technology and as a representative of innovative 'technology' the computer shapes our behaviour, scheduling and ways of engagement; the reality of computer technologies demand we conform to an entirely new set of social conventions revolving around computer etiquette. So, writing emails does not require the formality of a letter (or if it does, the letter is often included as an 'attachment') and email access ensures constant contact with colleagues, friends and family even as it dictates a presumption for an immediate response to those emails. How far can you resist the computers and associated technologies? Resistance as a form of engagement may be a choice, but as others in society integrate computer technologies such as emailing into everyday practice, one is forced to compromise. As social etiquette regarding how we respond to emails or produce letters and documents in electronic form changes, the technological reality of how we communicate also changes. Compliance regarding new technologies and practices – such as in computers and communication – are not just choices but necessities as they create social networks and tie us into knowledge systems.

Thus, the computer and its facilitating technologies are not only tools for disseminating social knowledge (writing research grants, preparing visual presentations, emailing and blogging) but they actually enable us to be social. In the end, the computer can be argued to have significant agency in social relationships – mediating them, facilitating them and even creating them. This active role for material culture is unproblematic if we draw on Latour's assertion regarding subject-object relationships as blurred and continuous (1991, 2000). From this perspective, the computer can be construed as an extension of the self (bodily and cognitively) just as with any other material culture-based technical, practical or performative engagement. But, would these engagements and knowledge systems have emerged if the computer had remained the enormous, complicated and overly specialised machines they were forty years ago? Forty years ago, people were forced to engaged with them very differently due to their physical presence as they filled entire rooms, required punch cards and spools located on their exterior (Fig. 11.1). This was a very different type of physical engagement; the materiality of the computer affected access to the computer and its technical knowledge. The size, complexity and mechanical aspects of the computer meant that very few people actually engaged and interacted with the machine, as it required specialist knowledge and specialist settings not accessible to wider society. This type of practical engagement involved mechanical, or sets of specialist technical knowledge, as well as the organisation of knowledge practice which we often use computers for today. However, as the technology expanded beyond the specialist engagements – meaning, as it became an increasingly open social technology – those overt technical attributes became masked and closed off, thereby affecting our practical and technical engagement with the computer. The open mechanics of the computer (i.e. spools, punch cards) were replaced with closed mechanics (i.e. microchips) hidden away in 'aesthetically' pleasing boxes, which should not be opened under serious penalty (your warranty will be void). In effect, although conceptual knowledge revolving around the use of computers is embedded in social practices, the technical aspects of computing still rest in the hands of specialists holding specific technological knowledge, even if its material and practical reality exists across a broad spectrum of social knowledge systems (see Woolgar 1991). The social use of the computer demonstrates how material culture, and our engagement with it, shapes broader social relations and conceptual knowledge systems: social and technical knowledge systems are enmeshed through continual and changing practical material engagements.

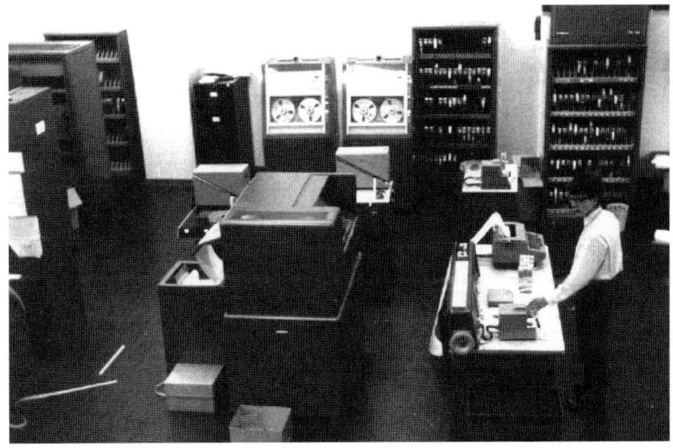

Fig. 11.1: TITAN computer of the 1960s–1970s (reproduced with permission from the Computer Laboratory, University of Cambridge).

Pottery technologies, variability and practical knowledge within a Copper Age community

Let us look now at a case study from prehistory – that of pottery production and the role of variability and conformity in materialising appropriate social meaning. Just as computers have become ubiquitous and everyday within our own society, pottery has been described in the same manner, being seen as a technology integral to a diverse set of everyday practices (i.e. Braun 1983). These practices are informed by the production event, but reside also in historically- and community-based styles, aesthetics and uses beyond just the technological choices involved in a vessel's production. In essence, pottery technologies represent exceptionally well how technologies of production are influenced by and materialise social and conceptual knowledge within society.

Here, I wish to highlight the importance of visual differences and technical choices within pottery assemblages. Visual attributes, unlike compositional or fabric characteristics, can be experienced directly by all producers and consumers of a vessel. As such, visual characteristics can be relatively easy to link to social knowledge regarding identity (who can use what vessels, who has made what vessels) and the meaningful use of particular vessels in appropriate contexts. As outlined above, multiple ideologies, institutional facts and shared conceptions of the world are all materialised within a single pottery-based engagement and a single finished vessel. The technological sequence, or the practical application of technological knowledge, provides a baseline for the implementation of pottery knowledge. It cements the rules that structure the life biography of the vessel, starting with the sequence of production, but also affecting consumption and depositional practices. It also continues to influence future production choices. These sequences do not just happen and the technology does not just mechanically reproduce the same pottery vessel over again. The potter is enmeshed in existing technological and social knowledge systems which reinforce and help shape the choices available – this is a learning and socialising environment which can be framed as a community of practice (per Wenger 1998). From an embodied individual to the community of practice in which they are inculcated, we can consider a potter (or any producer or user of material culture) as situated within this network and retaining various levels of potting skill and knowledge within a broader societal-based conceptual technological structure.

Our case study focuses on the 3rd to 2nd millennia BC Middle Guadiana Basin, which partially delineates the modern national borders between Spain and Portugal. The site and assemblage described here come from an enclosed Copper Age settlement on the Spanish side of the river named San Blas (Fig. 11.2). The University of Seville excavated the site over four seasons as part of the Alqueva Dam Project (see Hurtado 2004a, 2004b). San Blas had two phases of Copper Age settlement and it is the second, or later, phase pottery assemblage and associated technologies that will be discussed here. This later occupational phase of the site includes multiple dry-stone wall constructed houses – all approximately the same size and sharing very similar spatial organisation – an internally demarcated, ditched central zone and a large enclosing wall with bastioned entrance facing a sheltered location on the Guadiana River (Hurtado 2004a). Material culture is abundant at the site, indicative of the permanence and occupational focus within the community. Pottery is the dominant material recovered, but evidence for metallurgical, lithic and bone-working practices is also widespread. The ceramic assemblage indicates a diverse repertoire of attributes and styles present within the community, although the vessels can also be consolidated into general formal types such as bowls, deep storage vessels, fine wares and *campaniforme*, or Bell Beakers (Kohring 2007). These categories – and, more importantly, their variability – are the bases for exploring the interrelationship between object/human/knowledge as materialised in technological production practices. This materialisation of technological knowledge is held against a backdrop of context, social relationships, social realities and embodied technical practices.

The realm of pottery production decisions and engagements

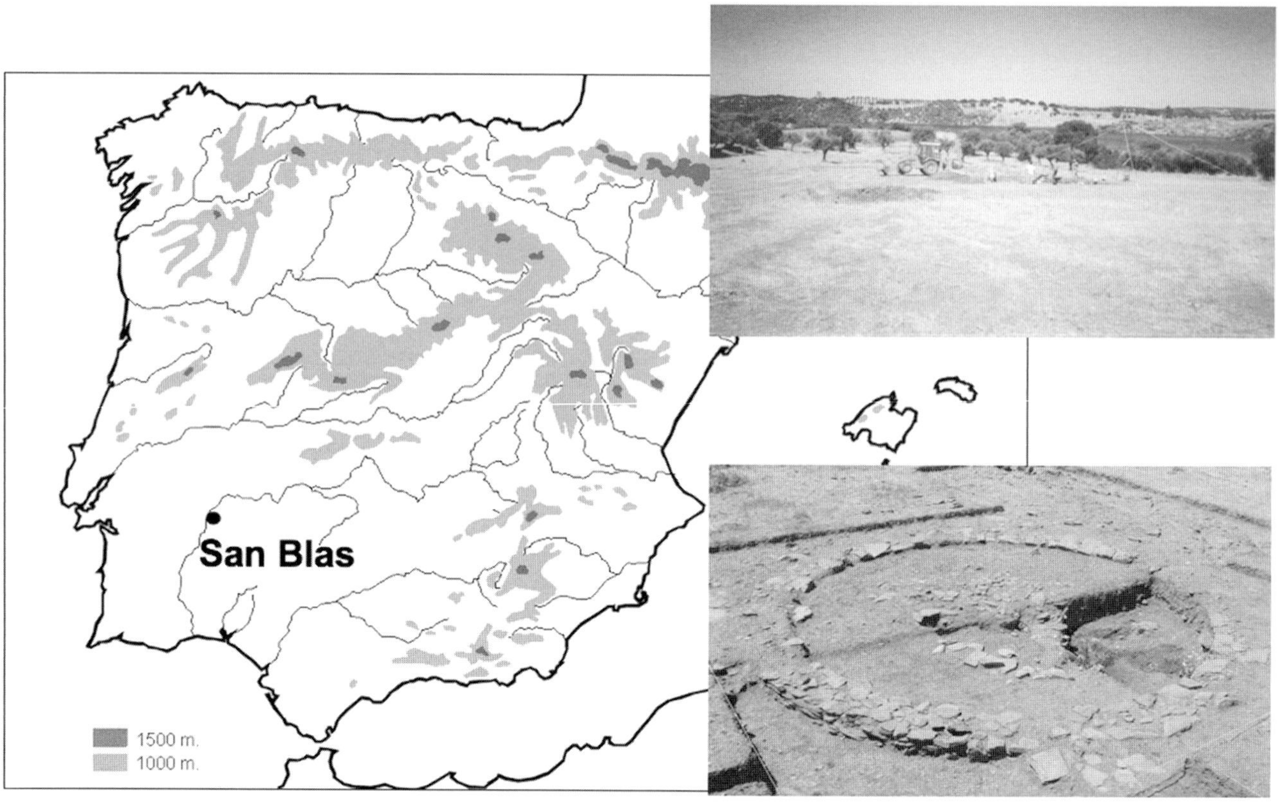

Fig. 11.2: The Site of San Blas, Cheles, Extremadura, Spain.

is an active arena in which social reproduction is continually being practically reworked and experienced in the context of this prehistoric Iberian community. Daily activities and social engagements are experienced through a suite of visual and material avenues in which producers and consumers make themselves simultaneous technical and social agents extending knowledge through the material forms. Those material forms spread into multiple knowledge systems and create new contexts for engagements, experiences and socialities. Knowledge in these engagements does not just apply to the symbolic and contextual aspects of the material culture (Hodder 1987, 1990) nor do the techniques of materialisation simply reference isochrestic variations in style (Sackett 1990). Instead, there is an extension of subject and object through the materialisation-of-knowledge process instigated and reproduced through the technological experience. As Latour (1991) states in his title, it reflects the reality that 'technology is society made durable'.

To set the context for our case study, I wish to briefly expand on the community wide ceramic assemblage during the Later Copper Age at San Blas. At the most general level, one word can describe the predominant vessel form used and produced in the community – bowls. The San Blas assemblage is marked by particularly overarching conventions regarding pottery forms and technical representations, especially regarding the predominance of bowls in every context examined in the study (Kohring 2007). Those contexts include several different structural or domestic settings, and several open, or public, areas. Yet, there is an intermingling of social/technical conventions and technological knowledge systems evident in the production and *chaîne opératoire* utilised by the community potters in the production of these bowls. While vessels are generally round-lipped and highly polished, the lips and surface finish can be seen as areas for individual expression and technical manipulation.

These two areas for the expression of social/technical knowledge can be discussed in terms of the techniques of production regarding surface finishing quality and those regarding lip and rim final modifications. Both technical aspects, as represented in the San Blas assemblage, are not difficult to manipulate nor do they require high degrees of specialised training to create visually distinct differences in these attributes. Surface polishing and burnishing are the most common finish for bowls in the assemblage and the finest burnished sheen is more a product of labour investment and controlled repetitive technique than requiring highly specialised training. Lip and rim manipulation can be highly variable, although within this community the lips are often rounded or slightly thickened. Despite this prevalence in rim production technique, more than sixty-five different variations

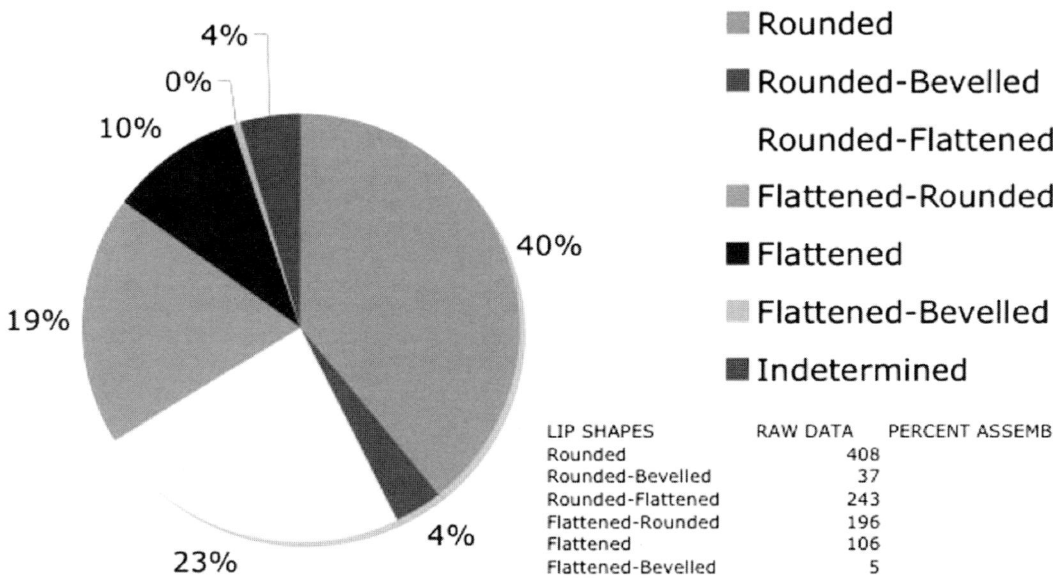

LIP SHAPES	RAW DATA	PERCENT ASSEMBLAGE
Rounded	408	39%
Rounded-Bevelled	37	4%
Rounded-Flattened	243	23%
Flattened-Rounded	196	19%
Flattened	106	10%
Flattened-Bevelled	5	0%
Indetermined	46	4%
ALL RIMS	1041	100%

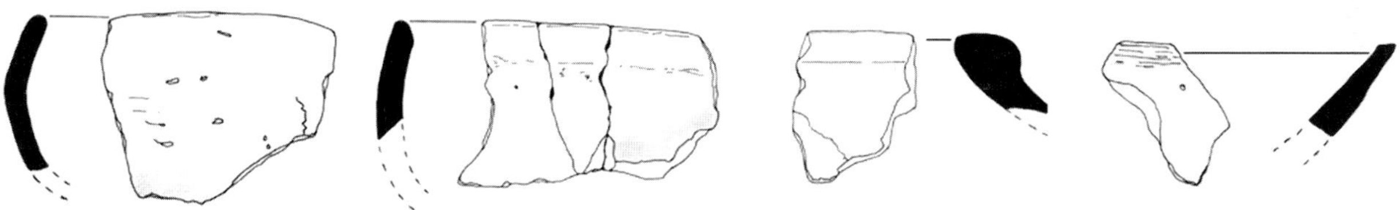

Fig. 11.3: Lip Shape Variability in the San Blas Assemblage. Percentage and raw data provided. Bottom demonstrates some of the continuum between rounded and flattened lip shapes suggesting individual, embodied practice and learning.

of rim and lip formation were recognised in the San Blas repertoire (Kohring 2007). Again, greater manipulation or elaboration in this area does not require significant specialised knowledge, but rather a knowledgeable competency in the media and techniques of production. If both surface finish and rim/lip production steps are easily learnt, manipulated and open to great variation, how do we explain or make sense of the patterning of visual characteristic combinations recognised in the San Blas assemblage?

The key to understanding choices in production techniques must be sought first in the socially learnt knowledge of what makes an appropriate pottery vessel in the San Blas community and then, how this is translated into the equally socially learnt bodily practices involved in making vessels. Both the techniques for producing surface finish and lip and rim form may be used as examples of how conceptual knowledge systems interlink with the technologies of pottery production through the social reproduction of these vessels, as both surface and lip attributes can have a significant visual impact. As discussed previously, visual characteristics create and inform socially understood conceptions of material culture appropriateness and as such can be firmly situated in the social realm. However, through social engagement processes, these visual characteristics are reproduced again in the technical realm of practical pottery making. In essence, social categories and meanings are reproduced in the act of producing.

The degree of conformity or variability in these visual characteristics becomes an important point of analysis as it reflects the boundaries surrounding social categories (which are created within the community and are applicable to all material culture). The categorisation process situates variability by placing it against a conceptual template, or aesthetic standard of acceptable material culture genres. At San Blas, the fact that the majority of sherds have a highly polished dark surface and rounded or thickened lip and rim is important as it indicates a certain level of conformity for a specific visual aesthetic for vessels and hence, a pertinent conceptual pottery category. But by also considering the variability in techniques, surface finish and lip and rim form, technical choices – often the product of embodied practices inculcated in the community of practice – can be imbued with more meaningful interpretations of social conceptual knowledge and categorical systems within society.

Taking the analysis down to micro-scalar variability, the range of minute differences in mouth orientation, rim shape, lip style and lip shape as separate variables make the lip and

Primary Permutations			
Lip Shape	**Rim Shape**	**Lip Style**	**Designation**
Rounded	Simple	Neutral	IA1
Rounded	Simple	Thinned	IA3
Rounded-Flattened	Simple	Neutral	IIIA1
Rounded-Flattened	Simple	Thinned	IIIA3
Flattened-Rounded	Simple	Neutral	IVA1
Flattened	Simple	Neutal	VA1

Permutation Cluster	Number in Cluster	Number of Examples	Percent Assemblage
Primary	6	598	57%
Secondary	5	113	11%
Tertiary	9	130	12%
Quaternary	21	118	11%
Unique Examples	24	36	3%
Unknown	1	46	4%
Total	66	1041	1

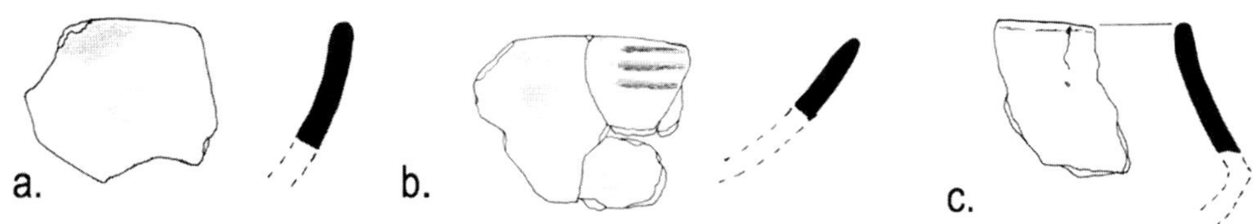

Fig. 11.4: Rim and Lip production permutations. Examples below include a. IA1, b. IA3 and c. IIIA1. Evidence of individual practice and community-based learning.

rim an interesting arena for exploring inculcated practice, individual expression and the individual materialisation of a broader social aesthetic. A combination of practical knowledge – such as how much pressure is exerted on the clay or how much smoothing is employed around the lip edges – and the lack of strongly instituted rules over lip shape within the technological repertoire results in a gradient of acceptable lip shapes between roundness and flatness (Fig. 11.3). The diversity of these permutations in the assemblage suggests that any of the ranges within this continuum were considered appropriate lip shapes. Importantly, the actual differentiation between variations may have been noticeable and meaningful within the community, but as they remained in keeping with the acceptable aesthetic for lip and rim form they were not categorically differentiated. These features were then combined in very flexible permutations with other 'finishing touches' around the rim and are applicable to the majority of vessel forms. Thus, in terms of lip shape we see the inculcation of practical and individually embodied knowledge affecting the materialisation of the technological knowledge systems. However, these same techniques and choices also have knock-on effects on the materialisation of other practical techniques of rim form production, some of which may be more rigorously structured, and which also contribute to the creation of conceptual knowledge categories.

If we then consider the practical employment of embodied techniques in relation to other rim and lip variables (such as orientation, which is also affected by shape categories), we see a set of linked choices, or permutations, consistently being used by the potters at San Blas. Immediately, and unsurprisingly, it is obvious that not all possible permutations were materialised by the community's potters and many potential permutations do not seem to have even been produced experimentally or as one-offs. Patterning suggests social rules regarding the practice of technological knowledge structured and channelled the potters' physical and embodied technique, style and aesthetic into appropriate forms and ways of making. Appropriate pottery was not homogenous in its visual characteristics, and variability and expression had their place. Variability and expression, however, existed within boundaries. An analysis of rim and lip attribute distribution frequencies demonstrated five different levels, or clusters, of acceptable permutation within what might be considered appropriate San Blas pottery. These permutations were then ranked based on their prevalence and repeated patterning in the technological repertoire.

Primary permutations were the most prevalent set of linked lip and rim visual characteristics (see Fig. 11.4). They were dominated by simple (where the rim follows the natural line of the vessel body) and thinned (where the lip is more than 2mm thinner at the lip than at the juncture with the neck of

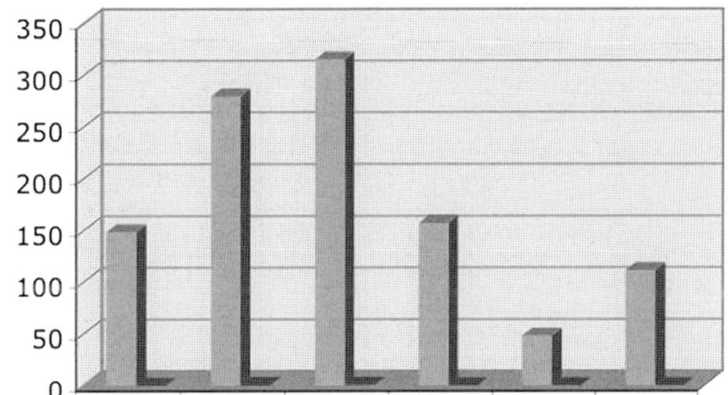

	I, II	III	IV	V	VI	VII
Total Number of Examples	148	279	315	157	48	111
Percent of Assemblage	14%	26%	30%	15%	5%	10%

Fig. 11.5: Surface Finish Quality. Evidence of technical choices for creating a community-based visually appropriate vessel.

the vessel) rim and lip forms. The repeated presence of some permutations suggests technical traditions, styles of making and learnt *habitus* within the community (see Dietler and Herbich 1989, 1998, Gosselain 1998). Non-existent permutations must also be considered in terms of materialised knowledge as their absence suggests the physical and technical restrictions within the technological knowledge repertoire of the potter due to their socially dictated practicality or logic; those permutations are simply not possible either as potential embodied practices or conceptual knowledge. Thus, both prevalence and presence of particular permutations is significant when considering the technological-social knowledge of pottery production and its materialisation.

All the variables and techniques employed in the creation of the rims and lips of the San Blas pottery embed within the fabric of social knowledge. This means that all of the variables and techniques present in the assemblage recursively inform the choices – unconscious and conscious – for pottery production in any given context. Thus, as an arena of high variability within the potting *chaîne opératoire*, the formation of lips and rims is complicated, dynamic and informed strongly by both unconsciously and consciously inculcated bodily and social practices. Can the same be said for surface finish?

Surface finishing procedures reflect technological knowledge in terms of the type and technique of finish produced and social knowledge in terms of knowing which finish is appropriate to which vessel and later post-production use context. Of course, these two realms of knowledge overlap and in a practical reality they cannot be separated. However, the knowledge employed in creating a specific surface finish, perhaps more than during lip formation, is often an overt decision even if the techniques and bodily actions are inculcated. Decisions such as time, labour, post-production use and other non-technical considerations affect whether surface finish is simply a smoothing of the exterior, whether it is burnished, or even, whether it is decorated with incised or painted motifs. For the purposes of this discussion, I want to make clear that surface finishing does not include decoration. Based on the *chaîne opératoire* evidence it appears that primary surface finishing and surface decoration were distinct procedural steps. The rationale for this separation was twofold: first, the majority of sherds in the study are not decorated (in terms of incised or painted motifs) and, secondly, all decorated sherds went through a primary surface finishing stage where they were smoothed, polished or burnished prior to the decoration process (see Kohring 2007 for further discussion). Thus, primary surface finish refers only to the production step in which the vessel was smoothed, polished or burnished and excludes secondary stages where the vessel surface may have been further decorated.

At San Blas, the most common *chaîne opératoire* procedure for surface finishing was intensive smoothing of the pottery body with the hand, fabric or a hard object. The polishing/burnishing process does not necessarily require specialised knowledge, but does reflect an investment in production activity within this realm of technical choices and a certain amount of bodily repetition depending on the intensity of the act and desired surface finish outcome – especially with highly burnished vessels. The intensity of the surface finishing investment and activity was placed on a continuum and rated from I/II (Intense/special finish, including burnishing until the texture is hard and slick to the touch and with a shiny visual nature) to VI (Low intensity finishing, such as basic smoothing and wiping of the surface with soft material leaving a slightly rough feel and matte visual look) (Fig. 11.5). All sherd surfaces (unless otherwise eroded or encrusted due to post-depositional erosion) indicated vessels were well smoothed, with 60% of the assemblage having been polished or burnished. While this process of polishing/burnishing had technical repercussions that might have been related to functional knowledge (i.e. making the surface impermeable) it also had social affects, creating a visually and texturally specific aesthetic, which might have had implications for potential use and display choices.

Thin-walled *paredes finas* and black bowls are the most commonly burnished forms and important, socially meaningful vessel categories at San Blas. They are found in public and domestic contexts, suggesting a display or consumption use-context in which their visual characteristics would have been important for establishing their appropriateness. Again, this is the interplay of social and technical knowledge co-informing each other. Socially constructed concepts such as style and aesthetic preferences intersect with use contexts and affect the *habitus* of the potter during the production of the surface finish. A surface aesthetic would have been achieved during the production process through visual cues (i.e. the increasing shine from the surface) and the feel and texture of the surface as it became increasingly slick or silky to the touch. A recursive relationship between social and technical would emerge with increasing skill and experience; as a potter's experience regarding differential surface treatments and their appropriate application grew in competency, they could assess the time and labour required to create the desired aesthetic and manipulate the entire *chaîne opératoire* accordingly. Unlike the variability of individual *habitus* or inculcated practices seen in the arena of lip and rim formation, surface finish represents a subtly different type of engagement with the technical and social knowledge. This type of technical engagement reinforces a social aesthetic as channelled by bodily knowledge of the visual and tactile experience; not only for the producer, but for all those engaging with these vessels in post-production settings and contexts.

When completed, the surface finish may be seen to materialise social knowledge within an understood framework of material culture, as conceptual categories of social use and meaning, while the lip and rim forms may materialise a slightly different set of social knowledge regarding community based relationships and preferences. Together, all these types of knowledge are materialised through practical engagements both in the technical process and social use. Looking at the production of pottery, we can see many conventions and heterogeneities. Lip and rim variability is highlighted, but other elements open for individual expression and preference include decoration and orifice size, while conventions are much more rigorously held for vessel form, surface finish and paste recipes (Kohring 2007). The creation of socially appropriate pottery affects both the production stages which show great consistency or even standardisation, as well as those stages that permit more variability and expression within the acceptable range. These visual characteristics are the key to the engagement, relationships and knowledge networks being materialised.

The habitus *and Community of Practice of the San Blas Potter*

The social 'situatedness', experience and learnt behaviour of the San Blas potters structured how they, as embodied and community-based individuals, linked up the different choices and potentials available in each pottery production event. Mechanical reproduction of techniques does not result in homogeneity and each vessel shows the subtle variability as event, technique and choice coalesce in slightly different ways with each production activity. Consistencies and patterns are, however, equally evident in the technical repertoire. Conventions of production, or the steps rigorously structured and held to great conformity, sway the technological knowledge system as much as the steps and attributes of production which are less constrained. The *chaîne opératoire* as such provides a conceptual framework about social appropriateness and a method for producing an appropriate vessel (see van der Leeuw 1993 on conceptualisations and executive functions). Yet, any *chaîne opératoire* must be inculcated through practice and this happens through learning within the community setting.

The San Blas potter, like all his/her contemporaries within the community, materialised the social realities, rules and relationships which structured the community of practice to which they belonged. Thus, while the potter had a choice in the creation of a unique vessel (albeit structured by the *chaîne opératoire*), these options were highly contingent on the inculcated social and technical knowledge of learnt practices. The learning of both technical skill and social aesthetic are channelled through the community of practice so that practices, even those that become inculcated or are non-discursive, are socially reproduced in meaningful cultural ways (see Wenger 1998). Let us consider embodied practice and community practice more dialectically. *Habitus* and structuration, which we may use to refer to the community of practice process, are crucial in appropriately executing technological knowledge because technological systems are intertwined with the materialisation of the social worlds in which producers/consumers are engaged. Shared technological systems, linked to contexts of social engagement, create communities of practice because they materialise the relationships between people and their world; the act of production is a materialisation of your social, and community, understanding of the world.

Furthermore, multiple meanings are always embedded in material culture during production and use, and the meanings and relationships being explicitly materialised often reflect the social context and personal relationships affecting any given event. Thus, different production and use patterns within a community may represent different access to technological and social knowledge, different social responsibilities and duties, and different social status among community members (see Gosselain 2006). The assemblage comparison among San Blas households supports differences in household groups. A look at the intensity of production (such as that of burnishing and surface finishing), found that in one household (H22) three different levels of technical investment could be seen in the assemblage (Kohring 2007). These three levels of investment were equally represented in production activity and within the domestic contexts and it may be that they represent multiple producers at work, perhaps relating to skill and the learning process within a microcosm of the San Blas potting community of practice.

The San Blas pottery assemblage and its distribution represents the conceptualisation of a pottery aesthetic and style as channelled through the community of practice and individual potters' embodied technical practice. By looking at the patterned range of variability employed within the numerous production sequences and individual events, we can access the palimpsest of individually 'situated potters' within their community of practice, enacting a technological and conceptual knowledge of production and aesthetics with each potting event and decision. The exploration of lip and rim variability and primary surface finishing provide two ways of understanding the situated potter's *habitus* and how the community of practice shapes the materialisation of this technical knowledge. While the community of practice maintained a *chaîne opératoire* which structured pottery production in highly patterned ways, certain elements were more open to individual reflexes, choices and contingencies. The consistent pattern of polishing and even burnishing the vessel surface indicated an aesthetic ideal regarding the visuality of appropriate San Blas pottery. On the other hand, the great diversity of minute lip variations represent an attribute that in some ways is difficult to control, as it often integrates a great deal of bodily technique, which was learnt within the community of practice.

Conclusions

This case study compares just two steps within the potting *chaîne opératoire* employed at San Blas. These two steps, however, demonstrate the inherent variability of production and show the materialisation of social and technical knowledge at work within a presumably close community of potters. Yet, by looking only at either the technical or the social aspects of pottery production, the societal wide conceptual knowledge of production and use is only partially accessible. Pottery, like all material culture, must be embedded into webs of socio-technical knowledge systems which crosscut both production and use and reflect a diversity of knowledge and meaning within society. Conceptual knowledge maintains an aesthetic regarding what makes an ideal San Blas vessel, but the practical reality within the assemblage suggests that this aesthetic permits variation within certain production choices and attributes. Based on the variation in rim and lip styles, it is argued these elements were open to potters for expressing their own unique styles and signatures on the vessels they produced. Choice, however, was still affected by individual potters' embodied knowledge, the community of practice in which they learnt their *chaîne opératoire* and the wider community of practice in which they learnt their own sense of aesthetic.

In comparison, other attributes may be more constrained by the aesthetic ideal. The individual choice of surface treatment, which could have been done in different ways, seems in practice to have been much more restricted. The community aesthetic seems to have required vessels with a well-polished finish. This has implications beyond the production event, as the visuality and materiality of the polished and burnished vessels provided an overall sense of unity and appropriateness to the San Blas pottery assemblage. Through use in daily social contexts, this visual aesthetic would have been an important attribute for defining the Copper Age San Blas community and hence seems to have been rigorously adhered to within the *chaîne opératoire* for producing pottery.

Communities of practice, as learning contexts and dynamic social networks, do not reproduce material culture mechanically. Each individual is a member of many overlapping communities and communities of practice and, in terms of any technical act, they bring with them many different experiences and skills. The enacting of many embodied materialisations of technical and social knowledge results in variability. The aesthetic ideal of what constitutes an appropriate vessel must incorporate differences to allow for individual stylistic expression, skill and choice as much as it must restrict variability in other attributes. This mediation between individual choices and wider aesthetic ideals is maintained by the potting community of practice and the teaching and inculcation of social knowledge through practical actions of the *chaîne opératoire*. The materiality and visual cues of the vessel, however, further extend this knowledge beyond production and into other social contexts, informing appropriate contextual use and engagement throughout the entire community. The sharing of social and technological knowledge is crucial for the reproduction of social systems within communities. Technology, as the materialisation of society through the production of appropriate material culture, becomes an integral avenue for exploring the convergence of inculcated practice, embodied knowledge and social knowledge systems.

Acknowledgements

I would like to thank Víctor Hurtado Pérez and John Robb for their support of this research and reading previous versions of this chapter. Thank you to Katharina Rebay-Salisbury and Marie Louise Stig Sørensen for their comments in both a theoretical and editorial vein. Funding for the original research was partially provided by EDIA Portugal, the Universidad de Seville and the University of Cambridge.

References

Barth, F. 2002. An anthropology of knowledge. *Current Anthropology* 43, 1: 1–18.

Bourdieu, P. 1977. *Outline of a theory of practice*. Cambridge: Cambridge University Press.

Braun, D. 1983. "Pots as tools," in J. A. Moore and A. S. Keene (eds) *Archaeological Hammers and Theories*. 107–134. New York: Academic Press.

DeMarrais, E. 2004. "The materialisation of culture," in E. DeMarrais, C. Gosden, and C. Renfrew (eds) *Rethinking Materiality: The Engagement of Mind with the Material World*. 11–22. Cambridge: McDonald Institute for Archaeological Research.

DeMarrais, E., L. J. Castillo, and T. K. Earle. 1996. Ideology, materialization, and power strategies. *Current Anthropology* 37, 1: 15–31.

Dietler, M., and I. Herbich. 1989. Tich Matek: The technology of Luo pottery production and the definition of ceramic style. *World Archaeology* 2, 1: 148–163.

Dietler, M., and I. Herbich. 1998. "*Habitus*, techniques, style: An integrated approach to the social understanding of material cultures and boundaries," in M. T. Stark (ed.) *The Archaeology of Social Boundaries*. 232–263. Washington, D.C.: Smithsonian Institute Press.

Dobres, M.-A. 2000. *Technology and Social Agency*. Oxford: Blackwell.

Gosselain, O. P. 1998. "Social and technical identity in a clay crystal ball," in M. T. Stark (ed.) *The Archaeology of Social Boudaries*. 78–106. Washington, D.C.: Smithsonian Institute Press.

Gosselain, O. P. 2000. Materialising identities: An African perspective. *Journal of Archaeological Method and Theory* 7, 3: 187–217.

Gosselain, O. P. 2006. "Heritage, adjustments and trickery in African pottery traditions and identities," in *Breaking the Mould: Prehistoric Ceramic Research Group Annual Conference*. Manchester.

Hodder, I. 1987. "The contextual analysis of symbolic meaning," in I. Hodder (ed.) *The Archaeology of Contextual Meanings*. 1–10. Cambridge: Cambridge University Press.

Hodder, I. 1990. "Style as historical quality," in M. W. Conkey and C. A. Hastorf (eds) *The Uses of Style in Archaeology*. 44–51. Cambridge: Cambridge University Press.

Hurtado, V. 2004a. El asentamiento fortificado de San Blas (Cheles, Badajoz). III milenio AC. *Trabajos de Prehistoria* 61, 1: 141–155.

Hurtado, V. 2004b. San Blas. The discovery of a large Chalcolithic settlement by the Guadiana River. *Journal of Iberian Archaeology* 6: 93–116.

Ingold, T. 1993. The temporality of the landscape. *World Archaeology* 25, 2: 152–174.

Ingold, T. 2000a. "Making culture and weaving the world," in P. M. Graves-Brown (ed.) *Matter, Materiality and Modern Culture*. 50–71. London: Routledge.

Ingold, T. 2000b. *The Perception of the Environment. Essays in Livelihood, Dwelling and Skill*. London and New York: Routledge.

Knappett, C. 2005. *Thinking Through Material Culture. An Interdisciplinary Perspective*. Philadelphia: University of Pennsylvania Press.

Kohring, S. 2007. Pottery Technologies and the Materialisation of Society: Late Copper Age Community Practices in Western Spain, University of Cambridge.

Latour, B. 1991. "Technology is society made durable," in J. Law (ed.) *A Sociology of Monsters: Essays on Power, Technology and Domination*. 103–131. London: Routledge.

Latour, B. 1993. "Ethnography of a "high-tech" case," in P. Lemonnier (ed.) *Technological Choices: Transformation in Material Culture Since the Neolithic*. 372–398. London: Routledge.

Latour, B. 2000. " The Berlin key: Or how to do words with things," in P. M. Graves-Brown (ed.) *Matter, Materiality and Modern Culture*. 10–21. London: Routledge.

Lechtman, H. 1977. "Style in technology: Some early thoughts," in H. Lechtman and R. S. Merrill (eds) *Material Culture: Styles, Organization, and Dynamics in Technology*. 3–20. St. Paul, MN: West Publishing.

Lemonnier, P. 1993. "Introduction," in P. Lemonnier (ed.) *Technological choices: Transformations in material cultures since the Neolithic*. 1–35. London: Routledge.

Mauss, M. 1979. "The notion of body techniques," in *Sociology and Psychology: Essays by Marcel Mauss (Trans. B. Brewster)*. 95–123. London: Routledge and Kegan Paul.

Miller, D. 2005. "Materiality: An introduction," in D. Miller (ed.) *Materiality*. 1–50. London: Duke University Press.

Pfaffenberger, B. 1992. Social anthropology of technology. *Annual Review of Anthropology* 21: 491–519.

Renfrew, C. 2001. "Symbol before concept: Material engagement and the early development of society," in I. Hodder (ed.) *Archaeological Theory Today*. 122–140. Cambridge: Polity Press.

Sackett, J. R. 1990. "Style and ethnicity in archaeology: The case for isochrestism," in M. W. Conkey and C. A. Hastorf (eds) *The Uses of Style in Archaeology*. 32–43. Cambridge: Cambridge University Press.

Thomas, J. 2004. *Archaeology and Modernity*. London: Routledge.

van der Leeuw, S. E. 1993. "Giving the potter a choice," in P. Lemonnier (ed.) *Technological choices: transformation in material cultures since the Neolithic*. 238–288. London: Routledge.

Wenger, E. 1998. *Communities of Practice: Learning, Meaning and Identity*. Cambridge: Cambridge University Press.

Woolgar, S. 1991. "Configuring the user: The case of usability trials," in J. Law (ed.) *A Sociology of Monsters: Essays on Power, Technology and Domination*. 57–99. London: Routledge.

12. Many hands make light work: potting and embodied knowledge at the Bronze Age tell at Százhalombatta, Hungary

Joanna Sofaer and Sandy Budden

The well-known idiom 'many hands make light work' refers to the power of a group working together to get things done faster or more easily. Conversely, 'too many cooks spoil the broth' suggests that there are occasions when the outcome of a process may be better, or more efficient, if only one or a few individuals participate. Following developments in other humanities and social science disciplines (Giddens 1986, Bourdieu 1977, Hacking 2004), the relationship between the individual and society has become a major theme in social archaeology through a focus on the relationship between agency and structure (Dobres and Robb 2000, Dobres 2000, Dobres and Hoffman 1994). Nonetheless, it has been relatively little explored in relation to prehistoric craft production, in particular the making of pottery. As with other crafts, there remains a highly individualistic approach implicit in discussions of pottery manufacture. In other words, as Patricia Crown has recently argued (Crown 2007), there is an assumption that a single individual is responsible for the production of a pot from start to finish.

This perception is problematic on two levels. First, an assumption that a pot is made by a single person hides the relationships required for the development of potting expertise. Not all potters are the same, since people have different skills, knowledge, strengths and weaknesses. This variability can be related both to individual aptitude (not everyone will be equally good at everything) and to potting experience. The latter involves the development of potting proficiency which requires the deployment of a range of learnt skills and is thus a product of the relationship between learners and more experienced potters (David 1990, Michelaki 2008, Arnold 1994, Arnold 2008, Crown 2001).

Second, Crown (2007) suggests that an individualistic approach to pottery production sits uneasily with the ethnographic evidence in which many different people may be involved in the making of finished pots. Recent explorations of ceramic *chaîne opératoires* show the range of stages and processes required to make a vessel, from sourcing, procurement and sorting of raw materials, through fetching water and preparing clays, to forming, decorating and firing the pot (Roux 1990, Roux 2003, Roux and Courty 2005, Gelbert 1999, Gosselain 1998, Gosselain 2000, Gosselain 2008, Gosselain and Livingstone Smith 2005, van der Leeuw 1999). Such potentially complex sequences may require not only the active co-ordinated participation of several people, but also negotiation with others who may not be directly involved in ceramic production. This may include, for example, negotiations regarding the timing and location of pottery, making activities vis a vis a range of daily or seasonal maintenance needs such as childcare, food processing or farming (c.f. Michelaki 2008, Sørensen 1996, Sofaer 2006a). The more people that are involved in pottery production, the more complex and wide-ranging the network of such negotiations may become. It has been suggested that increased task division in craft production is an indicator of workshop production (Arnold 1999) although in domestic settings the contributions of several individuals may still be involved (Crown 2007, David 1990). The distinction between different arenas of pottery production may therefore be more of a sliding scale than of absolute difference. Nonetheless, irrespective of whether pots are made in domestic settings or in workshops, the network of social relations involved in their creation impacts on household, kin, or wider social contacts since each participant in pottery production is ultimately part of a network involving people of different ages, gender, abilities and skills.

In this paper we examine a range of different vessel types from the Bronze Age tell of Százhalombatta in Hungary with a view both to exploring differences between potters and to identifying which objects are the products of many hands. Our method involves the identification of different levels of skill or embodied knowledge involved in the production of pots (Budden 2007, Budden 2008, Budden in press). By tracing these it becomes possible to explore the identities of makers, to infer where pots may have been the work of more than one person (c.f. Budden and Sofaer 2009, Crown 2007), and hence to explore the complexity of the networks involved in their production.

The Bronze Age Tell of Százhalombatta

The site of Százhalombatta is situated on the right bank of the Danube, 30 km south of Budapest. It is one of the largest and best preserved Bronze Age tell settlements in Central Europe, being about 200 m by 100 m in area, excluding the south and southwest parts of the site, which may represent up to one third of the original area and which have been destroyed during clay extraction by a local brick factory and erosion by the River Danube (Poroszlai 2000). The site has been the subject of three excavation campaigns (Kovács 1969, Poroszlai 1996, Poroszlai 2000, Poroszlai and Vicze 2000, Poroszlai and Vicze 2005), the most recent of which is still on-going. Excavation has focussed on domestic contexts and work is revealing a detailed picture of settlement structure and architecture; all the material discussed in this paper comes from domestic contexts, in particular the inside of houses. Finds from the site include pottery, metalwork, moulds, loom weights, phytoliths, worked wood, bone and stone, faunal and human remains. Of these, pottery is the most abundant, including many complete vessels, pot profiles and large sherds.

The majority of the rich ceramic assemblage dates from 2000–1500/1400 BC, spanning the Early Bronze Age classic Nagyrév (Szigetszentmiklós) to late Nagyrév (Kulcs) transition, through the following Middle Bronze Age Vatya tradition (Vatya I–III) and Vatya-Koszider phase at the end of the Middle Bronze Age, to the start of the late Bronze Age (equivalent to the Tumulus phase elsewhere in Europe). There follows a hiatus in occupation at the site until the Urnfield period. There is a wide range of vessels. While basic types such as cups, bowls, jugs and urns are present throughout and form the core of the assemblage, the range and elaboration of vessels within each of these categories changes through time, resulting in complex typological variation (Budden 2007, Vicze 2001). The transition from the Early to the Middle Bronze Age sees an increase in the range of forms within individual vessel types as well as the introduction of new types. This is traditionally understood as a result of the shifting cultural traditions of the Nagyrév and the influence of contemporary Kisapostag communities leading to the emergence of the Vatya tradition (Poroszlai 2003, Poroszlai 2000, Vicze 2001, Bóna 1992, Bóna 1975). Vessel forms include small cups, sieves, fish-dishes, deep domestic (cooking) bowls, cooking jar forms, storage vessels, small bowls (for eating), storage vessels, fineware bowls and jugs, urns, ember covers, and miniature forms that replicate the assemblage as a whole. In the Koszider phase at the end of the Middle Bronze Age the range of vessel forms decreases, but there is noticeable elaboration and exaggeration of existing forms (Budden 2007, Sofaer 2006a, Vicze 2001). Although there are some imported Middle Bronze Age ceramics at the site, notably from the neighbouring Incrusted Pottery Culture, these are clearly identifiable on the basis of their characteristic decoration. The Vatya pottery from Százhalombatta has a distinct quality and style, notably in its finewares (Budden 2007), and recent petrological and geochemical work has demonstrated that the overwhelming majority of the pottery at the site is locally made (Kreiter *et al.* 2007).

Tracing Embodied Knowledge at Százhalombatta

Potting is a very physical process that requires the potter to engage bodily with his or her clay and tools (Budden and Sofaer 2009). The production of different kinds of vessels requires different suites of body actions from the potter who has to engage with varying technical requirements and tools. The acquisition of potting skill is a matter of practice. In other words, it is a gradual process requiring the repeated performance of specific vessel-appropriate potting actions. It constitutes the development of embodied knowledge in the most literal sense, since the repetition of actions required to learn and to practice potting (or indeed any body action from knapping flints to writing with a pencil) has material repercussions for the human body by changing the neurology, musculature or skeleton (Dickens 2001, Keating and Miller 1999, Gardner 1993, Ingold 1993, Anderson 1983, Sofaer 2006b, Budden and Sofaer 2009, Mauss 1935).

Existing embodied knowledge can be overlain, or added to, with new or more advanced knowledge. It is, however, important to note that its physiological nature means that under usual circumstances it is impossible to 'unlearn' embodied knowledge, since it becomes cumulatively and more competently incorporated into practices of manufacture through time. Thus, even should the need arise for pots to be produced quickly, or with materials of a poorer quality, the skilled potter will almost certainly produce an article that demonstrates clear technical competence. Furthermore, elderly potters retain the knowledge for skilled action. Crown (2001:454) notes that elderly Pueblo potters continue to make high quality vessels until they die. Meanwhile, Kramer (1985) and Stark and Longacre (1993) note a tendency for older potters to produce more complex items because of an incremental increase in skill with age and experience. Shoji Hamada, one of the greatest potters of the 20th century, continued to make highly skilled pots well into his old age (Peterson 1974). It is also worth considering that it is not in the best interests of potters to produce poorly made vessels,

Vessel Type	Characteristics	Degree of technical complexity
Cups	*Small, ubiquitous vessels the majority of which have simple morphologies*	*Easiest* 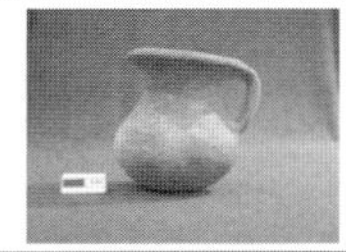
Domestic vessels	*Generally of moderate size with comparatively neutral shapes*	*Intermediate* 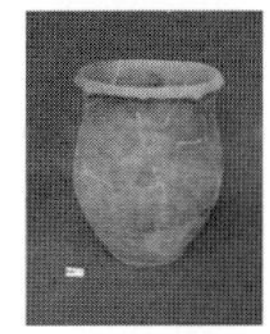
Urns	*Often very large with complex morphologies, additions and embellishments*	*Very technically demanding* 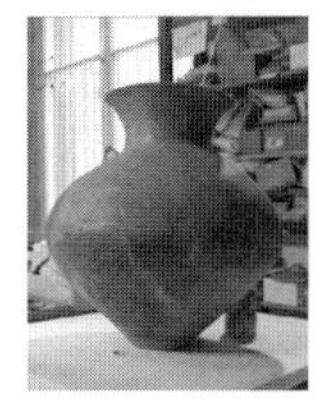
Fine ware	*Complex tripartite forms with highly exaggerated morphology, complex handles and embellishments, and very fine wall thickness*	*Very technically demanding* 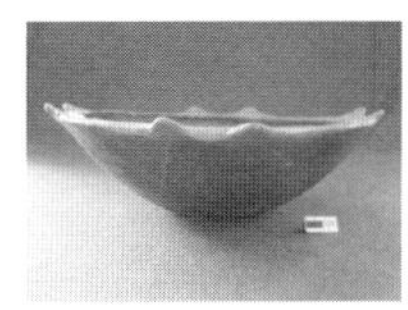

Fig. 12.1: Classification of technical difficulty in vessels from Százhalombatta, Hungary.

since it results in damage to their reputation (Longacre *et al* 2000). In a discussion surrounding reasons for poorly executed pottery, Crown (2001) notes that once potters can no longer satisfy the necessary criteria for production they simply stop potting.

Thus, even where a skilled potter or other craftsperson produces less than their 'best' work they will still incorporate a degree of knowledge not achievable by a practitioner who has not fully acquired the proper suite of actions to create a successful outcome (Budden 2008: 10). An example of this in hand-made pots, such as those found at Százhalombatta, would be the difference between a skilled potter producing a pot in a hurry but maintaining the skill to correctly 'bond' successive coils or slabs of clay and a novice potter who has not yet learnt to ensure that clay is of the correct consistency for this process to be successfully completed (Budden and Sofaer 2009). 'Poorly made' pots are not therefore just pots made in a hurry (Budden and Sofaer 2009, c.f. Crown 1999, Crown 2001). Nonetheless, the full acquisition of embodied knowledge can only happen where an aptitude exists. Not every potential participant will have the aptitude to acquire a specific skill; what one person can do, another cannot. Crown (1999), for example, argues that people who do not wish to pot or cannot pot, do not pot. Hodder (1991: 88) makes a similar point regarding the decoration of calabashes. Even professional potters may exhibit strengths and weaknesses in their work, although these will not be of the same order as differences in skill between novices and master potters.

The pottery from Százhalombatta is a complex assemblage but can be divided into four groups of vessels of contrasting technical difficulty relating to vessel size, and morphological characteristics (Fig. 12.1). Smaller, simpler forms require a smaller reservoir of embodied knowledge and less skill to produce than large complex ones (Arnold 1985, Caiger-Smith 1995, Rhodes 1957, Rice 1987, Rye 1981, Michelaki 2008). Thus, small simple cups represent the easiest vessels made at the site, domestic vessels of moderate size with comparatively neutral shapes are of intermediate difficulty, large urns with complex morphologies, additions and embellishments are very technically demanding, while fine wares with complex tripartite forms, highly exaggerated morphology, complex handles, embellishments and very fine wall thickness are the most technically demanding of all.

Since clay is a plastic additive medium, embodied knowledge can be traced in the ceramic record. The level of acquisition of embodied knowledge can be assessed by recording the degree of skill exercised by a potter for different elements of vessel manufacture against the expected outcome for a given vessel type (Budden 2007, Budden 2008, Budden and Sofaer 2009). For example, if like domestic wares, a particular vessel

Technological Variable	Cups (N)			Domestic Vessels (N)			Urns (N)			Fine Ware (N)			x^2	df	Sig. p≤ 0.05
	Poor	**Moderate**	**Good**	**Poor**	**Moderate**	**Good**	**Poor**	**Moderate**	**Good**	**Poor**	**Moderate**	**Good**			
Clay preparation	26	47	67	5	14	36	5	11	47	3	12	82	40.245	6	**.000**
Manufacturing	39	49	52	1	20	34	2	14	47	3	28	66	62.875	6	**.000**
Wall Thickness	43	34	58	10	19	26	9	14	40	7	25	63	28.284	6	**.000**
Additions	32	25	34	2	7	14	1	8	14	5	14	42	27.474	6	**.000***
Interior Surface Treatment	37	33	57	4	21	30	3	26	34	7	17	73	45.484	6	**.000**
Exterior Surface Treatment	34	48	55	5	23	27	8	11	44	6	11	79	54.850	6	**.000**
Decoration	9	1	14	7	14	13	1	10	20	4	15	44	27.651	6	**.000**
Rim Deviation on Horizontal Plane	27	49	14	4	32	10	2	18	14	7	28	35	37.640	6	**.000**
Rim Symmetry	5	36	43	6	17	25	3	12	25	5	16	59	13.035	6	**.042***
Handle Symmetry	13	30	46	1	4	20	1	3	18	4	7	47	19.599	6	**.004***
Profile Symmetry	27	43	45	5	10	9	4	5	17	4	9	45	27.174	6	**.000***
	Soft	**High**	**Good**	**Soft**	**High**	**Good**	**Soft**	**High**	**Good**	**Soft**	**High**	**Good**			
Firing	3	6	131	0	0	55	1	2	60	1	3	93	3.938	6	.685*

Fig. 12.2: Chi-squared tests for differences in skill in technological variables between vessel groups at Százhalombatta (= cell with less than 5).*

			Manufacturing			Total
			good	moderate	poor	
Clay preparation	good	N	45	19	3	67
		% of Total	32.1%	13.6%	2.1%	47.9%
	moderate	N	6	23	18	47
		% of Total	4.3%	16.4%	12.9%	33.6%
	poor	N	1	7	18	26
		% of Total	.7%	5.0%	12.9%	18.6%
Total		N	52	49	39	140
		% of Total	37.1%	35.0%	27.9%	100.0%

Fig. 12.3: Frequency of cups with contrasting degrees of skill in clay preparation and manufacturing.

	N	Spearman's Correlation	T	Sig.
clay prep - manufacturing	140	.659	10.302	**.000**
clay prep – wall thickness	135	.580	8.222	**.000**
clay prep – profile symmetry	115	.590	7.767	**.000**
clay prep – rim deviation on H. plane	90	.458	4.835	**.000**
clay prep – rim symmetry	84	.540	5.813	**.000**

Fig. 12.4: The relationship between skill in clay preparation and skill in variables related to the making of the body of cups: manufacturing, wall thickness, profile symmetry, rim deviation on the horizontal plane, rim symmetry.

type is repeatedly produced to have thick, even walls, then this is the expected or correct outcome and would be scored as good in the recording form. If, however, the wall thickness was extraordinarily variable or thin it would score as poor. Conversely, if the wall thickness of a small cup or fine ware Koszider jug was excessively thick, lumpy or uneven then this variable would score as 'poor' for that vessel form. Thus, vessel groups are not compared with each other and may follow different criteria. Furthermore, it is essential to stress that this recording system is not a measurement of aesthetics but of technical competence exercised for any given technological variable, and that these variables are not ranked in any hierarchy (Budden 2008). For example, decoration and wall thickness are evenly weighted. In combination, the range of technological variables represents the proposed outcome and characteristics that should be encompassed within a single vessel type. The more highly skilled the potter, the greater the acquisition of embodied knowledge, and the closer to the expected outcome for its particular type a vessel will be. Conversely, where poor execution of skill occurs, potters have not yet acquired the finesse of body actions characterised by skilfully made pots.

Hands at work

Given that cups are the least technically demanding form made at Százhalombatta one might expect that, if they were all made by experienced potters, they should show the least amount of error. Fine ware, being more complicated to make, should demonstrate the most errors. What we find, however, is the reverse (Fig. 12.2) (Budden 2007). In other words, cups are much more error-prone, with a great deal of variation in skill level and with an overall significantly lower investment of skill. By contrast, fine wares, the most technically complex vessels, are consistently good and show the fewest errors. The only exception to this is firing, which is consistently good for all vessel groups (Budden and Sofaer 2009).

We can interpret these patterns as a reflection of differences in the identity of potters (Budden and Sofaer 2009). In order to be able to make fine wares, potters have to go through the process of learning to make less complex vessels such as cups (Budden and Sofaer 2009, Singleton 1998). Some cups may therefore represent the work of apprentice or learner potters. Domestic vessels are the work of moderately skilled potters, and urns of skilled potters, while the fact that fine wares show relatively little error overall reflects the accumulated embodied knowledge of experienced professionals.

Unpicking this data in terms of the relationship between variables relating to different aspects of the pot-making *chaîne opératoire* for individual vessels, in particular cups, which are the vessels showing the greatest variability in skill, allows us to further explore differences between potters in embodied knowledge and to consider the issue of vessels as collaborative work. Individual pots are frequently remarkably consistent

	N	Spearman's Correlation	T	Sig.
decoration - manufacturing	24	.710	4.724	**.000**
decoration – wall thickness	23	.443	2.262	**.034**
decoration – profile symmetry	21	.775	5.343	**.000**
decoration – rim deviation on H. plane	12	.252	.825	.429
decoration – rim symmetry	13	.149	.501	.626

Fig. 12.5: The relationship between skill in decoration and skill in variables relating to the production of the body of cups (note: results for rim deviation and rim symmetry may be affected by small N).

	N	Spearman's Correlation	T	Sig.
exterior surface treatment - manufacturing	137	.642	9.727	**.000**
exterior surface treatment – wall thickness	132	.625	9.133	**.000**
exterior surface treatment – profile symmetry	112	.512	6.252	**.000**
exterior surface treatment – rim deviation on H. plane	88	.424	4.340	**.000**
exterior surface treatment – rim symmetry	82	.411	4.036	**.000**

Fig. 12.6: The relationship between skill in exterior surface treatment and skill in variables relating to the production of the body of cups.

	N	Spearman's Correlation	T	Sig.
additions - manufacturing	91	.684	8.851	**.000**
additions – wall thickness	88	.566	6.368	**.000**
additions – profile symmetry	80	.600	6.620	**.000**
additions – rim deviation on H. plane	70	.543	5.331	**.000**
additions – rim symmetry	69	.508	4.824	**.000**

Fig. 12.7: The relationship between skill in additions and skill in variables relating to the production of the body of cups.

in skill level. Thus, cups which are made from well-prepared clay are well or moderately manufactured but rarely poorly manufactured, while those that are made from poorly prepared clay are poorly or moderately manufactured but rarely well manufactured (Fig. 12.3). A similar relationship exists between clay preparation and all variables related to the making of the body of the vessel (wall thickness, profile symmetry, rim deviation on the horizontal plane, rim symmetry) (Fig. 12.4). It therefore seems that well-prepared clays were reserved for more highly skilled potters. This practice has been observed in modern potters where apprentices are not allowed to work with high value resources (Herzfeld 2004, Singleton 1998). It is of note that there is almost no poor clay preparation for the highly skilled production of fine wares (Fig. 12 2).

Cups that display little or no error in the production of the vessel body also tend to have well-executed decoration, those with poorly made bodies have poor decoration, and those that are of moderately skilled execution display a full range of skill in decoration from good though to poor (Fig. 12.5). Thus, in most cases the level of skill exercised for both variables is consistent within the same vessel. A similar relationship exists between variables reflecting the production of the vessel body and the skill displayed for external surface treatments such as burnishing (Fig. 12.6). Crown (2007) suggests that where there is task division in pottery manufacture, decoration and surface treatments are frequently carried out by a different person to that making the vessel body. If this was the case at Százhalombatta, one might expect skill in decoration and surface finish to be consistently good across all cups, including those that are poorly made. At Százhalombatta, however, although a reservoir of skill in decoration clearly exists, it is not deployed for poorly constructed vessels. Thus, although decoration and vessel formation require different kinds of motor skills, our results do not suggest that decoration or surface treatment were usually performed by different hands to those making the pot; at Százhalombatta most pots may well have been decorated by the same person who made the body of the vessel. Where pots show good vessel formation and moderate decoration or surface treatment, or poor vessel formation and moderate decoration or surface treatment, differences of this order in the level of skill can be attributed to the particular strengths and weaknesses of individual potters and the technical complexity of particular tasks. For example, maintaining the correct wall thickness for cups may have been particularly tricky to master and this may be why it is more error-prone than other variables (Fig. 12.2).

The attachment of handles, lugs, foot-rings or other additions to a vessel is another task that can potentially be carried out by a different person to that making the vessel body or decorating the vessel. Here again, however, we see that well-made cups tend to have good well-attached additions, while those that are poorly made tend to have poor ill-bonded

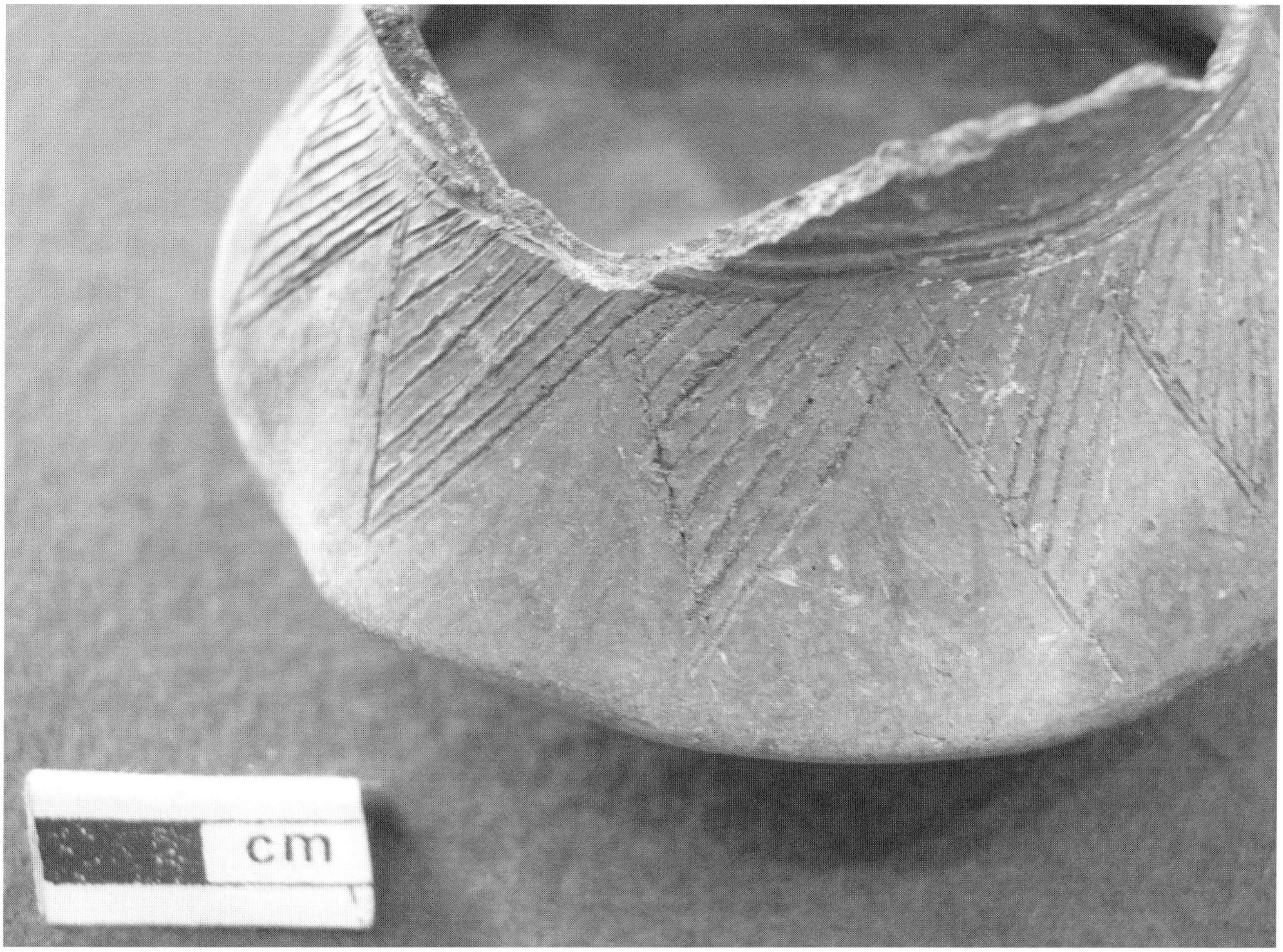

Fig. 12.8: A 'mixed message pot' with good skill in wall thickness but poor skill in decoration.

	Cups	Domestic Vessels	Urns	Fine Wares
Number of mixed message vessels	45	13	15	15
% of mixed message vessels	32%	24%	23%	16%
Total number of vessels	140	55	65	97

Fig. 12.9: The frequency of 'mixed message' pots in each vessel group (vessels which have one or more technological variables scored as good and one or more technological variables scored as poor, excluding firing).

additions (Fig. 12.7). This suggests that overall the execution of skill for each of these variables is consistent within a given vessel. Again, there is no evidence for many hands at work on the same pot. It is likely that most additions were made by the same potter who was responsible for the other aspects of the vessel.

Given that many cups are internally consistent in skill level, it is possible to suggest that minor differences in skill are the result of individual potters' strengths and weaknesses. For example, some potters may be more skilled at decoration, while others are better at creating a symmetrical vessel profile. Nonetheless, there are two areas where it is possible to identify the input of more than one person in making pots. First, the overwhelmingly accurate firing of cups (and indeed of all vessel types) is not consistent with the number and range of errors found for other technological variables (Fig. 12.2). This strongly suggests that there were skilled practitioners who took on the firing of different vessel types, perhaps in a centrally organised manner. Second, there are a few vessels which buck the general trends described above and which seem to send 'mixed messages', displaying contradictory skill levels for different elements of the potting process (Fig. 12.8). These 'mixed message pots' stand out for their inconsistency in skill against a general background of vessels that are otherwise relatively internally consistent in execution of skill. It may be that such pots reveal a relationship between novice and experienced potters, both contributing to the making of individual vessels (cf. Crown 2007). Such collaborations appear to have been *ad hoc* since, apart from firing, there does not appear to be any clear direction in the data regarding which processes were more frequently carried out by novices or experienced potters.

'Mixed message' vessels can be more frequently identified in cups than other vessel groups. They decline in frequency as vessel complexity and overall skill levels increase (Fig. 12.9). Thus, cups, although technically more simple, potentially involve more hands than other vessel types. If these vessels represent collaborations between different potters with contrasting degrees of skill, then they may reflect the kind of learning process within which potters learnt their craft, and thus the nature of relationships at play between individual learners and more experienced craftspeople. There are two main kinds of informal learning methods: scaffolded learning and relatively independent trial and error (Greenfield 2000, Greenfield, Maynard and Childs 2000, Wallaert-Pêtre 2001, Gosselain 1992, Gosselain 1998). In scaffolded learning, learners are guided by teachers who provide practical help, models and verbal direction in accordance with the developmental level of the learner (Greenfield 2000). Advice is freely available and learners observe and follow models presented by more skilled practitioners. In strongly scaffolded situations, help is given before learners have the opportunity to make a serious error and learners have little chance to make a mistake, innovate or experiment. Thus, the transmission of traditional ways of doing things is ensured (Greenfield 2000). By contrast, in trial and error learning, learners embark on a process of discovery with very little guidance. This leads to a much higher rate of errors but also encourages innovation and experimentation (Greenfield 2000, Greenfield, Maynard and Childs 2000).

At Százhalombatta, the evidence suggests that the learning method lay somewhere between these two extremes. If we take 'mixed message' pots to be collaborative work, then these vessels indicate that more experienced help was available to guide learner potters; 'mixed message' pots could be interpreted as vessels where learner potters are either assisted through the input of more experienced helpers, or where learners are encouraged to add to their skills by working on pots made by more experienced potters. Certainly, both the 'mixed message' and internally consistent vessels point to a range of competencies – a situation which might be expected when beginners or apprentice potters, who have not yet acquired a full range of skills work alongside more experienced ones. Nonetheless, the existence of errors in less technically complex vessel forms suggests that potters were allowed to make their own mistakes. Furthermore, synchronic and diachronic variation in vessel form suggest that innovation was possible, albeit within strict rules surrounding the 'correct' culturally acceptable way to make a pot (Budden and Sofaer 2009). Learning methods at Százhalombatta may therefore have been lightly or moderately scaffolded; novice potters frequently worked on their own but acted within a wider environment where they were able to draw on the help and support of others. The nature of the wider social and economic setting in which potters learnt was, however, highly-structured. Potters of all skill levels from novice to master worked within a system of progression where skill was not only required, but also rewarded, by moving on to the making of more complex vessels and access to better quality resources once relevant skills had been acquired.

Potting, Embodied Knowledge and Social Relations at Százhalombatta

Learning to be a potter requires the development of embodied knowledge and the education of potters generates identity in relation to their communities of practice (Singleton 1998, Rogoff 1990, Michelaki 2008). At Százhalombatta it is possible to identify potters with different degrees of skill (in other words with different degrees of embodied knowledge) and thus different identities ranging from novices to highly skilled masters. Our data therefore fragment the notion of 'potter' to show tighter categories of individual craftspeople at work.

The development of such a range of identities was, however, set within a wider social milieu. While many vessels seem to have been formed and decorated by one person and then fired centrally, some appear to have involved more hands. The latter tend to be technically easier vessels, in particular cups. Thus, while such vessels are simple to make, they are, in fact, socially complex; the greater the number of people involved, the more social engagement and negotiation is required. Conversely, the most technically difficult vessels – fine wares – can, in this respect, be considered socially more simple, since it appears that they were made wholly by potters whose high skill level suggests an oft-repeated engagement with embodied knowledge, in other words, master potters. They do not have the same potential to involve many people since they are not the focus of learning and teaching communities, although in practice it is likely that these different networks would have been interrelated.

Given the contrasting nature of the social networks surrounding the production of different vessel types it may be that the production of cups was particularly vulnerable to strain. Complex networks may be put under strain when one of the supporting links falls down (for example childcare falling through), or when the output of craftspeople presents increased demands for labour either directly or indirectly related to pottery production (for example, a need for increased manpower to extract and sort clays) (Crown 2007). At Százhalombatta, connections may also have extended outside pottery production to include metalwork and woodworking since technical relationships between pottery and other crafts described at the site (Sofaer 2006a) must have involved communication between craftspeople (Sofaer 2006a, Budden and Sofaer 2009). In hierarchical societies, as in the Bronze Age, links between craftspeople are frequently based around close relations between kin (Arnold 1985, Kramer and Douglas 1992). This provides strong support for complex networks, thereby reducing potential strains, as well as providing opportunities for guided learning. Many ethnographic studies of scaffolded learning highlight the role of close kin (mothers, fathers, uncles and aunts) in the transfer of knowledge (Greenfield 2000).

To fully understand the development of embodied knowledge at Százhalombatta and its link to social relations, however, it is also important to understand why it was 'socially permitted' for novice potters to learn on cups and not on fine wares. Although technically simpler vessels were clearly more suited to learners, the use of cups as 'practice pieces' was also made possible because they did not fulfil the same social role as fine wares. Fine wares were high status objects strongly implicated in social display and close visual scrutiny (Budden and Sofaer 2009, Sofaer 2006a, Sofaer 2010, Vicze 2001). High-quality clay, good manufacture, decoration and firing of fine wares were therefore critical to their social function. It is likely that cups, although implicated in ritual events linked to family and kinship, (including hoarding and burial) (Budden 2007:242, Budden in press) did not have the same prestige role as fine ware. Hence, they would be considered acceptable as long as they fitted into general notions regarding culturally acceptable shape and decoration. It was not therefore crucial for the same consistently high standards of manufacture to be applied to cups, allowing their production to be part of a learning process.

Concluding Remarks

In this paper we have used the investigation of embodied knowledge as a means of exploring the relationship between individuals and society. At Százhalombatta, many pots appear to have been primarily the work of a single craftsperson and to reflect variations in individual levels of skill and aptitude. Differences in embodied knowledge created distinctions between different kinds of potters, such as between masters and apprentices. However, an individualistic approach to pottery production cannot be wholly sustained. The evidence speaks to a structured system of apprenticeship which required a series of social relationships to support the development of potting expertise. These relationships include those constructed between novice and master potters visible in the 'mixed message pots', as well as potentially complex extended support networks that underpinned ceramic production. Furthermore, firing of all vessels, including poorly-made cups formed by novices, appears to have been carried out by people who had the skilled knowledge to avoid the pitfalls inherent in firing (c.f. Barley 1997, Rye 1981, Fournier 1973), suggesting they were well practiced specialists at this task, and adding another layer to the making of pottery at the site. Thus, while large-scale social structures may be implicated in the relationship between individual and society, it is clear that embodied knowledge also acts as a means of informing group membership at a scale that lies between the individual and society at large.

References

Anderson, J. R. 1983. *The Architecture of Cognition*. Cambridge, Massachusetts: Harvard University Press.

Arnold, D. 1985. *Ceramic Theory and Cultural Process*. Cambridge: Cambridge University Press.

Arnold, D. 1994. "Patterns of learning: residence and decent among potters in Ticul, Yucatan, Mexico," in S. Shennan (ed.) *Archaeological Approaches to Cultural Identity*. London: Routledge.

Arnold, D. 1999. "Advantages and Disadvantages of Vertical-Half Molding Technology: Implications for production organization," in J. M. Skibo and G. M. Feinman (eds) *Pottery and People*. 59–80. Salt Lake City: University of Utah Press.

Arnold, D. 2008. *Social Change and the Evolution of Ceramic Production and Distribution in a Maya Community*. Colorado: University Press of Colorado.

Barley, N. 1997. "Traditional Rural Potting in West Africa," in I. Freestone and D. Gaimster (eds) *Pottery in the Making: World Ceramic Traditions*. 140–145. London: British Museum Press.

Bóna, I. 1975. *Die mittlere Bronzezeit Ungarns und ihre südöstlichen Beziehungen. Archaeologia Hungarica 49*. Budapest: Akadémiai Kiadó.

Bóna, I. 1992. "Bronzezeitliche Tell-Kulturen in Ungarn," in W. Meier-Arendt (ed.) *Bronzezeit in Ungarn. Forschungen in Tell Siedlungen an Donau und Theiss*. 9–39. Frankfurt: Pytheas.

Bourdieu, P. 1977. *Outline of a theory of practice*. Cambridge: Cambridge University Press.

Budden, S. 2007. *Renewal and Reinvention: The role of learning strategies in the Early to Late Middle Bronze Age of the Carpathian Basin*. PhD dissertation, Southampton University.

Budden, S. 2008. "Skill Amongst the Sherds: Understanding the role of skill in the Early to Late Middle Bronze Age in Hungary," in I. Berg (ed.) *Breaking the Mould: challenging the past through pottery*. 1–17. Oxford: Archaeopress.

Budden, S., and J. Sofaer. 2009. Non-Discursive Knowledge and the Construction of Identity. Potters, Potting and Performance at the Bronze Age Tell of Százhalombatta, Hungary. *Cambridge Archaeological Journal* 19, 2: 203–220.

Budden, S. A. in press. Arenas of Skill: Learning strategies, family, kinship and specialisation in the Early to Late Middle Bronze Age in Hungary. *Journal of Family History (special edition edited by Duncan Sayer)*.

Caiger-Smith, A. 1995. "Pottery, People and Time: A Workshop in Action. Shepton Beauchamp: Richard Dennis.

Chalkin, S., and J. Lave (eds) 1996. *Understanding Practice. Perspectives on Activity and Context*. Cambridge: Cambridge University Press.

Crown, P. 2001. Learning to Make Pottery in the Prehispanic American Southwest. *Journal of Anthropological Research* 57: 451–469.

Crown, P. 2007. Life Histories of Pots and Potters: Situating the Individual in Archaeology. *American Antiquity* 72: 677–690.

Crown, P. L. 1999. "Socialisation in American Southwest Pottery Decoration," in J. M. Skibo and G. M. Feinman (eds) *Pottery and People*. 25–43. Salt Lake City: University of Utah Press.

David, N. 1990. *Vessels of The Spirit: pots and people in North Cameroon (Video)*. Calgary: University of Calgary.

Dickens, P. 2001. Linking the Social and Natural Sciences: Is capital modifying biology in its own image? *Sociology* 35, 1: 93–110.

Dobres, M.-A. 2000. *Technology and Social Agency*. Oxford: Blackwell.

Dobres, M. A., and C. R. Hoffman. 1994. Social Agency and the Dynamics of Prehistoric Technology. *Journal of Archaeological Method and Theory* 1, 3: 211–258.

Fournier, R. 1973. *Illustrated Dictionary of Practical Pottery*. London: Van Nostrad Reinhold Company.

Gardner, H. 1993. *Frames of Mind: The Theory of Multiple Intelligences*. London: Fontana Press.

Gelbert, A. 1999. "Technological and Stylistic Borrowings between Ceramic Traditions: A case study from Northeastern Senegal," in L. R. Owen and M. Porr (eds) *Ethno-Analogy and Reconstruction of Prehistoric Artefact Use and Production, Urgeschichtliche Materialhefte14*. 201–224. Tübingen: Mo Vince Verlag.

Giddens, A. 1986. *The Constitution of Society: Outline of a Theory of Structuration*. Cambridge: Polity Press.

Gosselain, O. P. 1992. Technology and Style: Potters and Pottery among the Bafia of Cameroon. *Man* 27: 559–586.

Gosselain, O. P. 1998. "Social and Technical Identity in a Clay Crystal Ball," in M.T. Stark (ed.) *The Archaeology of Social Boundaries*. Washington: Smithsonian Institute Press.

Gosselain, O. P. 2000. Materializing Identities: An African Perspective. *Journal of Archaeological Method and Theory* 7, 3: 187–217.

Gosselain, O. P. 2008. "Thoughts and Adjustments in the Potter's Backyard," in I. Berg (ed.) *Breaking the Mould: challenging the past through pottery*. 67–79. Oxford: Archaeopress.

Gosselain, O. P., and A. Livingstone Smith. 2005. "The Source Clay Selection and Processing Practices in Sub-Sharan Africa," in A. L. Smith, D. Bosquet, and R. Martineau (eds) *Pottery Manufacturing Processes: Reconstruction and Interpretation, British Archaeological Reports International Series 1349*. 33–47. Oxford: Archaeopress.

Greenfield, P. 2000. "Children, material culture and weaving: historical change and developmental change," in J. S. Derevenski (ed.) *Children and Material Culture*. 72–86. London: Routledge.

Greenfield, P. M., A. E. Maynard, and C. P. Childs. 2000. History, Culture, Learning, and Development. *Cross Cultural Research* 34, 4: 351–374.

Hacking, I. 2004. Between Michel Foucault and Erving Goffman: between discourse in the abstract and face-to-face interaction. *Economy and Society* 33, 3: 277–302.

Herzfeld, M. 2004. *The Body Impolitic. Artisans and Artifice in the Global Hierarchy of Value*. Chicago: The University of Chicago Press.

Hodder, I. 1991. "The Decoration of Containers: an Ethnographic and Historical Study," in W. Longacre (ed.) *Ceramic Ethnoarchaeology*. Tucson: University of Arizona Press.

Ingold, T. 1993. "Technology, language, intelligence: a reconsideration of basic concepts," in K. R. Gibson and T. Ingold (eds) *Tools, Language and Cognition in Human Evolution*. 449–472. Cambridge: Cambridge University Press.

Ingold, T. 1998. From complimentary to obviation: on dissolving the boundaries between social and biological anthropology, archaeology and psychology. *Zeitschrift für Ethnologie* 123: 21–52.

Ingold, T. 2001. "Beyond art and technology: the anthropology of skill," in M. B. Schiffer (ed.) *Anthropological Perspectives on Technology*. 17–31. Albuquerque: University of New Mexico Press.

Kamp, K. A. 2001. Prehistoric Children Working and Playing: a Southwestern Case Study in Learning Ceramics. *Journal of Anthropological Research* 57: 427–450.

Keating, D., and F. Miller. 1999. "Individual pathways in competence and coping: from regulatory systems to habits of mind," in D. Keating and C. Hertzman (eds) *Developmental Health and the Wealth of Nations*. 220–234. New York: Guildford.

Kovács, T. 1969. A százhalombattai bronzkori telep (The Bronze Age Settlement at Százhalombatta). *Archaeologiai Értesítő* 96: 161–169.

Kramer, C. 1985. Ceramic Ethnoarchaeology. *Annual review of Anthropology* 14: 77–102.

Kramer, C., and J. E. Douglas. 1992. Ceramics, caste and kin: Spatial relations in Rajasthan, India. *Journal of Anthropological Archaeology* 11, 2: 187–201.

Kreiter, A., B. Bajnóczi, P. Sipos, G. Szakmány, and M. Tóth. 2007. Archaeometric Examination of Early and Middle Bronze Age Ceramics from Százhalombatta-Földvár, Hungary. *Archeometriai Műhely* 2: 33–46.

Longacre, W. A., J. Xia, and Y. Tao. 2000. I Want to Buy a Black Pot. *Journal of Archaeological Method and Theory* 7, 4: 273–293.

Mauss, M. 2006 [1935]. "Les techniques du corps," in N. Schlanger (ed.) *Marcel Mauss: Techniques, Technology and Civilisation*. 77–95. Oxford: Durkheim Press.

Michelaki, K. 2008. Making Pots and Potters in the Bronze Age Maros Villages of Kiszombor-Új-Élet and Kláraflava-Hajdova. *Cambridge Archaeological Journal* 18, 3: 327–252.

Minar, C. J., and P. Crown. 2001. Learning and Craft Production: An Introduction. *Journal of Anthropological Research* 57: 369–380.

Peterson, S. 1974. *Shoji Hamada: A Potter's Way and Work*. Tokyo and New York: Kodansha International.

Poroszlai, I. 1996. *Excavations at Százhalombatta 1989–1995*. Százhalombatta: Matrica Museum.

Poroszlai, I. 2000. "Excavation Campaigns at the Bronze Age site at Százhalambatta-Földvár," in I. Poroszlai and M. Vicze (eds) *Százhalombatta Archaeological Expedition SAX Annual Report 1*. 13–73. Százhalombatta: Matrica Museum.

Poroszlai, I. 2003a. "Fortified Centres along the Danube," in Z. Visy (ed.) *Hungarian Archaeology at the Turn of the Millennium*. 151–155. Budapest: Ministry of National Cultural Heritage.

Poroszlai, I. 2003b. "The Koszider Period," in Z. Visy (ed.) *Hungarian Archaeology at the Turn of the Millenium*. 161. Budapest: Ministry of National Cultural Heritage.

Poroszlai, I., and M. Vicze (eds) 2000. *SAX. Szazhalombatta Archaeological Expedition. Annual report 1*. Százhalombatta: Matrica Museum.

Poroszlai, I., and M. Vicze (eds) 2005. *SAX. Szazhalombatta Archaeological Expedition. Annual report 2*. Százhalombatta: Matrica Museum.

Rhodes, D. 1957. *Clay and Glazes for the Potter*. Philadelphia: Chilton Book Company.

Rice, P. M. 1987. *Pottery Analysis: A Sourcebook*. Chicago: The University of Chicago Press.

Rogoff, B. 1990. *Apprenticeship in Thinking: Cognitive Development in Social Context*. Oxford: Oxford University Press.

Roux, V. 1990. The Psychological Analysis of Technical Activities: a contribution to the study of craft specialisation. *Archaeological Review from Cambridge* 9, 1: 142–151.

Roux, V. 2003. A Dynamic Systems Framework for Studying Technological Change: Application to the Emergence of the Potter's Wheel in the Southern Levant. *Journal of Archaeological Method and Theory* 10, 1: 1–30.

Roux, V., and M. A. Courty. 2005. "Identifying Social Entities at Macro-Regional Level: Chalcolithic Ceramics of South Levant as a Case Study," in A. L. Smith, D. Bosquet, and R. Martineau (eds) *Pottery Manufacturing Processes: Reconstruction and Interpretation, British Archaeological Reports International Series 1349*. 201–214. Oxford: Archaeopress.

Rye, O. 1981. *Pottery Technology: principles and reconstruction*. Washington: Taraxum.

Singleton, J. 1989. "Japanese folkcraft pottery apprenticeship: Cultural patterns of an educational practice," in M. Coy (ed.) *Apprenticeship: From Theory to Method and Back Again*. 8–15. Albany: State University of New York Press.

Singleton, J. (ed.) 1998. *Learning in Likely Places. Varieties of Apprenticeship in Japan*. Cambridge: Cambridge University Press.

Singleton, J. 1998 "Craft and art education in Mashiko pottery workshops," in J. Singleton (ed.) *Learning in Likely Places. Varieties of Apprenticeship in Japan*. 122–133. Cambridge: Cambridge University Press.

Sofaer, J., *et al.* 2010. "Technology and Craft," in T. Earle and K. Kristiansen (eds) *European Society in Late Prehistory: A Comparative Approach*. Cambridge: Cambridge University Press.

Sofaer, J., and M. Vicze. in prep. *The Late Bronze Age and Middle Bronze Age Koszider Pottery from Százhalombatta, Hungary*.

Sofaer, J. R. 2006a. *The Body as Material Culture. A Theoretical Osteoarchaeology*. Cambridge: Cambridge University Press.

Sofaer, J. R. 2006b. Pots, Houses and Metal: Technological Relations at the Bronze Age Tell at Százhalombatta, Hungary. *Oxford Journal of Archaeology* 25, 2: 127–147.

Sørensen, M. L. S. 1996. "Women As/And Metalworkers," in A. Devonshire and B. Wood (eds) *Women in Industry and Technology: from prehistory to the present. Current research and the museum experience*. 45–52. London: Museum of London.

Stark, M. T., and W. A. Longacre. 1993. "Kalinga Ceramics and New Technologies: Social and Cultural Contexts of Ceramic Change," in W. D. Kingery (ed.) *The Social and Cultural Contexts of New Ceramic Technologies*. Westerville: American Ceramic Society.

Toren, C. 1999. *Mind, Materiality and History. Explorations in Fijian Ethnography*. London: Routledge.

van der Leeuw, S. E. 1999. "Exchange and trade in ceramics: some notes from a potter's point of view," in J. P. Creilard, V. Stissi, and G.v. Wrjngaarden (eds) *The Complex Past of Pottery: production, circulation and consumption of Mycenaen and Greek pottery*. 115–136. Amsterdam: Gieben.

Vicze, M. 2001. Dunaújváros-Dunadűlő. The Early and Middle Bronze Age Cemetery of Dunaújváros-Kosziderpadlás. Ph.D. Thesis, Eötvös Loránd University.

Wallaert-Petre, H. 2001. Learning how to make the right pots: apprenticeship strategies and material culture, a case study in handmade pottery from Cameroon. *Journal of Anthropological Research* 57: 471–492.

13. Spinning faith

Lise Bender Jørgensen

In Antiquity, spindle spinning was a basic craft that went on more or less continuously. Considerable lengths of yarn were needed for making textiles; one square metre of average quality cloth contains about two km of yarn. It takes practice to spin evenly and efficiently, but once learnt it becomes second nature – a simple, subconscious movement, almost a slight-of-hand. Yarn can be spun in two directions – clockwise or anti-clockwise, termed z and s respectively (Fig. 13.1). Once the spinning process has begun it is essential to keep twisting in the same direction; if not, the fibres will unwind, causing the yarn to disintegrate. In most societies, including Western Europe,[1] one direction of twist is the norm. The other direction is described as wrong, in some cases even seen as the direction of evil (Lysenko and Komarova 1992: 1, Stenberg-Tyrefors 1982: 11–12). A spinner in Bzou, Morocco, reacted in a similar way when asked to reverse her twist direction from clockwise to anti-clockwise by visiting Scandinavian textile scholars in November 2001.

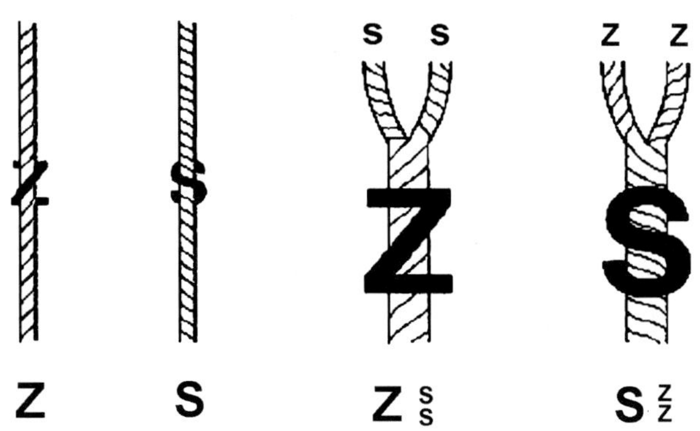

Fig. 13.1: z- and s-twist. When two or more threads are plied together they are twisted the opposite direction (illustration by Catharina Oksen, copyright Lise Bender Jørgensen).

In most of Europe, clockwise twisted yarn (z) has been the rule ever since the end of the Bronze Age, except for special effects such as spin patterns, where alternating groups of s- and z-twisted yarns are used to create subtle patterns, or for weft yarns if a fabric with a smooth surface was desired. Scandinavians deviated from the norm, preferring anti-clockwise twisted yarn (s) since the beginning of the Bronze Age and throughout the 1st millennium BC. They changed their minds around 200 AD to conform to continental Europe. During the Pre-Roman and the first part of the Roman Iron Age, what looks like a 'border' between the two directions of twist may be perceived across Denmark (Bender Jørgensen 1984, 1986a: 134, 343ff, 1992: 116ff). In Pharaonic Egypt, yarns were usually twisted anticlockwise (s) (Barber 1991: 66, Hall 1986: 12, Vogelsang-Eastwood 1999: 22); this continued throughout the Hellenistic, Roman and Byzantine Periods, and even after the Arab conquest in 641 AD (van't Hooft *et al.* 1994: 14, 143). Textiles dated between the 10th and 13th Century seem to have been made of clockwise twisted yarns (z) (van't Hooft *et al.* 1994: 189), and textiles from Mamluk Egypt (1250–1517 AD) are almost exclusively twisted clockwise (Vogelsang-Eastwood 2004, 47–48, Table 3). The Fellahin and Bedouin of modern Egypt and northern Sudan, however, still twist their yarns anti-clockwise (s), contrary to the practice of inhabitants of the Arabian Peninsula (Crocker Jones 1989: 27) and – presumably – the Berbers of North Africa. Handspun yarns and a range of hand-woven fabrics bought by the author in Morocco 2001 are all z-spun. According to Grace M. Crowfoot, several other features seem to have remained unchanged in Egypt since the Pharaonic period, such as a preference for spindles with whorls mounted on the top, for rotating them on the thigh rather than twirling them with the fingers in the air, and the absence of a distaff (Crowfoot 1931: 31–32, 36).

The simple movement of the spinner's hand is much more than just a technique. Long-standing traditions of performing in exactly the same way over millennia, as has been observed in Europe and Egypt, show that it constitutes embodied cultural practice, reflecting the close relationship between knowledge, beliefs and technology. The daily chore of spinning represents complex knowledge on what to do and how to do it, on the raw materials available and their properties, which tools to use, and on various treatments such as the application of heat or oil to facilitate fibre processing. In the following, ethnographic, archaeological and experimental evidence on the labour, learning and rituals of textile making are explored, followed by a discussion of spinning in Bronze Age Scandinavia and what the direction of twist may reflect in terms of technology and beliefs.

Textile Making: Labour and Learning

In pre-industrial societies, textile production was a major undertaking, involving most members of the household. Textiles were needed for clothing, bedding, soft furnishings and other types of household goods, and a myriad of other items such as sacking, animal trappings, etc. Textile production included a wide range of individual processes, from the production of fibre through sheep husbandry or the cultivation of flax and other fibre plants, the harvesting and processing of the fibres, followed by their preparation for spinning, the spinning process itself, weaving and various forms of finishing processes. Some tasks were carried out by groups of people, others were individual undertakings (for detailed descriptions of processes, tasks, tools, etc see, for example, Cardon 1999, Højrup 1967: 177ff or Walton Rogers 2007: 9ff). Textile crafts are usually considered women's work (but see Crocker Jones 1989: 14, 23, 25, Crowfoot 1931: 36, 39, Hansen 1947: 41, Højrup 1967: 194, Picton and Mack 1993: 19–20); some of the processes were, however, usually performed by men, e.g. the cultivation of flax or wool shearing. Children also helped, learning the various aspects of the textile crafts as they grew and gained dexterity and skill.

Spinning is one of the most time-consuming parts of textile making. Recent experiments indicate that an experienced spinner can produce 30–50 m of yarn per hour on a hand spindle, with variations depending on the type of fibre and thickness of yarn (Mårtensson *et al.* 2006: 8, Nørgaard 1999). For clothing of medium weight, i.e. some 10 threads/cm in warp as well as weft,[2] more than 2 km of yarn is needed per square metre. Finer or denser fabrics require much more. The weatherproof, very densely woven wool twills used for military cloaks in Roman Egypt are a good example of this. One of the finest had 17 warp threads and no less than 72 weft threads/cm, adding up to almost 9 km yarn per square metre (Bender Jørgensen 2004: 96). Similarly densely woven wool fabrics have been found in Viking Age graves in Scandinavia (Bender Jørgensen 1986: 94ff, Geijer 1938: 22ff). Royal pharaonic linens reportedly have up to 100 warp threads/cm (Vogelsang-Eastwood 1999: 22). This meant consuming still more highly skilled spinners' time – or splicers', as yarn for the finest pharaonic linens was spliced, rather than spun (Barber 1991: 47, Kemp and Vogelsang-Eastwood 2001: 76). Around the Mediterranean, many garments have been constructed from lengths of cloth of about 3x1 m. This applies to Antiquity as well as recent pre-industrial societies (Rabaté and Sorber 2007, Vogelsang-Eastwood 1992). Each of these would demand substantial amounts of yarn; for the examples mentioned above, some 6 km of yarn would have been needed for a garment of medium weight, 30 km of yarn for a weatherproof cloak of the type found in Roman Egypt or for the fine textiles of the Vikings, i.e. 120–600 hours of spinning at 50 m/hour. For the finest Pharaonic linens, the hours would be even longer.

It takes long practice to spin evenly and efficiently. Ethnographic records, for example those of G.M. Crowfoot (for Egypt and the Sudan) or G. Chantréaux (for Algeria) indicate a long apprenticeship in these areas, starting at the age of 5 or 6 by learning to spin weft yarn. Warp yarns are much more difficult as the warp has to sustain tension and wear during the weaving process; some apprentices never get beyond the spinning of weft (Chantréaux 1941–42: 98, Crowfoot 1931: 42). Once learnt, however, spinning can be performed while walking, sitting, standing, waiting at the village fountain, looking after children or animals, chatting, and during any other unoccupied moments. A range of different methods may be used, such as by the hands alone; twisting a hooked stick; by grasped spindle; rotation of spindle in the hand, or rolling it down the thigh (Fig. 13.2); and by supported or suspended hand spindle (Fig. 13.3) (see Crowfoot 1931 for an overview of methods and tools). Spindles may, or may not, be provided with a spindle whorl; the position of the whorl may be on the top of the spindle, at the bottom, or in fact at almost any position on the stem. They often have a thread notch, cut in the top of the stem, or a metal hook, but may also be used without either. Similarly, spinners may, or may not, use a distaff. Each method can be employed for making high quality yarns, but some have certain advantages over others, such as greater speed, greater length of spin, etc, and in several cases only one of them is used by a certain group of people (Crowfoot 1931: 44–45). Often, however, different methods are employed for different types of yarn. Suspended spindles are often used for warp yarns, supported spindles for weft yarns (Chantréaux 1941–42: 96ff).

Weaving, too, takes long years of apprenticeship to master. It includes making a warp, mounting it on the loom, the knitting of heddles, inserting the weft, and – when the web is finished – cutting it off and dismantling the loom (Fig. 13.4). In traditional societies of North Africa, girls started at the age of eight. First the girl watched her mother weaving; later she practised on simple tasks, e.g. repairing broken warp threads

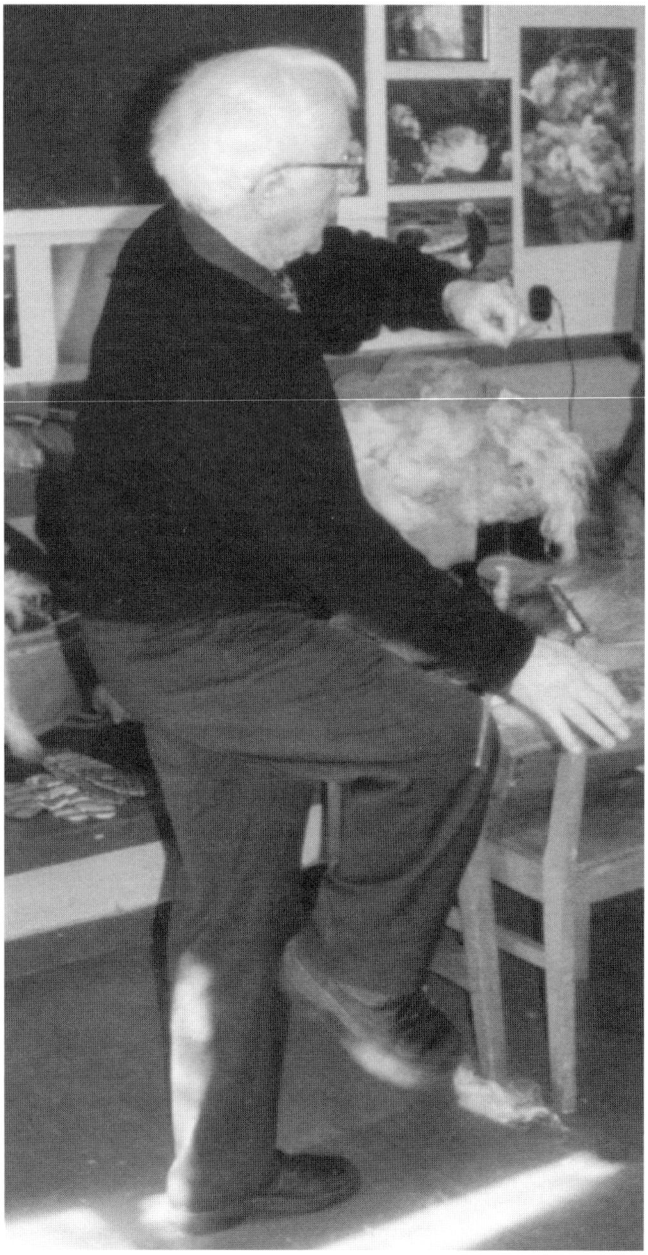

Fig. 13.2: Spinning by rolling the spindle down the thigh (photo by Lise Bender Jørgensen 1999).

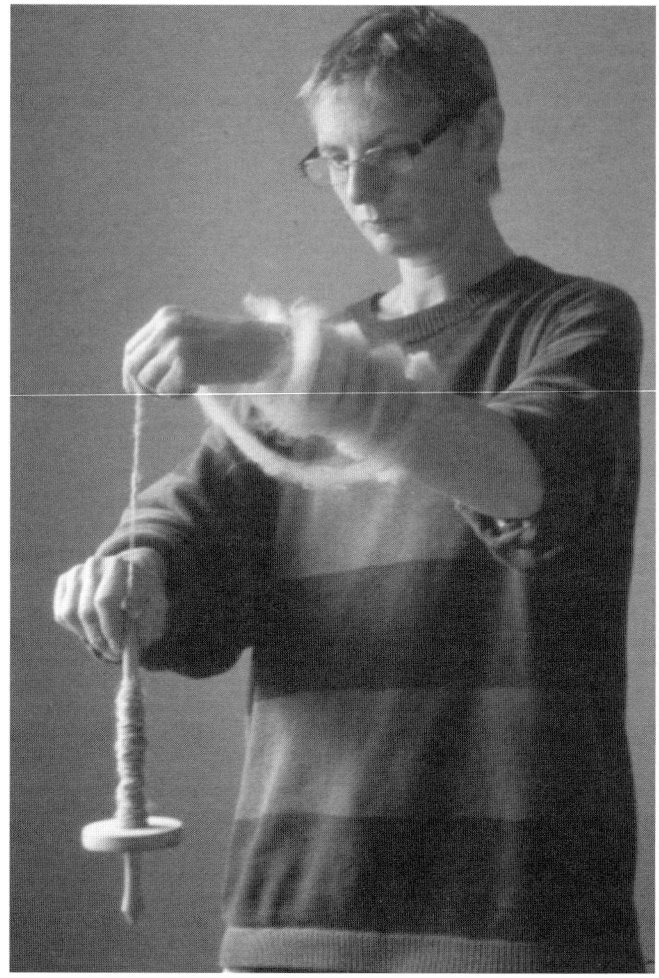

Fig. 13.3: Suspended or drop spindle; (photo Oddmunn Farbregd 1999).

Fig. 13.4: (right) The basic structure and workings of a loom, illustrated by a two-beam loom. This is the type still used in North Africa. The warp is the threads held under tension between the two beams at each end. The weft is the threads that are inserted to form the web. In order to do that sheds, dividing the warp threads in two layers, are necessary. This is done by heddles, loops attached to a heddle rod across the loom. It makes it possible to separate the warp threads by lifting part of them and make a path for the weft. The heddle rod is placed in the middle of the loom and is lifted (a) or lowered (b) to form sheds (illustration by Poul Wöhlicke, copyright Lise Bender Jørgensen).

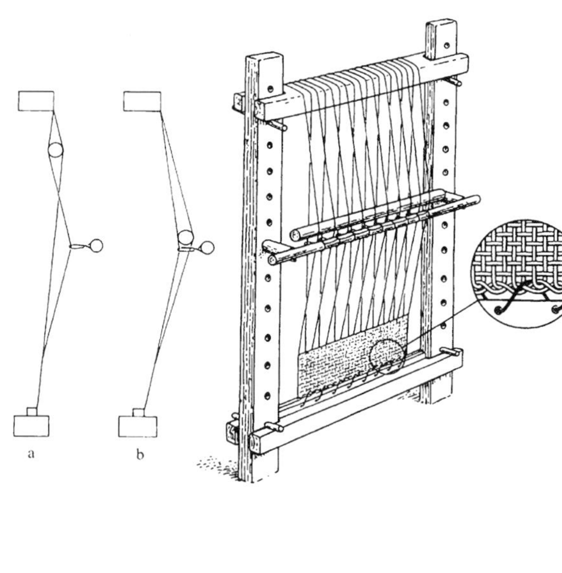

and making braids. Sitting beside her mother, assisting her, the girl slowly absorbed the knowledge. When she had mastered the basic technique of plain weave, her mother could begin to teach her the more complicated techniques, such as making motifs. Weaving was mostly carried out in the colder months of the year, and held a rhythm of its own, accompanied by the thump, or tinkling, of the weaving comb beating down weft (Chantréaux 1941–42: 227, Messick 1987: 219, Reswick 1985: 70ff).

The technology of weaving involves principles of mechanics and mathematics that belies the simplicity of loom construction. Looms of pre-industrial societies consists of a few beams and rods, held together by cords; a skilled weaver can work those pieces of wood and string to make anything from coarse rugs or sacking to the finest, elaborately decorated fabrics. S/he also knows how to manipulate them so that woven patterns are created with mathematical precision. To do so requires complex knowledge, dexterity and skill, and long training. The proficient weaver is indeed an accomplished designer who knows exactly how to obtain a specific type of fabric from the raw material (fibres and yarn), techniques and equipment available.

Rites of textile production

Ethnographic records of the 20th century indicate that among the Berbers of North Africa, all stages of textile production are steeped in notions of belief. The transformation of wool to clothing and other textile items is perceived as a creative process, comparable to giving birth (Chantreaux 1941–42, Laoust-Chantréaux 1990, Makilam 2007: 77ff., Messick 1987, Picton and Mack 1993: 24–25, Reswick 1985: 33ff.). The freshly shorn wool is taken into the house, washed, dried and then left, covered, in a dark room for a week, to rest and presumably 'grow'. When rested, the wool is sorted, combed or carded; these processes are perceived as altering the life of the wool by submitting it to a new linearity (Makilam 2007: 84). The next stage, spinning, is thought to be able to change the lives of men and women due to the rotational movement. Different spindles are used for warp and weft; the small drop spindle for warp is considered feminine, the larger, leg twirled spindle for weft as masculine (Messick 1987: 212). The tools, especially the distaff, are considered to have magical powers, and cannot be lent to women from another house; in fact, they used to be buried with the owner (Makilam 2007: 85). Similarly, the warp is considered as male, the weft female. When a warp is ready to be mounted on the loom, the weaver raises her skirts, straddles the warp and beams and walk over them, symbolically giving birth. The weaver's relationship to her warp has been described as similar to that of mother and son. The mounted loom is perceived as 'alive' (Messick 1987: 212–213). According to Rabaté and Sorber, the loom is regarded as a powerful, living creature, reborn every time a new warp turns beams and sticks into an entity. It has to be fed, with the skilled work of the weaver but also with salt or other substances; as long as the web is unfinished, no thread may be cut with knife or scissors. However, a live thread may be torn out to heal a sick person. When the fabric is finished and cut off, the loom dies. Weaving is thus perceived as a metaphor of human existence; knots used in the loom's construction serve as metaphors for moral concepts (Rabaté and Sorber 2007: 44).

Oleg Lysenko and Svetlana Komarova have investigated ethnographic records collected mainly in the period 1902–1914 and now in the State Ethnographical Museum in St. Petersburg. They show that in Slavonic folk traditions the spindle was perceived as the axis of the world. Much as the earth began with a twisting movement, the spindle creates order out of a chaotic mass of fibre. The processes of textile production were perceived as having important cosmological meaning, especially at key life passages such as birth, marriage and death (Lysenko and Komarova, 1992: 7, 12f). Similarly, in the Norse world, textile crafts seem to have been perceived as something more than inconspicuous, functional daily actions (but see Bek-Pedersen 2007). The First Lay of Helgi Hundingsbane (The Poetic Edda, *Helgakviða Hundingsbana I*, 2–4) tells how the *nornir*, goddesses of fate, arrives at Helgi's birth to spin and weave the web of his life. In a famous passage in Njal's saga, the *Darraðarljóð*, twelve valkyries are described as directing the outcome of the battle of Clontarf by weaving on a loom made of human body parts (Njals saga, 157). Eldar Heide (2006 a, b) has used these and other examples to argue that the Norse term *seið*, which is used to refer to magical processes carried out by women, means spinning, and that witchcraft was carried out by the intermingling of braids, threads, spinning and wind. He contends that this word/meaning can be traced etymologically in old Germanic and other Indo-European languages, as well as Finno-Ugric languages including Sámi (Heide 2006b: 235ff, 304–5).

Ethnographic studies and records of traditional textile work in 19th and early 20th century Scandinavia also indicate that the processes involved in textile making were circumscribed by rituals. From Sweden it has been reported that small stones symbolising food were thrown into the water after washing the sheep; that shearing was best done at a new moon and that each sheep was thanked and given a grain of salt afterwards (Kjellberg 1943: 613ff). In Denmark, flax fields were birched, i.e. elder twigs were stuck in the ground at midsummer (Højrup 1967: 201–202). Each stage of production had its proper time of the year; the processing of wool started after Michaelmas, and – spinning included – had to be completed before Christmas. After Christmas, it was time for weaving and for the spinning of flax (Højrup 1967: 236, Kjellberg 1943:625, Skougaard and Hansen 1983). Yarn should be wound clockwise, and the warping and mounting of a loom started at a new moon so that the web would grow. It was important to shield the warping process from the evil eye of strangers, so it had to be carried out in the course of a single day – if not, weft yarn supposedly would become sparse. A good clear shed, i.e. the establishment of two layers of warp so that the weft can be inserted, is crucial for successful weaving; anyone who came into a room where warping or mounting of a loom was in progress had to lift one leg, symbolising the shed, and pronounce a well-wishing formula (Grenander-Nyberg 1974: 309ff). As in North Africa, the loom in Scandinavia could be perceived as a living being, and the process of weaving associated with notions of pregnancy and giving birth (Grenander-Nyberg 1974: 306ff).

Notions like these may be perceived as embedded beliefs, reflecting the order in which things must be done. Although couched in superstition, many of them represent good sense and indeed factual knowledge, based on the experience of many generations. An interrupted warping process, for example, results in uneven tension on the loom, followed by difficulties in the weaving and irregularities in the web. Still, the ritualisation of the routines of textile making is more than common sense. It may be perceived as a way of making sure that things are done in the way they should be done, according to the worldview of the community in question. Rituals and routines constitute a way to express this in a non-discursive way, as craft, or technology in the sense transmitted by the Greek words *techne* that emphasise that art, skill, craft, methods, knowledge, understanding and awareness cannot be separated and that ancient technologies materialise prevailing worldviews (Dobres 2010: 106).

Spinning in Prehistoric Europe

If we bring an awareness of the complexity of textile making both in terms of labour, learning and rituals to the study of textiles from Prehistoric Europe, it is possible to begin engaging with the material in new ways. Why did people in prehistoric Scandinavia develop a preference for s-twisted yarns during the Bronze Age, contrary to those in other parts of Europe, and stick to it throughout the first Millennium BC and into the early Roman Iron Age, while their neighbours in what is now northern Germany twisted their yarns the opposite way? Difference in twist does not cause different textures. A close look, and preferably a microscope or magnifying glass, is needed to tell the difference between textiles made from s- and z-twisted yarns. The crafting process, the act of spinning itself would constitute the main difference, as the spinning of s- and z-twisted yarns require different movements. A change requires spinners to exchange an ingrained set of movements with another, in addition to learning to use a different tool.

Sunwise and withershins

As we have seen, the direction of spinning is perceived as 'right' or 'wrong' in many traditional or pre-industrial societies. In English, the modern designation of 'clockwise' may be replaced by terms like sunwise, sunward, or in the Shetlands *sungaets*, 'anti-clockwise' by 'withershins' or 'widdershins', all words that hold notions on the right or wrong way to move that supposedly go a long way back. According to the *Encyclopedia Britannica,* studies of Native American dance show that tribes that used to be hunters move sunwise, while agriculturalists move withershins. Lysenko and Kamerova (1992: 12) describe beliefs among traditional Russian peasants that access to the underworld could be obtained by moving against the sun, and that wizards could be identified by the way they moved around a table, a house or a herd. Could related notions of right and wrong, sunwise or withershins, or similar notions of right and wrong, have been held by the Bronze Age and Early Iron Age inhabitants of Scandinavia, and be embedded in reasoning about the spinning direction?

In his studies of Bronze Age religion, Flemming Kaul has shown that depictions of ships on razors illustrate the movements of the sun, sailing to the right during the day and to the left during the night (Kaul 1998, 2004). He further perceives the wheel cross as a symbolic expression of the sun's cyclic movement (Kaul 2004: 250–251, fig. 67), and argues that the sun chariot from Trundholm, Denmark, and various wheeled cauldrons are all expressions of what he calls the myth of the eternal cycle of the sun (Fig. 13.5). A wheeled cauldron from Bujero in Rumania carrying waterbirds is a case in point: the wheels are mounted with further small figures of waterbirds that will move around when the wheels are turned (Kaul 2004: 252–253, fig. 68). The Bujero cauldron may, however, be moved in two directions. The waterbirds – both those on top of the wagon and those on the wheels – are facing opposite directions. This suggests that wheels embodied the conception of opposite directions. Kaul also draws attention

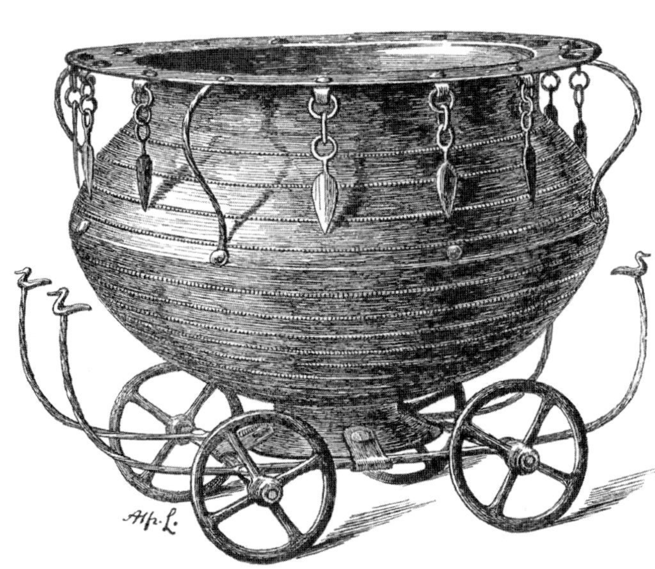

Fig. 13.5: Wheeled cauldron from Skallerup, Denmark (reconstructed). The bird figures look both ways (after Broholm 1943: 170, fig. 67).

to a motif he terms the double horse – animals with head and sometimes forelegs at both ends – that he explains as sun horses travelling with the sun both during the day and during the night. (Kaul 2004: 254, figs 70–72). Kaul's right-left logic awards the 'right' direction, symbolising the day, a higher value than the left, and the left direction, symbolising night, is much less frequent. The dual nature of the double horse and of the Bujero wheeled cauldron does, however, suggest that both directions were considered of equal importance and that wheels served as a symbol of this. Kaul suggests that this might reflect influences from the Central European Urnfield culture where symmetry appears to have been an essential feature in symbolic art to a greater extent than in Scandinavia (Kaul 2004: 255).

At first glance, Kaul's image of the eternal cycle of the sun seems to move the wrong way, as he depicts the direction we know as South at the top, with North at the bottom. It takes a moment to realise, and although the logic is clear enough it keeps puzzling the modern observer and makes one realise how strongly we are conditioned to perceive North at the top of a map, South at the bottom. Still, the sun turns sunwise, regardless of the way its journey is depicted. The wheel, however, is a different matter. It can turn forwards as well as backwards. Whether this is perceived as sunwise or withershins depends on the location of the observer. Can this be translated into terms that capture and relate to the twisting of yarn? To the spinner, the rotation of the spindle by rolling in on the thigh, or by twirling it between the finger and the thumb would appear equally as a wheel turning forwards. The former would result in s-twisted yarn (withershins), the latter in z-twisted (sunwise). Could this mean that both directions may have appeared 'right'?

Textiles from Early/Middle[3] Bronze Age Scandinavia and northern Germany mostly have s-twisted warp but z-twisted weft. Did this, perhaps, reflect notions of both directions being of equal value, and that both were necessary to make a fabric complete? Other fabrics are made of s-twisted yarns in both systems, and this undoubtedly was the main way of twisting yarn. Would this also have been perceived as the 'right' way? Inga Hägg has noted that among the Bronze Age textiles found in Denmark, those made entirely of s-spun yarns were only used for three purposes: cloaks, rugs and foot coverings (Hägg 1995: 40). Investigating contemporary textiles from Schleswig-Holstein, Solveig Ehlers found that Hägg's observation held up, and added gender differences (Ehlers 1998: 220ff). In the Early Bronze Age (Montelius Period II, c. 1500–1300 BC), textiles made solely from s- twisted yarns are more common in male graves than in female graves; this is reversed in Period III (c. 1300–1100 BC). Ehlers interprets this as influences from the Lüneburger area (Ehlers 1998: 282ff), but could it also – or instead – be that the making and selection of yarns for specific, gendered garments have been steeped in notions of right or wrong, or sunwise or withershins?

Twisting the Bronze Age

Spirals were much beloved in the Bronze Age, both as a motif and as a form. Interlinked spirals are a characteristic ornament on bronzes in Scandinavia, particularly in Montelius Period II. They twirl both ways – first in, then out and then onwards to form another spiral; but they usually twirl sunwise first. Arm rings, finger rings, and what may have been ear or hair rings were often shaped as spiral bands or spiral wires. Here, the spiral is a single, three-dimensional form (Fig. 13.6); many spiral rings appear in pairs, sometimes twisted opposite ways.[4] Other types of jewellery, including neck rings, fibulas, and various forms of arm- or leg-guards have spiral-shaped ends (Fig. 13.7). These usually twist the same way, sunwise, unless each end of the item split, in duplets, quadruples or even sextets of spirals, turning opposite ways. Some arm and neck rings are shaped as cords of twisted bronze or gold; the metal itself appear as 'spun'. Fibula bows may appear as twisted too, probably imitating a 'real' thread or cord. The fibula 'cords' are almost always s-twisted (Fig. 13.8). This is also the most common direction of the twisted or corded neck and arm rings. A number of z-twisted rings show that this direction was perfectly acceptable too, but in much lower numbers. The spiral-shaped arm and finger rings also show regional variation

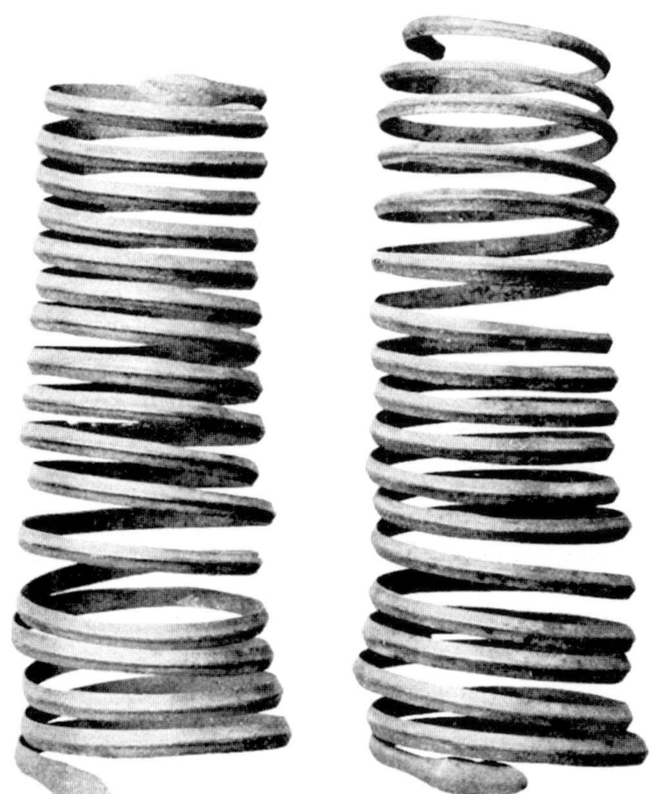

Fig. 13.6: Spiral-shaped bronze armrings twisted withershins from Hundslev, Kjølstrup Sogn, Denmark (after Broholm 1945: 188, M. 36).

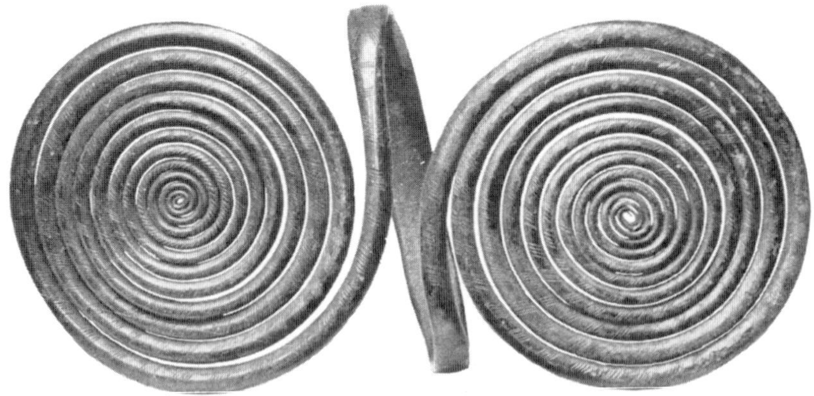

Fig. 13.7: Bronze armring with spiral-shaped end from Simons Mose, Denmark. Both spirals are twisted sunwise (after Broholm 1943: pl. 21, fig. 14).

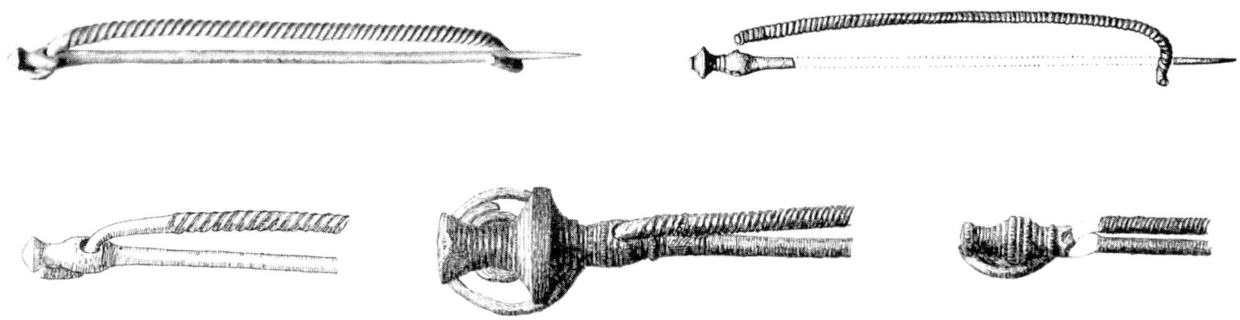

Fig. 13.8: Bronze fibulas with bows shaped as cord, from Toppehøj, Resen Sogn, Kallerup Sogn, Haverslev Sogn, Hagested (Denmark). The cord is twisted withershins (after Broholm 1943: pl. 23, fig. 1, 4, 9, 22, 20).

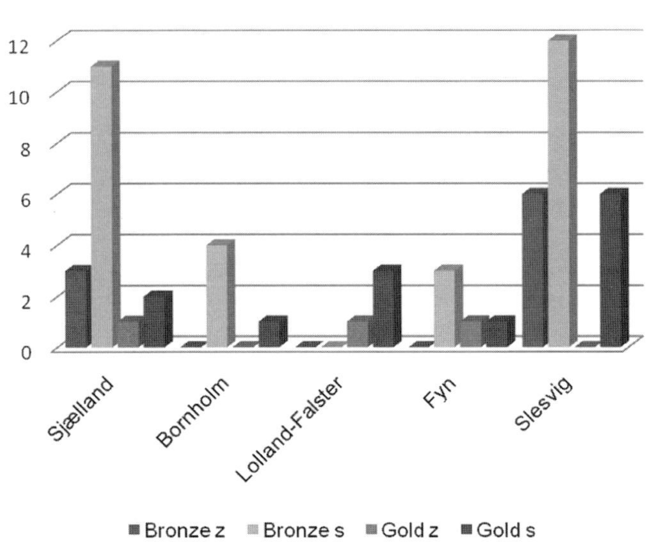

Fig. 13.9: Twisted bronze and gold rings from the Danish isles of Sjælland (Zealand), Bornholm, Lolland-Falster and Funen, and from the former Duchy of Slesvig, now the southernmost districts of Jutland (Denmark) and the northernmost part of Schleswig-Holstein (Germany). Z stands for sunwise twisted rings, S for withershins.

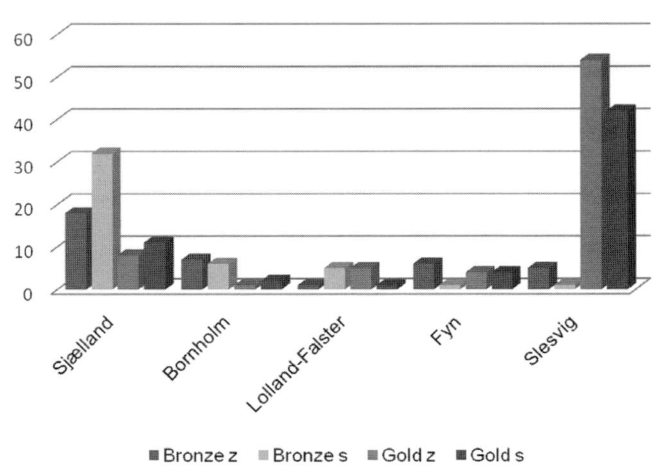

Fig. 13.10: Bronze and gold spiral rings from the Danish isles of Sjælland (Zealand), Bornholm, Lolland-Falster and Funen, and from the former Duchy of Slesvig, now the southernmost districts of Jutland (Denmark) and the northernmost part of Schleswig-Holstein (Germany). Z stands for sunwise twisted rings, S for withershins.

with a marked preference for z-twisted gold rings in south Jutland and Schleswig-Holstein compared to the Danish Isles (Fig. 13.9 and 13.10).[5] This indicates that directions of twist did hold meaning, although not necessarily the same for the twist applied to wool yarns as to bronze or gold. The bronze and gold cords require entirely different work procedures than that of the spinner. Where the spinner would roll her spindle down her thigh to obtain yarn that appear as twisted withershins, the metal smith presumably would twist his rod of gold or bronze upwards – sunwise – although the resulting ring would turn out with exactly the same twist as the yarn.

In North African textile crafts, the warp is, as we have seen, considered male, the weft female. This is further emphasised by the yarns being made on spindles with the opposite connotations. Similar notions were held in ancient Greece, where the crossing of the masculine warp and the feminine weft was perceived as a marriage, and a textile was seen as *being* marriage (Scheid and Svenbro 1996: 29, 68). The most common combination of twist in Bronze Age textiles is that of s-twisted warp and z-twisted weft, something that would fit well to an idea of a fabric constructed from opposite but complementary elements. Would that, perhaps, indicate that the fabrics made solely of s-spun yarns signified the wholly male? Or did it signify the state of being (or becoming by death) unmarried? As Solveig Ehlers found, those dated to Montelius Period II were mainly found in male graves. In the following period III, however, this is reversed; the few finds from the Late Bronze Age suggest that the tendency observed in Period III persisted. By the Early (Pre-Roman) Iron Age (c. 550/530 BC), woven textiles seem to have been made from one type of yarn only, s-twisted north of the 'border' across Denmark, z-twisted south of it. This may reflect a shift from a worldview where complementarities were important to one of different values. If the idea of symmetry that Flemming Kaul views as an influence from the Urnfield culture means that both directions are of equal value, it does, perhaps, not include the sense of complementarities.

We know little of the textile tools of Bronze Age Scandinavia; they obviously were not made of non-perishable materials such as stone or burnt clay. If spindles had spindle whorls, or looms loom weights, they must have been made of materials like wood, bone or unfired clay. The hardware of technology is missing. Our knowledge of the textile technology of the Scandinavian Bronze Age has been studied solely from its products, the textiles themselves. Technology is, however, far more than the hardware of tools. It also consists of software – the knowledge, skill, techniques and movements of the craftsman or – woman, and includes beliefs, and a sense of what is right and wrong in the community where it is enacted.

Notes

1 Clockwise twisted yarn (z) is the norm currently, at least in Western Europe (Bender Jørgensen 1992b, 32).

2 Warp is the system of threads kept under tension in the loom, and runs from back to front or vertically on the loom. Individual warp threads may also be termed 'ends'. Weft is the system of threads that runs from side to side in the loom. Individual weft threads may also be termed 'picks' (Walton and Eastwood 1988). See also fig. 4.

3 Terminologies of Bronze Age chronologies differ in various parts of Europe. In Scandinavia, the Early Bronze Age comprise Montelius periods I-III; in Central Europe, Montelius periods II and III would be the Middle Bronze Age.

4 Assessments of twist directions are based on drawings in Aner and Kersten 1973, volumes 1-7. Designations of '2 Expl' has been taken as two identical rings twisted the same way.

5 Data from Aner and Kersten 1973, volumes 1-7. No distinction between Montelius phases I, II and III have been made.

References

Aner, E., and K. Kersten, 1973ff. *Die Funde der älteren Bronzezeit in Dänemark, Schleswig-Holstein und Niedersachsen*. Neumünster: Wachholtz Verlag.

Barber, E. J. W. 1991. *Prehistoric Textiles: the development of cloth in the Neolithic and Bronze Ages with special reference to the Aegean*. Princeton: Princeton University Press.

Bek-Pedersen, K. 2007. Are the spinning *nornir* just a yarn? *Viking and Medieval Scandinavia* 3: 1–10.

Bender Jørgensen, L. 1984. North European Textile Production and Trade in the 1st Milennium AD – a Research Project. *Journal of Danish Archaeology* 3: 124–134.

Bender Jørgensen, L. 1986. *Forhistoriske textile i Skandinavien. Prehistoric Scandinavian Textiles*. København: Det Kongelige Nordiske Oldskriftselskab.

Bender Jørgensen, L. 1992a. *North European Textiles until AD 1000*. Aarhus: Aarhus University Press.

Bender Jørgensen, L. 1992b. I Knud den Stores klæ'r. *Nationalmuserts arbejdsmark* 1992: 32–42.

Bender Jørgensen, L. 2004. A Matter of Material: Changes in Textiles from Roman Sites in Egypt's Eastern Desert. *Antiquité Tardive* 12: 87–99.

Broholm, H. C. 1943. *Danmarks Bronzealder II. Kultur og folk i den ældre bronzealder*. Købnhavn: Nyt Nordisk Forlag. Arnold Busck.

Broholm, H. C. 1945. *Danmarks Bronzealder III. Samlede fund fra den yngre bronzealder*. Købnhavn: Nyt Nordisk Forlag. Arnold Busck.

Cardon, D. 1999. *La Draperie au Moyen Âge. Essor d'une grande industrie européenne*. Paris: CNRS Éditions.

Chantréaux, G. 1941–42. Le Tissage sur Métier de haute lisse à Aït-Hichem et dans le Haut-Sebaou. *Revue Africaine* 85: 78–115, 86: 212–229.

Crocker Jones, G. 1989. *Traditional Spinning and Weaving in the Sultanate of Oman*. Ruwi: The Historical Association of Oman.

Crowfoot, G. M. 1931. *Methods of Hand Spinning in Egypt and the Sudan*. Halifax: Bankfield Museum Notes Second Series, No. 12.

Dobres, M.-A. 2010. Archaeologies of technology. *Cambridge Journal of Economics* 34: 103–114.

Ehlers, S. 1998. *Bronzezeitliche Textilien aus Schleswig-Holstein. Eine technische Analyse und Funktionsbestimmung*. Unpublished PhD thesis. University of Kiel.

Geijer, A. 1938. *Birka III: Die Textilfunde aus den Gräbern*. Uppsala: KVHAA.

Grenander Nyberg, G. 1974. *Lanthemmens vävstolar. Studier av äldre redskap för husbehovsvävning*. Stockholm: Nordiska museet.

Hall, R. 1986. *Egyptian Textiles*. Aylesbury: Shire Publications.

Hansen, H. P. 1947. *Spind og Bind. Binde-hosens- Bindestuens og Hosekræmmerens Saga*. København: Munksgaards Forlag.

Heide, E. 2006a. "Spinning seiðr," in A. Andrén, K. Jennbert and C. Raudvere (eds) *Old Norse Religion in Long-Term Perspectives*. 164–170. Lund: Nordic Academic Press.

Heide, E. 2006b. *Gand, seid og åndevind*. Unpublished PhD thesis. University of Bergen.

Højrup, O. 1967. *Landbokvinden. Rok og Kærne. Grovbrød og Vadmel*. København: Nationalmuseet.

Hägg, I. 1995. Rezension zu Lise Bender Jørgensen, North European Textiles until AD 1000 (Kopenhagen 1992). *Praehistorische Zeitschrift* 70, 1: 136–144.

Kaul, F. 1998. *Ships on Bronzes. A study in Bronze Age Religion and Iconography. Publications from the National Museum, Studies in Archaeology and History 3*. Copenhagen: The National Museum of Denmark.

Kaul, F. 2004. *Bronzealderens religion. Studier af den nordiske bronzealdersa ikonografi. Nordiske Fortidsminder Serie B, 22*. København: Det Kongelige Nordiske Oldskriftselskab.

Kemp, B. J. and G. Vogelsang-Eastwood 2001: *The Ancient Textile Industry at Amarna*. London: Egypt Exploration Society.

Kjellberg, S. T. 1943. *Ull och ylle. Bidrag till den svenska yllemanufakturens historia*. Unpublished PhD thesis. University of Lund.

Laoust-Chantréaux, G. 1990. *Kabylie coté femmes. La vie féminine à Aït Hichem, 1937–1939. Notes d'ethnographie*. La Calade, Aix-en-Provence: Édisud.

Lysenko, O. and S. Komarova. 1992. *Fabric – Ritual – Man. Weaving Traditions of the East Europe Slavs*. St. Oetersburg: Fortis.

Makilam 2007. *The Magical Life of Berber Women in Kabylia*. New York: Francophone Cultures and Literatures, Peter Lang.

Messick, B. 1987. Subordinate discourse: women, weaving, and gender relations in North Africa. *American Ethnologist* 14, 2: 210–225.

Mårtensson, L., E. Andersson, M.-L. Nosch and A. Batzer. 2006. *Technical Report, Experimental Archaeology Part 2:2: Whorl or Bead?* Copenhagen: The Danish National Research Foundation's Centre for Textile Research.

Nørgård, A. 1999. *Weaving samples of sailcloth on a warp-weighted loom. Experiments carried out at the Viking Ship Museum in Roskilde, Denmark in 1999*. Roskilde: Vikingeskibsmuseet.

Picton, J. and J. Mack. 1993. *African Textiles*. London: British Museum Press.

Rabaté, M.-R. and F. Sorber. 2007. *Costumes berbères du Maroc: Décors traditionnels*. Paris: ACR Édition Internationale.

Reswick, I. 1985. *Traditional Textiles of Tunisia and related North African weavings*. Los Angeles: Craft and Folk Art Museum.

Scheid, J. and J. Svenbro. 1996. *The Craft of Zeus. Myths of Weaving and Fabric*. Cambridge/Massachusetts: Harvard University Press.

Skougaard, M. and H. T. Hansen. 1983. *Uld og hør. Tekstilfremstilling i 1800-tallets bondesamfund*. København: Nationalmuseet.

Stenberg-Tyrefors, B. 1982. *Höger eller vänstervriden tråd? En beskrivning av den betydelse vridningsriktningen kan ha haft vid förindustriell framställning av tråd*. Essay for the BA Degree at Stockholm University, Department of Ethnology.

Van't Hooft, Ph. P. M., M. J. Raven, E. H. C. van Rooij and G. M. Vogelsang-Eastwood. 1994. *Pharaonic and Early Medieval Egyptian Textiles. Collections of the National Museum of Antiquities at Leiden 8*. Leiden: Rijksmuseum van Oudheden.

Vogelsang-Eastwood, G. M. 1992. *Patterns for Ancient Egyptian Clothing*. Leiden: Museum for Ethnology.

Vogelsang-Eastwood, G. M. 1999. *Tutankhamun's Wardrobe. Garments from the tomb of Tutankhamun*. Rotterdam: Barjesteh van Waalwijk van Doorn and Co.

Vogelsang-Eastwood, G. M. 2004. *The Quseir al-Qadim textiles 1978–1982*. Leiden: Textile Research Centre.

Vogt, E. 1937. *Geflechte und Gewebe der Steinzeit*. Basel: Verlag E. Birkhäuser.

Rogers, P. 2007. *Cloth and Clothing in Early Anglo-Saxon England, AD 450–700. CBA Research Report 145*. York: Council for British Archaeology.

Walton, P. and G. Eastwood. 1988. *A Brief Guide to the Cataloguing of Archaeological Textiles*. London: Institute of Archaeology Publications.

14. The sound of fire, taste of copper, feel of bronze, and colours of the cast: sensory aspects of metalworking technology

Maikel Henricus Gerardus Kuijpers

> We might wish to simplify and rationalize skills [...] but this is not possible because we are complex organisms (Sennett 2009: 238).

Ask any craftsperson about their practice and how s/he knows that something is done right and they will mention the senses. Touch, feel, smell, sound or sight can all tell the craftsperson that s/he is doing a good job. It is what some people may call 'finger-feeling'. It is part of what Aristotle defined as *tekne* (Aristotle 2008). It may also be referred to as *technique* in contrast to technology (Ingold 1990: 7). In response to such observations, this paper will explore how we may add sensory aspects of technology and bodily engagement with the material to develop a more nuanced understanding of *skill*. These reflections will be used to think about metalworking in the

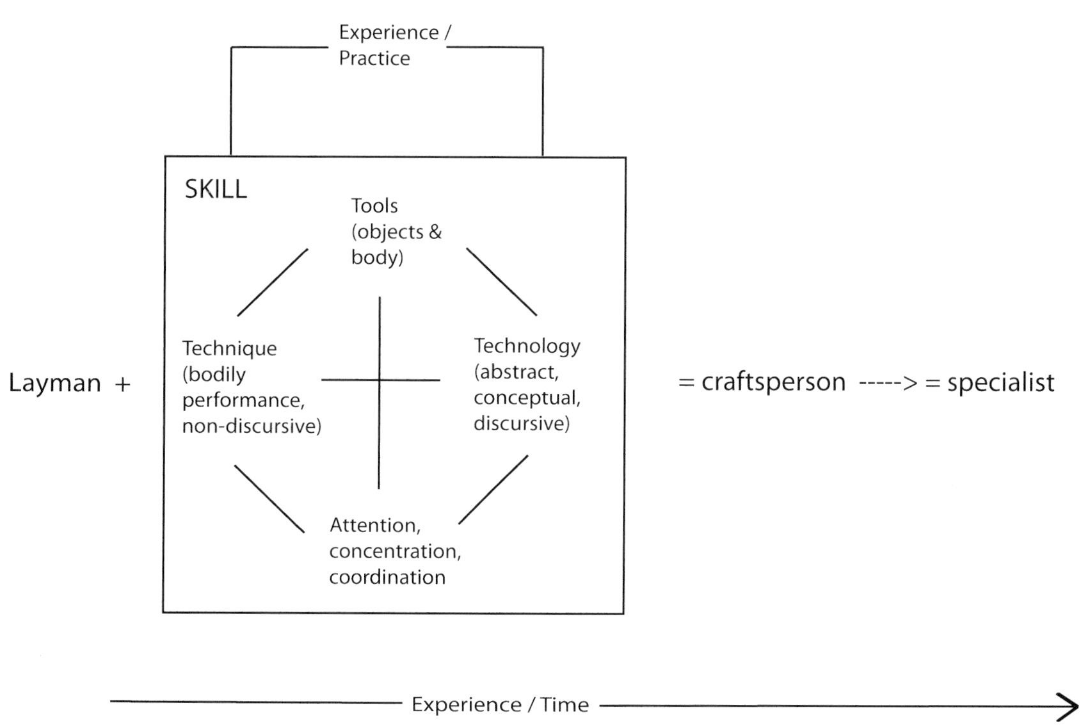

Fig. 14.1: The transformation from layman to specialist; learning skill is not necessarily a linear process.

European Bronze Age, offering new avenues and a different perspective for investigating metallurgy. How does one become a metalworker? Which skills are needed, and is there a way in which we can analyse what transforms a layperson into a specialist (Fig. 14.1)? The focus is on basic metalworking skills, since, in my opinion, the impact of metallurgy on Bronze Age society is more likely to be understood from the spread and production of simple artefacts, such as axes, rather than only from a focus on more specialised and complex objects (cf. Fontijn 2002, fig. 6.15, Kienlin 2010: 171).

Within Bronze Age studies the role of metallurgy has been a central topic. As Harding (2000: 239) has observed 'the elements of skill and restricted knowledge that the smith's art involves, and the processes by which this knowledge could be passed on to others, have often been stressed'. In existing research, *skill* is often used as an argument to explain why metalworking was a more specialised technology, and the advent of metallurgy is used to argue that the Bronze Age constitutes a threshold in European prehistory. In this paper I will argue that the supposed skills conceived for metalworking have hardly been adequately explored to account for the way in which they are stressed (e.g. Bridgeford 2002: 124, Earle 2004: 161, Goldhahn and Oestigaard 2007, Kristiansen and Larsson 2005: 52–53).

DISCURSIVE (knowledge)	**NON-DISCURSIVE (know-how)**
technology	technique
brain-knowledge/theoretic /explicit memory	body-knowledge/muscle/ unconscious memory
top-down approach (abstract)	bottom-up approach
what we expect	what they did
structure	agency
objective (universal)	subjective (context specific)
purpose orientated	action orientated
static	fluid
explaining/words /descriptive	acting/experience /tacit
etic	emic

Fig. 14.2: Dichotomies related to the academic analyses of discursive and non-discursive knowledge in archaeology (after Apel 2008, Budden and Sofaer 2009, Ingold 2000, Dobres 2010, Pelegrin 1990).

To properly understand metalworking technology in the Bronze Age, or any technology in prehistory, it is essential to study skill in a more nuanced manner, however difficult it may be to determine exactly what it is and how it can be measured or valued (Bleed 2008, Budden and Sofaer 2009, Dobres 2006, Hurcombe 2000, Ingold 2000, Sennett 2009). In the first half of this paper, I will discuss the way in which prehistoric metallurgy has been researched and interpreted. I shall show how most research has privileged discursive knowledge, which has led to the current interpretation that specialised Bronze Age smiths had a close relationship with the ruling class (e.g. Kristiansen 1987, Kristiansen and Larsson 2005, Randsborg 1986, Winghart 1998, Vandkilde 2007, 2010). Responding to the limitations of these approaches, the second part of the paper explores an archaeology of the senses as a promising method to understand the relation between the body and knowledge, which will allow researchers not only to enquire what skill is and how it is acquired, but also what craft is and what makes a craftsperson. All of these concepts are only vaguely understood in our modern society in which many tasks of the hand have been taken over by computers and machines (Sennet 2009). Finally, the argument is made that the integration of both discursive and non-discursive knowledge, which are two sides of the same coin, may be better understood through the study of craftsmanship and skill, rather than through the concept of technology.

Discursive versus non-discursive knowledge

Current analysis often places skill with non-discursive knowledge, residing in the body rather than the brain. Skill, however, defies the mind-body dichotomy, as it seems to be both discursive and non-discursive. It refers to the acquisition and experienced operation of certain motor skills ('muscle memory'), but also involves understanding what is possible, what needs to be done, and what is the *right* way of doing a particular action, which may be conveyed as discursive knowledge. Thus, skill draws on cognitive *and* motor activities and all skill involves bodily engagement (Bleed 2008: 157, Mauss 1973, Olausson 2008: 36, Sennett 2009: 10).

Fig. 14.2 shows a list of characteristics that are directly or indirectly linked to either discursive or non-discursive knowledge in current archaeological analyses, sometimes also referred to as the distinction between knowledge and know-how. They are posed as a dichotomy, opposites in a theoretical framework for studying technology and in many ways they reflect Post-Enlightment dualisms, such as mind/body, or modern concerns about the relationship between structure and agency (Budden and Sofaer 2009: 2, Hacking 2004).

The polarisation of these different kinds of knowledge may not be helpful for any attempt at understanding and analysing technology, yet, for now, the differences help clarify my point and explain the current problem we are facing in exploring metalworking technology in the Bronze Age.

Within archaeology, Dobres (2010) in particular has discussed such dichotomies. The discursive epistemology is part of what Dobres has termed '*practical reason ontology*', whereas the non-discursive shares many values with what she has labelled a '*cultural reason approach*'. Dobres argues that the practical reason ontology perceives technology as a medium *between* nature and culture, existing thus both 'outside' society as well as at its very core. It underlies and shapes society, and in this position it therefore gives room to technological determinism, which, it can be argued, is still strongly embedded in many interpretations of Bronze Age society. In contrast, the cultural reason ontology perceives technology as being mediated only *by* culture, thus making people the ontological starting point (Dobres 2010: 104–106). In the next section I will explore some of the core dichotomies outlined above with regards to Bronze Age metalworking.

De-constructing metalworking technology

The most common works on prehistoric metallurgy are encompassing in nature and provide a general overview and explanation of metalworking technology (e.g. Coghlan 1975, Craddock 1995, Ottaway 1994, Tylecote 1987). What these main studies of metalworking technology have in common is a descriptive approach and an economical emphasis. Additionally, material sciences provide the archaeometallurgical discourse with highly detailed analyses of metal artefacts, only strengthening the attention towards the knowable facts or 'hard' aspects of technology at the expense of cultural context and the *actors* who would actually be manufacturing these objects. Many of these studies refrain from making interpretations about the meaning or organisation of the craft as a practice, limiting themselves to a supposedly objective explanation of technology. The supposed straightforward understanding of technology (cf. Hawkes 1954) leads to a top-down approach. From the general structure of technological practice (*viz.* physics / universalities) towards the specifics of (prehistoric) cases and agency (cf. Bray in press), scientific data on metalworking creates a structure from which it is subsequently inferred what to expect. Thus, explanations of metalworking technology seem to be applicable to wide areas, as if the technology would have been the same everywhere. The process is clearly perceived as purpose oriented, meaning that every step in the technological *chaîne opératoire* needs to have purpose, which is the creation of the actual artefact. It is furthermore taken for granted that these purposes are rational and efficient in terms of modern epistemologies and ontologies. Hence, it is all too easily assumed that the aim of the metalworker was to achieve the best mechanical properties in terms of functionality. This might be the reason why possible ritual acts are difficult to integrate in the technological process; to us, with our modern understanding of technology, they appear not to serve the purpose of creating an artefact, but to relate to ways of reasoning that sit outside the metallurgical process (see Kuijpers 2008: 27–30, 59–65 for a discussion). Hence, these aspects either are left aside[1] or are seen as the 'soft' aspects of technology, which might be addressed *after* the knowable aspects have been studied (e.g. Budd and Taylor 1995).

Useful as the approaches described in the last section may be, they are extremely technological in nature and do not answer the questions *who crafts* and *what is craft*. Furthermore, they also refrain from studying skill and the bodily engagement with the material. Both research and interpretation into prehistoric metallurgy have been predominantly scientific, emphasising discursive knowledge. This strong focus on the rational aspects of technology was also observed by Budd and Taylor when they argued that assumptions about the special nature of metalworking technology is based on the idea 'that it is qualitatively different from, and fundamentally more difficult and complex than, preceding technologies (or crafts) such as basketry, flint-knapping or potting, which might all be done "unscientifically"'(1995: 134). Budd and Taylor, however, used this to advocate a more symbolist approach (see below). As the archaeological tradition has decided to perceive the change from stone to bronze as a break between two eras in prehistory, there has been a strong focus on discontinuity, which has also affected understandings and approaches to the two materials (cf. Sofaer-Derevenski and Sørensen in press). Not only do we adhere to the idea that metal was the motor behind this discontinuity, but the idea that metalworking technology saw a break from former technologies is also implicitly embedded in this discontinuity. We think of ourselves as having achieved 'power' over nature through the rational application of scientific knowledge (Ingold 1990: 5) and searching for the roots of that achievement in the past, metallurgy is presented as one of the decisive moments in the history of mankind (Childe 1944, Piggot 1965: 71). We regard it as a step forward, towards the rational and discursive application of knowledge. I suggest that one of the reasons for the disproportional strong focus on the rational elements of metalworking technology is that the technology itself has been regarded as objective. Clearly, the approach described above follows the 'practical reason ontology' described by Dobres (2010). It holds constant the material and applies a modern definition of technology in which the dichotomy between discursive and non-discursive is silently present. The technology is knowable because it is reconstructed from 'pure' scientific data and therefore it does not need a social context – this is a distinctly etic approach. In the past decades, however, it has become clear that technologies and technological systems are, at least, *also* profoundly social.[2]

The idea of metalworking as highly specialised (discursive) knowledge, leading to specialisation, has profoundly affected thinking about prehistoric metalworking technology and has led to the proposition that metalworking facilitated (or at least catalysed) the emergence of elites. Although the argued correspondence between the spread of metalworking technology and the rise of what seems to be an elite is frequently presented as a strong inductive argument, it is based on theoretical assumptions about the relation between specialisation and complex societies (see Costin 2001 for a discussion). Though some alternatives are being proposed (e.g. Bartelheim 2009, Kienlin 2010, Kienlin and Stöllner 2009, Kienlin and Zimmermann in press), this relationship has largely gone unchallenged. Elsewhere I have argued that some of the interpretations and assumptions that underlie the idea of a specialist metalworker are in dire need of further critical exploration (Kuijpers in press, 2008: 51–67). I do not necessarily disagree with the assumptions made, and the following is not primarily intended as an argument against these interpretations. The reflections are provided to argue in favour of the methodology proposed in the second part of this paper. The point I wish to make is that our cognitive and theoretical frameworks for the study of metalworking technology (innovation, adaptation and effects) may be responsible for the very distinct line of reasoning in which the assumptions made (see below) appear to be confirmed, thereby strengthening the axiom. It is, for instance, an interesting inconsistency that metalworking is regarded as sacred or esoteric, specialist knowledge while other crafts are simply perceived as everyday activities (cf. Sofaer 2006: 139 for an interesting view on this inconsistent ranking of Bronze Age crafts).

> Compared to flint, bronze technology was both more demanding and exclusive (Kristiansen 1987: 33).

> ... the smith occupies a central role by manufacturing the aristocrats new weapons, whose power relies on a combination of technical skills and secret magic. As master of these transformative skills, the smiths were in many cultures linked to occult sciences (magic, healing and shamanism) and to the art of song, dance and poetry. The smith-gods' weapon was the thunderbolt, being gods of thunder and lightning. This relates to another important skill of smithing: pyrotechnics and the control of fire (Kristiansen and Larsson 2005: 53).

Kristiansen's groundbreaking work on the grand narratives of the European Bronze Age triggered my research into skill and made me question why exactly metalworking technology would be more demanding and exclusive than other technologies. Why can it only be done by a specialist and why does this specialist seem to be so much more important than any other craftsperson in the Bronze Age (see also: Bertemes 2004, Bridgeford 2002, Budd and Taylor 1995, Childe 1930, Earle 2004, Goldhahn and Oestigaard 2007, 2008, Harding 2000, Winghart 1998, Vandkilde 2010)? Moreover, and most importantly, what are these smithing skills? In comparison to the enormous amount of work done on the meaning and typology of metal objects, the interest in production appears to have become limited to the highly specialised sub-field of archaeometallurgy. Systematic research into the social organisation and characteristics of the craft and the metalworkers is still lacking in details (but see Goldhahn 2007, Jantzen 2008, Jockenhövel 1982, Kienlin 2007, 2008, 2010, Kuijpers 2008, Meurkens 2004, Williamson 1990). The axiom described above seems to ignore that we know so very little about these workers and the practices behind the objects. Yet, in societies where craft production is intimately linked with social structure, the meaning and function of objects can only be understood fully if the manufacturing and manufacturers of the objects are also taken into account. Who made what, how and why (Costin 2001: 275, Hurcombe 2007: 538)?

Although unfair to simplify it to this extent, following the 'metallurgy leads to elites' argument through reveals a series of assumptions on which it is built:

metal = rare
rare objects = prestige and status
specialist knowledge / skill = control
prestige / status / control = power
power = elites
→
metal = elites

Questioned here is the emphasis on *skill* and *specialist knowledge* used in favour of the above argument without actually exploring what they are, how they are acquired and how they relate to technology. Clearly, the presence of non-discursive knowledge is acknowledged here as skill, yet the contemporary archaeological interest lies mostly in the *social value* of skill, which is usually measured in capitalist terms. The idea is that doing something well adds value to the object and a form of power to the producer. However, skill – the ability to carry out a task and to do it well – is a culturally contingent term and dependent on the cultural reference of what is considered 'quality'. Hence, what Dobres has termed 'wow! Ooooh Ahhh' intuitive judgements (Dobres 2006: Fig. 1) should not be simply taken as valid arguments. It is not hard to find examples of such judgements and a closer look reveals that such intuitive argumentation has not changed in the last 80 years:

> The change in properties of copper by heat is really startling, it is distinctively more dramatic than the effect of baking upon potter's clay (Childe 1930: 4).

> The fantastic transformation of raw copper into finished objects may further have invited both myths and secrecy, thus being another possible medium for gaining control (Vandkilde 2010: 5).

It is these kinds of assumptions, found throughout scholarly literature as well as more popular media[3] (e.g. Budd and Taylor 1995, Clarke et. al. 1985, Goldhahn and Oestigaard 2007,

2008), that are of interest. They clearly feed into assumptions about experiencing the sensations of metalworking, but are, apart from our own perception and some anthropological analogies, entirely speculative. Clearly, our present day concepts of materiality affect our ability to perceive materiality in past cultures (Hurcombe 2007: 537). The metallurgical skills appear to be admired for their visual enchantment and the cognitive (and supposed esoteric) knowledge involved. However, my view is that this cognitive 'knowledge' is the fetish of our academic tradition, and may not be the same kind of knowledge that a craftsperson appreciates or prehistoric metalworker would have had. Instead of simply accepting the 'wow' arguments given above, which build on perception and experience, a more analytical and structured exploration into the experiencing of metalworking technologies is needed in order to define more precisely the skill involved in metalworking, how this skill was acquired and how this subsequently builds an identity of being a metalworker.

Recent theories, such as material engagement theory, focus on the relationship between humans and the material world (Renfrew 2004: 23) and recognise the value of phenomenology in this exploration (ibid. 24–25). Yet, materiality and material engagement theory seem to be more concerned with spheres of consumption than with production (Ingold 2007: 9), and thus there is little enquiry into the practices of the craftspersons. Furthermore, the question posed by Ingold, as a critique towards the philosophical and theoretical ponderings of scholars dealing with materiality, is also worth exploring as part of a research into technology. Would an engagement – working practically *with* material – offer a more powerful procedure of analyses than the approach bent on the abstract analyses of things already made (ibid. 3)? This argument, I think, is especially a matter of concern for archaeologists, because the prehistoric agents manufacturing the objects we are trying to understand would have made sense of the world through engagement and experience, not analytical abstract science. The 'origins' of cognitive knowledge and science, however, may lie in exactly these experiences, as explained by Renfrew: he argues that, for instance, the *experience* of balancing two different kinds of materials may have led to the concept of weight (2004: 26). The (non-discursive) motor skill of balancing and the senses telling that they are different consequently seem to lead to a (discursive) understanding of weight.

One of the challenges for the study of technology is thus to overcome the dichotomy between discursive and non-discursive knowledge, because technology is made up of both. 'The construction of conceptual knowledge systems, or ontologies, is immediately and inextricably linked to practical engagements' (Kohring, this volume). Accordingly, researching skill, which links these two together, is a vital part of understanding metalworking (or any) technology. It is by no means a way to fully explain prehistoric metalworking technology, and it is not an approach leading to a clear-cut conclusion, but it is a necessary perspective to be taken and one that might lead to the integration of discursive and non-discursive knowledge and related dualisms.

From material to specialist metalworker

As mentioned before, the role of the metallurgist as an important figure in society and a skilful (ritual) specialist seems to be readily acknowledged in general readings of the European Bronze Age, either explicitly (e.g. Goldhahn and Oestigaard 2007, Kristiansen and Larsson 2005: 32ff, Childe 1930) or with a little more caution (Barber 2003: 134, Harding 2000: 239–240, Cunliffe 2001, 2008: 155–157). Nonetheless, direct evidence for the presence of metalworkers as a recognised social class in prehistory, such as graves assigned to metalworkers based on their content (e.g. Bertemes and Heyd 2002, Butler and Van der Waals 1966, Jockenhövel 1982, Randsborg 1986, Sperber 2000), is still extremely limited. As such, the metalworkers remain an elusive group, and their actions and societal impact are clouded in theoretical discourse. Assumptions on the ritual nature of metalworking and the social position of the smith are seldom corroborated by archaeological data. Metalworking evidence is also scarce in some areas due to problems of recognition (Kuijpers 2008: 105). Interpretation of the societal impact of Bronze Age metalworking is therefore traditionally achieved through investigation of the finished objects and inferred from ethnographic and anthropological analogies. Hence, the bronze smith – commonly envisioned as a highly specialised male person who controlled the 'secret' and 'magical' knowledge of metalworking – and his supposed itinerancy (Childe 1930, Kristiansen and Larsson 2005: 58) remains largely a theoretical construct (Sørensen 1996: 46, Kuijpers 2008: 51ff).

In the wake of post-processual archaeology, attention turned towards agency and the symbolic, focussing more on the meaning of metal and questions such as *who crafts?* Notably, Budd and Taylor made a strong argument suggesting that the way in which metalworking technology had been studied so far was very anachronistic. Although they themselves were rather nuanced and made clear that their remarks were to be taken as suggestive (Budd and Taylor 1995: 141), the effect of this seminal paper was a major shift in how we perceive metalworking technology. The 'soft' aspects of technology (i.e. the symbolic meaning of metal, the technology and its producers, beliefs etc.) were now being discussed more intensely and interpretations clearly favoured explanations in the realm of ritual, power, symbolism and elite-driven metallurgy. These studies are, however, often focused on the consumption of metal and are thus more often than not what I theoretically would categorise as cause-effect studies, rather than being concerned with *understanding* the technology and manufacturing itself. Furthermore, as mentioned before, they do not explore the skills involved in metalworking but rather the social value of these skills. Though interesting, such approaches do not lead to a better understanding of the manufacturing process or a clearer image of the producers of metalwork. On the contrary, they place the metalworker outside the realm of 'simple' craftsmanship, or, as Randsborg (1986: 188) puts it 'we can scarcely speak of "craftsmen"'. Interpretations go directly from the objects to their significance in Bronze Age society and from there it is inferred back that the producers *must* have played an important role, even though their identity is uncertain (e.g. Goldhahn 2007: 91, Harding 2000: 239–240). In this manner, within the social interpretation of metalworking and with regards to the social organisation and characteristics of the craft and metalworkers, a short-cut is being made around the actual *making* of the object, which still remains unexplored, assumed rather than investigated, and we are left with the ritual, chiefly, specialist bronze smith – the 'master of mysteries' (Childe 1930: 4).

Through de-constructing current lines of reasoning and critically assessing some of the points made, I hope to have shown that there has been a disproportionate focus on the discursive aspects of metalworking technology. Although amends have been made by addressing the symbolic and ritual nature of the craft, these ideas were built on the underlying idea of a technology in which discursive knowledge is most important. I think such an interpretation stems from an overemphasis on the scientific results that archaeometallurgy has given us. By no means am I saying that these results are not extremely useful, or that we should stop this line of research. I am arguing that, although we were optimistic, these methods alone cannot answer the simple question posed: *who crafts?* If we are to understand the craft of metalworking and what makes one a metalworker, we also have to understand the non-discursive aspect, even if we cannot capture it in measurements and transform it to quantitative data (but see Budden and Sofaer 2009). It involves understanding what exactly *skill* is rather than discussing the social value or social consequences of skilled production.

The scientific knowledge provided by material sciences offers us a framework within which we need to work because both in cultural as well as in 'practical reason ontologies' the framework of artefact physics applies (cf. 'real-world factors' Dobres 2010: 106). Yet, in order not to lose the actors, we have to explore the way in which the prehistoric metalworker was aware of and used the physical rules of their craft. Technology and technological change is simply a far too complex research topic to be studied from a single perspective. Moreover, the technology not only needs to be studied from both a scientific and a social perspective, these perspectives also desperately need to be integrated, as they do not operate apart from each other. This call for integration between metallurgy as a human action and material science has been advocated several times now (Goodway 1991, Ottaway 1994, 2001, 2002, Thornton 2009) albeit without actually coming to terms with *how* this should be done. The second part of this paper explores such a methodology.

Re-constructing metalworking craftsmanship

Recent neuropsychological research into tool making skills is providing us with clear evidence that various parts of the brain are involved, placing the senses, rather than cognitive strategic action and planning, much more at the forefront of learning a skill (Stout and Chaminade 2007: 1091). An interesting case shows that one does not even need explicit cognitive knowledge in order to acquire certain skills. Patient H.M., a man suffering from severe anterograde amnesia (due to surgery), is incapable of creating new memories. He was asked to perform a mirror-draw task and although he had no recollection of having done this before on previous days, he nonetheless showed improvement (Corkin 1968: 255). Further tests on motor skills led the researchers to conclude that one is able to acquire skill, despite impairment in learning (ibid. 262). Skilled learning was obviously not lost, although the patient did not have the capability to draw on explicit cognitive knowledge. This is not to say that skilled learning does not take place in the brain – it does – but in a different part of the brain, and it therefore is tacit knowledge.[4] Many bodily gestures are examples of such tacit knowledge (Mauss 1973). The examples described above give us a clear motivation that in order to understand skill, we must explore the relationship between knowledge and the human senses.

The skill ascribed to metalworkers had to be acquired by experiencing their material and engaging with it. But how can we possibly study, describe and understand this skill, and hence metalworking craftsmanship, without researching the experiences and cognitive understanding that prehistoric metalworkers may have had with their material? In contrast to how a prehistoric metalworker would have dealt with metal, current material sciences focus strongly on discursive or 'hard aspects' of this technology, which can usually only be reached through detailed technical and chemical analyses. This distant and rational approach is unable to come to terms with the social aspects and skills that are involved. We can safely assume that the prehistoric metalworker did not hold the discursive knowledge that copper melts at exactly 1084.5 °C, nor that a reduction in thickness of 60% leads to an increase in hardness from 100 HV to 235 HV in a 10% Sn tin-bronze alloy.[5] Nor would they have known about the *exact* composition of their material. The scientific analyses that are possible for metal are useful for determining exact composition, which is measurable objective knowledge, and classifications are made based on these analyses, possibly showing patterns which are of interest to us for provenancing or determining groups of metal (e.g. Butler and van der Waals 1966, Merkl 2010, Needham 2002). However, they lack explanatory force as to the characteristics of the craft and the metalworker. Prehistoric metalworkers most likely had other matters of concern. Regarding these concerns, archaeometallurgical results need interpretation and need to be explained in their social context (Kienlin 2010: 1–2).

Although prehistoric metalworkers seem to be competent

in the recognition and manipulation of some compositional differences (*viz.* alloys), compositional analysis overshoots its goal with regards to this skill because it looks at variables of which they would have been unaware, and in such detail that it does not refer to human choice anymore. It is extremely unlikely that the prehistoric metalworker ever knew that his/her product contained 0.3% nickel and 0.6% iron, or that the difference in 6.5% and 7% tin had any deliberate meaning or function. The question now, in my opinion, is therefore not what the exact composition is of an object, but how the prehistoric metalworker perceived his/her product and what kind of 'composition' s/he was aiming at, how this was done and for what purpose. This was not based on assessments or knowledge similar to our scientific analyses, but on the basis of a skill in recognising colour, hardness, ores, smell, malleability and other factors, all of which may have given them information about the material properties or 'metalleity'.[6] Their technology is a sentient and sensual material practice (Dobres 2010: 109). Therefore, the way forward for archaeometallurgy, I believe, is the study of craftsmanship and associated skills, which involves dealing with at least two questions:

1) What exactly did the prehistoric metalworker know about his/her technology (what mattered to him/her)?
2) How did s/he go about working with these matters of concern (= engagement)?

Whereas the first question has ties with cognitive archaeology, and has recently been addressed by Kienlin (2007, 2008, 2010, Kienlin *et al.* 2006), for the latter question the heuristics of an archaeology of the senses is most likely to provide answers. As argued above, the prehistoric metalworker would have understood his/her craft through their own body and their senses. This experience, through which a craftsperson is able to see and feel when his/her product is perfect without formal measurements and without being able to explain it exactly (which would make it cognitive knowledge), is what I define as *embodied knowledge*. Knowledge that, without the involvement of *both* the body and the mind, cannot be gained. This rather tacit form of knowledge is difficult to come to terms with verbally and in writing, but is nonetheless an important part of any technology, as 'the portion of technical knowledge that people can verbalize represents only the tip of the iceberg' (Pfaffenberger 1992: 508). It is from this bodily engagement with material that more formal and abstract concepts of measurements and knowledge may have evolved (*cf.* Renfrew 2004). Because of this, we should be aware that using these concepts of measurement to explain a world in which they were not (yet) present is a distinct *etic* approach leading to a distant and outside perspective of the culture and technology studied. Understanding prehistoric metalworking skills may also tell us something about the inception of ideas such as 'composition' or 'ore', which we should not take for granted. Concepts such as 'ore', for instance, had to be invented, and the idea of what is the *right* rock to use differs among societies (Bray in press). To know exactly when the fire is hot enough, without using a temperature gauge, is a matter of skill, repetition and the use of the senses. Without any means of measuring it besides your senses, the only way to gain this knowledge is to *experience* it. It is non-discursive, embodied, fluid, dependent on the context and thus the outcome is subjective knowledge. Focussing on this knowledge may lead to a more nuanced interpretation of the craft of metalworking in the Bronze Age.

Hence, this paper explores a methodology that aims to come to terms with a more emic view on how the prehistoric metalworker could possibly have experienced and dealt with his or her craft – measurement by experience. It presents a technological understanding in which the clear line between discursive and non-discursive knowledge was not present and where Descartes had not yet torn apart the relation between knowledge and the sensual body. Simply opening a textbook on mineralogy written before Descartes beautifully describes the kind of 'technological' understanding I am hinting at:

> Color, taste, odor, and qualities of minerals which can be perceived by touch are most widely known because they are more easily recognized by the physical senses than qualities such as strength or weakness. A great many of these qualities are not known to everyone although those qualities which are learned through experience are widely known (Agricola 1955: 5).

Understanding skill through the senses

Having made clear that skill needs to be incorporated in technological research, *how* to do so remains a difficult issue. There is undeniably an appreciation of skill in Bronze Age research, shown, for example, in the interest in objects indicating high skill levels. Whether there is also an academic understanding and interest in what exactly these skills are, and how they come into being and are played out, is less obvious. In lithic studies 'skill matters' (Bleed 2008) and has been incorporated mostly by experimental reconstruction of the technology by modern flintknappers (e.g. Apel 2008). Here, experimental researchers use their personal experience to make general claims about the technology. Since many academics do not have this experience, it is impossible to dismiss this 'data', yet it seems equally impossible to incorporate, as it is 'only' a personal claim, made by a single agent in our contemporary society (which of course influences how they deal with the technology). This approach thus tends to lead to a polarisation of theoretical debates and practitioners' experimental data – an unacknowledged paradigm war according to Dobres (2006). Nonetheless, it does clearly lead to more nuanced questioning about the role of craftsmanship in societies and how this may be used in power struggles (cf. Olausson 2008).

How to study skill then? In order to understand the skills involved in metalworking, we have to try to understand the relationship between the craftsperson and the material on which skill is built. Since all experience of the world outside is mediated by the senses (bodily engagement) I am proposing an archaeology of the senses (Day forthcoming, Skaetes 2010) as a methodology to understand the skills involved in prehistoric (metalworking) craftsmanship. So far, only phenomenological archaeology has made an attempt at understanding prehistoric people through embodied experience, which has been a welcome change of perspective in contrast to the 'abstract models which prioritize the role of the mind in human cognition' (Brück 2005: 46). Although an archaeology of the senses is closely related to phenomenology,[7] it applies a different methodology and has a different goal. This is not the place to go into a discussion on the (mis)interpretation of phenomenology within archaeology, but briefly outlining the differences with an archaeology of the senses is needed in order to prevent confusion.

An archaeology of the senses

In the social sciences and humanities an interesting new paradigm has been evolving since the 1990s as part of the 'sensory turn'. This has led to the introduction of sensual culture studies in fields ranging from anthropology to architecture and philosophy to psychology. I advocate that these sensual studies may provide us with the necessary perspectives and new methodologies to further archaeological research on prehistoric technologies. As this field is still in its infancy, there has been no widely accepted methodology on its use yet, and within archaeology only few have made suggestions as to its implementation (e.g. Houston and Taube 2000, Hurcombe 2007, Skeates 2010, Day forthcoming).

Within anthropology there has been somewhat more consideration of the concepts (Classen 1997, Classen and Howes 1996, Howes 2003, XIII with references, Goody 2002) and a distinction seems to be made between the senses, sensation and emotion. All three are associated with each other, although some important differences are present. A physical experience (using your senses) may lead to a sensation, which in turn relates to emotions. Sensation is therefore both related to physical experience (feelings 1) as well as to emotions, sentiments and passions (feelings 2) (Goody 2002: 22, Harris and Sørensen 2010). The emotional side of sensation thus refers to the inward reflection to which experiences give rise. Since the senses are already both physiologically and culturally constituted, and since different cultures put different emphasis on particular senses ('sensory models', Classen 1997), the sensation emerging from experiences is distinctly culture specific. Research in the sensory perceptions of peoples in diverse societies have, however, shown that there are only small differences in the physiological variables (Goody 2002: 18, see below). Concerning the phenomenological approach, Tilley's (2004: 201) point that 'we and the people of the past share carnal bodies' therefore seems to hold some ground. Yet, how the sensual perception or sensation is experienced (feelings 2) is a completely different matter. In phenomenological archaeology, the researcher becomes a methodological 'tool', and this is a problematic proposition. The sensation, emotions and thus meaning emerging from the experience are defined not only culturally, but also individually.[8] Their attempts at reconstructing past experiences by modern bodily experiments as a way to understand past perception and meaning is therefore rightly criticised as highly subjective speculations (Brück 2005, Fleming 2006).

With an archaeology of the senses it is, however, not my intention to look for the sensation nor meaning of metalworking, but rather the ways in which the prehistoric metalworker could have 'measured' and understood his/her craft through the senses. Therefore, the methodology proposed here is also based on experiencing, but is limited by the senses and artefact physics, as will be explained below. Surely, symbolic meaning, and therefore ritual, is an aspect of metalworking craftsmanship too, but they are related to the practice of metalworking. As I have argued elsewhere (Kuijpers in press), I believe that interpretations regarding these sensations, at this moment, actually form an intellectual barrier in our understanding of prehistoric metalworking craftsmanship. Before addressing questions relating to these matters, I first and foremost want to explore how the prehistoric metalworker understood, experienced and played out his/her craft. I am aware that teasing these aspects apart may again create a non-existent dichotomy. Nonetheless, I separate them at this point because I first want to explore the framework in which the prehistoric metalworker was working. In a nutshell, I am interested in the (physiological) senses and how sensual perception (experiencing) relates to skill and knowledge.

Some thoughts on the methodology of an archaeology of the senses

As with the phenomenological approach, an archaeology of the senses needs to overcome two problematic issues. Firstly, the diversity of human experiences needs to be taken into account, without befalling to extreme relativism and subjectivity (cf. Bruck 2005: 57). This problem relates to the authority given to personal experience by researchers practicing experimental archaeology with regard to technology (cf. Dobres 2006). Secondly, the relationship identified in the present has to be proven significant in the past, too. The following paragraph will argue that the study of skill through the senses, especially considering technology, is well suited to overcome these problems.

The challenge is to analyse qualitative data (examples given below) in such a way that it will lead to a set of rules that apply to metalworking craftsmanship. These rules are not universalities, as people will use their senses in different manners. Nonetheless, even though the act of perception is culturally defined, there are limitations to what our senses can and cannot do, known as Weber fractions (e.g. Eerkens 2000).[9] In addition, there are limitations to what is possible with the material. In this manner I believe we might be able to combine two different perspectives towards technology: the distant, formalised, rational approach of the material sciences and the social, subjective, cultural approach that have the actors as its ontological starting point. In so doing, it challenges the gap that is present in archaeometallurgy between metallurgy as a material science and as a human practice. The distinct discursive approach of archaeometallurgical science has shown us the possibilities and limitations of the material, through artefact physics. Hence, it has created the 'material framework' to which the craft of metalworking is confined. The next step, I think, is to create a second embedded framework, relating to the actual possibilities and limitations of prehistoric metalworking (Fig. 14.3). Since the prehistoric metalworker was not able to go about his/her craft at the abstract level the sciences do, this 'human' or 'experience' framework is smaller and sits within the material framework. The limits of this framework are set by both prehistoric cognition and the senses, making them disputable but not unknowable. Studying the subjective, personal experience of craftsmen in a manner that tries to find the objective rules that underlie their experiences (cf. Apel 2006) avoids the problem of describing the sensational experience of a single agent, thus circumventing one of the pitfalls in the framework of the phenomenological methodology. Yet, it does provide us with the 'rules' that apply working with the body, engaging with the material. This I see as a step forward in our understanding of prehistoric metalworking craftsmanship.

The psychophysical framework sets another range of 'limitations', in close interplay with the social factors that constitute a technology. Here, Dobres' arguments for the 'cultural reason ontology' come into play. Technological practice is guided by a number of cross-cutting factors, ranging from gender, what is 'proper' raw material, rituals, *faux pas* to tradition (Dobres 2010: 109). Within the limitations of the material and the physical and cognitive capabilities of the metalworker, the craft was formed by personal *choice* and cultural dogma, which in turn affect the psychophysical framework. Thus, technology is in essence a social phenomenon (Pfaffenberger 1988, 1992, Killick 2004) and hence completely dependent on the social context in which it plays out. The third framework is the most difficult to reconstruct, as it includes factors such as symbolic meaning, the rituals and taboos involved, the meaning and organisation of the craft, and the social value of metalworking skills. For archaeologists, who cannot ask the craftsperson for his/her reasons, nor observe them, there appears to be no other option than to make inferences in order to form interpretations regarding these aspects of craftsmanship.

The use of experiments to learn about skill

Crafters, experiments and ethnographic analogies may help us in a more holistic understanding of prehistoric metalworking. They can teach us how fire was dealt with, how and on what basis certain choices were made and what kind of knowledge and skills may have been present. Furthermore, experimental archaeology is a way of opening up possibilities and questions. Following are a few examples that describe the kind of sensual experiences that I believe deserve more analytical attention.

My first example highlights the taste of copper, which is one of the ways in which a former copper prospector on the Fiji islands was able to find copper deposits. By looking for small streams and tasting the water, the distinctness of the copper taste led him to rich deposits (pers. comm. Rupert Grey 2010). The taste of minerals and metals is mentioned several times by Agricola as a way of determination, for instance: 'the taste of copper is very bitter and unpleasant, that of iron less so while the taste of tin is the weakest of all metals' (1955: 8). Second, the smell of processing arsenic-rich ores produces toxic

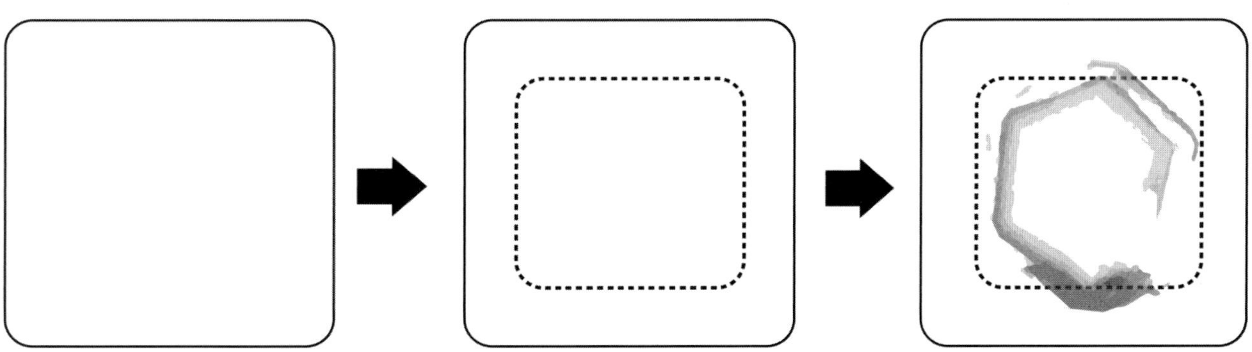

Fig. 14.3: The first framework of artefacts physics, a second framework of what was possible in prehistory – the psychophysical framework – and the third and most difficult to reconstruct, the social character of craftsmanship (drawing by the author).

white smoke and a strong garlicky odour (Charles 1967: 26, Harper 1987: 654, Nriagu 2001: 2). Besides visual markers of arsenic in the ore (silvery-white, turning greyish when exposed to air (Bray in press, Charles 1967: 25), the smell may have been the most obvious way to determine whether the ore was arsenic-rich. Again, historical documents seem to confirm this manner of identification (cf. Agricola 1950: 113, Feuchtwanger 1867: 123). The prehistoric metalworker may well have made a causal link between a certain kind of greyish ore, producing a garlicky smell and white smoke and the possibility of manufacturing an artefact from this, which then had a distinct white-silvery colour and was notably harder than objects different in colour, produced from different kind of rocks. This brings us to the third example, the colours of the cast, which may have been one of the most characteristic elements of the kind of bronze the prehistoric metalworker was dealing with and one of the main concerns for the prehistoric metalworker. Generally, an object rich in copper is reddish, while tin-bronze has a more gold-like colour. J. Zuiderwijk, an experienced bronze caster I have been working with, is able to distinguish differences in copper-tin composition in order of approximately 2% steps, based only on the colour of the artefact.[10] Following Renfrew's (2004) line of reasoning, it can be argued that seeing and experiencing these differences in colour could possibly have led the prehistoric metalworker to a basic cognitive understanding of 'composition' or alloys. We have strong indicators that in prehistory choices were also made on the basis of colour, be it for aesthetic, technological or symbolic reasons (Lehoërff 2008, Hosler 1995, Jones and MacGregor 2002, Juleff and Bray 2007, Pearce 2007, Stevens 2008). Thus, understanding the engagement of the Bronze Age metalworker with the material world is fundamental to our understanding of the craft as technological choices were made on the basis of these engagements.

Experimental and ethno-archaeology may therefore be useful as they provide insights into the skills used in metalworking craftsmanship. The analytical aim is that of recognising the way in which one can deal with technological knowledge through bodily engagement with the material, such as looking at the fire and knowing from the colour when it has reached the required temperature, or stirring in the bronze, feeling, and thus knowing, that it is ready to be poured. Experiments are especially important here since they allow us to measure exactly those aspects that seem to help the craftsperson understand their craft. In contrast to Reynolds' (1999) definition of what makes a good archaeological experiment, arguing that experiencing a past technology is interesting but not useful in scientific research, I advocate that it is exactly these experiences that we should try and understand (cf. Kuijpers 2008: 24–26). From this, we may gain a basic, more defined understanding of skill. Based on the reasonable assumption that there is only a limited number of ways to, for instance, assess/guess the heat of fire without using a temperature gauge, the focus of the methodology proposed here is to understand the sensual measurement made. We should ask how the fire looks, how it feels, sounds and burns and to what extent this leads to an understanding for the craftsperson involved. As argued above, there are limitations to what our senses are capable of and there are the constraints of (artefact) physics. This makes it possible to study the relationship between them within certain limitations. In addition, because both can be (re)tested, I believe this approach to be more in accordance with what we acknowledge as being sound research. It is, however, phenomenological in the sense that it tries to understand how sensual perception and conscious experience relates to skill and knowledge.

Conclusion: there was no such thing as metalworking technology

Whereas studies of other technologies such as flintworking (Pelegrin 1990, Apel 2006, 2008, Olaufson 2008) have had a strong focus on non-discursive knowledge, I perceive the opposite to be true in studies of metalworking, which is problematic. A strong appreciation of the discursive knowledge involved in metalworking has led to a distinct line of reasoning in which the non-discursive has remained largely unexplored. It also confirms Hurcombe's (2007) idea that we need to discuss explicitly the fact that different materials allow different possibilities of understanding, without projecting this onto past minds. While studies of technologies such as flint and stone working have been influenced by the introduction of the concept of a *chaîne opératoire*, a similar methodological approach rarely seems to be applied to metalworking, where detailed mechanical analyses are perceived as the method *par excellence* to understand the technology. This might have something to do with the fact that, in contrast to stone working, metalworking is not a linear process, hence complicating the *chaîne opératoire*. The focus of material science is on the knowledge trapped in the object. Results thus lie in what can be read from the object, mostly without asking which parts of that knowledge would also have been available and of interest to the prehistoric metalworker.

The methodology explored in this paper is based on one of the core arguments of phenomenology: that the world around us is experienced and understood from the perspective of the embedded and sensual human body. I have argued that Bronze Age metalworking technology was, in the first place, experienced rather than rationally understood. That is not a denial of rational knowledge in prehistoric metallurgy, but rather a necessary change of perspective towards craft, craftsmanship and skill. Although skill has often been used in arguments regarding the meaning and importance of metalworking, it is not a properly understood concept and exploring the skills that are actually involved in metalworking is a rather neglected field in metallurgical research. One of the reasons for this might be the highly qualitative and elusive nature of this 'data'. Nonetheless, there are ways to quantify skill, even archaeologically, as for example done by Budden and Sofaer (2009). Although this may be useful for certain analyses, I do not feel that such a distinctively quantitative approach is always necessary to incorporate skill in archaeological research. The methodology proposed here, an archaeology of the senses, may further our understanding of the relationships between the body, materials and knowledge, which underpins skill. The heuristic power of a sensory archaeology cannot yet be assessed, as it is still in its initial stages. It does lead, however, to a different way of exploring the data and hopefully a better and more holistic understanding of prehistoric metalworkers and their craft. Furthermore, it may challenge the dichotomy between discursive and non-discursive knowledge, for which it has been argued that this opposition is too absolute.

What has become clear is that deeply embedded in the intellectual heritage of technology studies are the dualisms we are now seeking to overcome. Even when we deconstruct the concept of technology, as Ingold (1990) for instance tried to do, the term inherently seems to carry a modern Western burden, while not being able to capture the original meaning of *tekne*. Integral in the concept of *tekne* was the idea of skill and the engagement with material with both the mind and the senses. Discursive knowledge (*episteme*) is the fetish of our academic tradition and is incapable of describing the way in which craftspeople understand and work with their material. Therefore, I am inclined to follow Ingold's (1990: 6) daring statement on technology and argue that there was no such thing as metalworking technology in the Bronze Age. Craft and craftsmanship captures the meaning of *tekne* far better. Yet, these concepts are poorly understood as they have been given only little attention (but see Needleman 1979, Pye 1968), but have become the subject of renewed interest recently (Adamson 2010, Ingold 1990, 2000, Sennett 2009). Surely archaeology, with its prime focus on things, is in a position to explore such concepts. Thus, the first step towards a more balanced exploration of metalworking in the Bronze Age is to study metalworking *craftsmanship* in contrast to metalworking technology. The study of craft and craftsmanship challenges the dualisms mind/body, discursive/non-discursive and therefore also the gap that is so apparent in metallurgical studies.

Acknowledgements

I would like to thank my supervisor Marie Louise Sørensen for the opportunity to contribute to this volume. She and Katharina Rebay-Salisbury are also thanked for their useful comments on earlier drafts. I have benefited greatly from discussions with Dacia Viejo Rose, Tim Flohr Sørensen, Jo Appleby, Frank Stevens and several of my fellow PhDs at Cambridge University, exploring the points made and commenting on my work. Jeroen Zuiderwijk and Neil Burridge are thanked for trying to put into words exactly the kind of knowledge which is hard to capture in language. The research leading to these results has received funding from the European Union Seventh Framework Programme (FP7/2007–2013) under grant agreement n° 212402.

Notes

1 There is almost no mentioning of the words 'ritual' or 'symbolic' in some of the standard works on metalworking technology (e.g. Craddock 1995, Tylecote 1987, Ottaway 1994).

2 Some would argue that they are fully social (Science and Technology Studies).

3 Museums in general acknowledge the idea, because of the visual attraction of metalworking and esoteric and mysterious knowledge involved.

4 I am indebted to Dacia Viejo Rose for making me aware of this research.

5 Which does not mean that there was no understanding of the idea that hammering leads to increased hardness and that copper should be at a certain temperature (*viz.* hot enough) to be poured. The difference between these kinds of knowledge, although in essence they describe similar things, is the manner in which they are gained and used. Whereas the former is strictly discursive and reflects the manner in which the academic tradition understands knowledge, prehistoric knowledge of the same principles, in contrast, would be based on experience.

6 An old term referring to the properties of metal (Huxam 1753, 859), re-introduced by Peter Bray as a useful term to describe the idea of metal as a package of attributes (properties and potentials) which are available to human society.

7 It is sometimes seen as a forerunner of sensory archaeology.

8 I am not arguing that it is impossible to understand other people's sensations or emotions – which is the field of affect theory. To do so, however, one must be able to observe the other. It is not something which can be reconstructed through contemporary social embodied experiments, which seem to be part of Tilley's methodology.

9 Psychophysical studies as to the limits of human perceptual abilities – the inception of which is attributed to Weber in the 1830s.

10 Up to 12% tin, after which it becomes harder to distinguish (pers. comm. Jeroen Zuiderwijk).

References

Adamson, G. (ed.) 2010. *The Craft Reader*. Oxford: Berg.

Agricola, G. 1950 [1546]. *De re metallica*. New York: Dover Publications.

Agricola, G. 1955 [1546]. "De Natura Fossilium (Textbook of Mineralogy)," in *The Geological Society of America Special Paper* 63. New York: Geological society of America.

Apel, J. 2006. "Skill and experimental archaeology," in J. Apel and K. Knutsson (eds) *Skilled production and Social Reproduction*. Uppsala: Societas Archaeologica Upsaliensis.

Apel, J. 2008. Knowledge, Know-how and Raw Material – The Production of Late Neolithic Flint Daggers in Scandinavia. *Journal of Archaeological Method and Theory* 15, 1: 91–111.

Aristotle. 2008. *De eerst filosofie*. Amsterdam: Muntinga B.V., Uitgevery Maarten.

Barber, M. 2003. *Bronze and the Bronze Age. Metalwork and society in Britain c.2500–800 BC*. Stroud: Tempus Publishing.

Bartelheim, M. 2009. "Elites and metals in the Central European Early Bronze Age" in B. Roberts and T. Kienlin (eds) *Metals and Societies. Studies in honour of Barbara S. Ottaway, Universitätsforschungen zur prähistorischen Archäologie 169*. 34–46. Bonn: Rudolf Habelt.

Bertemes, F. 2004. "Frühe Metallurgen in der Spätkupfer- und Frühbronzezeit," in H. Meller (ed.) *Der geschmiedete Himmel. Die weite Welt im Herzen Europas vor 3600 Jahren*. 144–149. Landesmuseum für Vorgeschichte Halle. Stuttgart: Konrad Theiss.

Bertemes, F., and V. Heyd. 2002. "Der Übergang Kupferzeit/Frühbronzezeit am Nordwestrand des Karpatenbeckens – Kulturgeschichtliche und paläometallurgische Betrachtungen," in M. Bartelheim, E. Pernicka, and R. Krause (eds) *Die Anfänge der Metallurgie in der Alten Welt. Forschungen zur Archäometrie und Altertumswissenschaft 1*. 185–229. Rahden: Marie Leidorf.

Bleed, P. 2008. Skill Matters. *Journal of Archaeological Method and Theory* 15, 1: 154–166.

Bray, P. in press. *Before ^{29}Cu became copper: tracing the recognition and invention of metalleity in Britain and Ireland during the third millennium BC*. Oxford: Prehistoric Society.

Bridgeford, S. D. 2002. "Bronze and the First Arms Race – Cause, Effect or Coincidence?" in B. S. Ottaway and E. C. Wager (eds) *Metals and Society. Papers from a session held at the European Association of Archaeologists Sixth Annual Meeting in Lisbon 2000, British Archaeological Reports International Series* 1061. 123–133. Oxford: Archaeopress.

Brück, J. 2005. Experiencing the past? The development of a phenomenological archaeology in British prehistory. *Archaeological Dialogues*, 12: 45–72.

Budd, P., and T. Taylor. 1995. The faerie smith meets the bronze industry: magic versus science in the interpretation of prehistoric metal-making. *World Archaeology* 27, 1: 133–143.

Budden, S., and J. R. Sofaer. 2009. Non-Discursive Knowledge and the Construction of Identity. Potters, Potting and Performance at the Bronze Age Tell of Százhalombatta, Hungary. *Archaeological Review from Cambridge* 19, 2: 203–220.

Butler, J. J., and J. D. v. d. Waals. 1966. Bell Beakers and early metal-working in the Netherlands. *Palaeohistoria* 12: 44–139.

Charles, J. A. 1967. Early Arsenical Bronzes – A Metallurgical View. *American Journal of Archaeology* 71, 1: 21–26.

Childe, V. G. 1930. *The Bronze Age*. Cambridge: Cambridge University Press.

Childe, V. G. 1944. Archaeological Ages as Technological Stages. *Journal of the Royal Anthropological Institute* 74: 7–24.

Clarke, D. V., T. G. Cowie, and A. Foxon. 1985. *Symbols of power at the time of Stonehenge*. Edinburgh: National Museum of Antiquities of Scotland.

Classen, C. 1997. Foundations for an anthropology of the senses. *International Social Science Journal* 49, 153: 401–412.

Classen, C., and D. Howes. 1996. Making Sense of Culture: Anthropology as a Sensual Experience. *Etnofoor* 9, 2: 86–96.

Coghlan, H. H. 1975 [1951]. *Notes on the prehistoric metallurgy of copper and bronze in the Old World. Occasional papers on technology*. Oxford: Oxford University Press.

Corkin, S. 1968. Acquisition of motor skill after bilateral medial temporal-lobe excision. *Neuropsychologia* 6, 3: 255–265.

Costin, C. L. 2001. "Craft Production Systems," in G. M. Feinman and D. Price (eds) *Archaeology at the Millennium. A sourcebook*. 273–329. New York: Kluwer Academic /Plenum Publishers.

Craddock, P. T. 1995. *Early Metal Mining and Production*. Edinburgh: Edinburgh University Press.

Cunliffe, B. W. 2001. *Facing the Ocean: the Atlantic and its peoples, 8000 BC–AD 1500*. Oxford: Oxford University Press.

Cunliffe, B. W. 2008. *Europe between the Oceans: themes and variations, 9000 BC–AD 1000*. New Haven/London: Yale University Press.

Day, J. forthcoming. *Making Senses of the Past: Toward a Sensory Archaeology*.

Dobres, M. A. 2006. "Skilled Production and Social Reproduction in prehistory *and* contemporary archaeology: a personal exegis on dominant themes and their psychosocial influences," in J. Apel and K. Knutsson (eds) *Skilled Production and Social Reproduction*. 25–33. Uppsala: Societas Archaeologica Upsaliensis.

Dobres, M.-A. 2010. Archaeologies of technology. *Cambridge Journal of Economics* 34, 1: 103–114.

Earle, T. 2004. "Culture Matters: Why Symbolic Objects Change," in E. DeMarrais, C. Gosden, and C. Renfrew (eds) *Rethinking materiality. The engagement of mind with the material world. McDonald Institute Monographs*. 153–165. Cambridge: McDonald Institute for Archaeological Research.

Eerkens, J. W. 2000. Practice Makes Within 5% of Perfect: Visual Perception, Motor Skills, and Memory in Artifact Variation. *Current Anthropology* 41, 4: 663–668.

Feuchtwanger, L. 1867. *A popular treatise on gems in reference to their scientific value: a guide for the teacher of natural sciences, the lapidary, jeweller, and amateur: together with a description of the elements of mineralogy and all ornamental and architectural material*. New York.

Fleming, A. 2006. Post-processual Landscape Archaeology: a Critique. *Cambridge Archaeological Journal* 16, 3: 267–280.

Fontijn, D. R. 2002. *Sacrificial landscapes. Cultural biographies of persons, objects and 'natural' places in the Bronze Age of the Southern Netherlands, c. 2300–600BC. Analecta Praehistorica Leidensia*. Leiden: Leiden University, Institute of Prehistory.

Goldhahn, J. 2007. *Dödens hand – en essä om brons- och hällsmed. Gotarc Serie C*. Gothenburg: Göteborg University.

Goldhahn, J., and T. Oestigaard. 2007. "Rituelle spesialister i bronse- og jernalderen. En introduksjon," in J. Goldhahn (ed.) *Dödens hand – en essä om brons- och hällsmed. Rituelle spesialister i bronse- og jernalderen 1. Gotarc Series C, 65*. 1–18. Gothenburg: Göteborg University.

Goldhahn, J., and T. Oestigaard. 2008. "Smith and death – cremations in furnaces in Bronze and Iron Age Scandinavia," in C. Prescott, K. Chilidis, and J. Lund (eds) *Facets of Archaeology. Essays in Honour of Lotte Hedeager on her 60th Birthday*. 215–242. Oslo: Oslo Archaeological Series.

Goodway, M. 1991. "Archaeometallurgy: Evidence of a Paradigm Shift?" in P. B. Vandiver, J. Druzik, and G. S. Wheeler (eds) *Material Issues in Art and Archaeology II*. Materials Research Society Symposium Proceedings 185: 704–712. Materials Research Society.

Goody, J. 2002. The Anthropology of the Senses and Sensations. *La Ricerca Folklorica*, 45: 17–28.

Hacking, I. 2004. Between Michel Foucault and Erving Goffman: between discourse in the abstract and face-to-face interaction. *Economy and Society* 33, 3: 277–302.

Harding, A. F. 2000. *European Societies in the Bronze Age. Cambridge World Archaeology*. Cambridge: Cambridge University Press.
Harper, M. 1987. Possible toxic metal exposure of prehistoric bronze workers. *British Journal of Industrial Medicine* 44: 652–656.
Harris, O. J. T., and T. F. Sørensen. 2010. Rethinking Emotion and Material Culture. *Archaeological Dialogues* 17, 2: 145–163.
Hawkes, C. 1954. Archeological Theory and Method: Some Suggestions from the Old World. *American Anthropologist* 56, 2: 155–168.
Hosler, D. 1995. Sound, Color and Meaning in the Metallurgy of Ancient West Mexico. *World Archaeology* 27, 1: 100–115.
Houston, S., and K. Taube. 2000. An Archaeology of the Senses: Perception and Cultural Expression in Ancient Mesoamerica. *Cambridge Archaeological Journal* 10, 2: 261–294.
Howes, D. 2003. *Sensual Relations: Engaging the Senses in Culture and Social Theory*. Ann Arbor: University of Michigan Press.
Hurcombe, L. 2007. A sense of materials and sensory perception in concepts of materiality. *World Archaeology* 39, 4: 532–545.
Hurcombe, L. M. 2000. "Time, skill and craft specialisation as gender relations," in M. Donald and L. M. Hurcombe (eds) *Gender and Material Culture in Archaeological Perspective*. 88–109. London: Macmillan.
Huxham, J. 1753–1754. Medical and chemical observations upon antimony. *Philosophical Transactions* 48: 832–869.
Ingold, T. 1990. Society, nature and the concept of technology. *Archaeological Review from Cambridge* 9, 1: 5–17.
Ingold, T. 2000. *The Perception of the Environment: essays on livelihood, dwelling and skill.* London: Routledge.
Ingold, T. 2007. Materials against materiality. *Archaeological Dialogues* 14, 1: 1–16.
Jantzen, D. 2008. *Quellen zur Metallverarbeitung im Nordischen Kreis der Bronzezeit. Prähistorische Bronzefunde 19 (2)*. Stuttgart: Franz Steiner.
Jockenhövel, A. 1982. Zeugnisse der primären Metallurgie in Gräbern der Bronze- und Alteisenzeit Mitteleuropas. *Archeologia Polski* 27: 293–301.
Jones, A., and G. MacGregor. 2002. *Colouring the Past: the significance of colour in archaeological research*. Oxford/New York: Berg.
Juleff, G., and L. Bray. 2007. Minerals, Metal, Colours and Landscape: Exmoor's Roman Lode in the Early Bronze Age. *Cambridge Archaeological Journal* 17,3: 285–296.
Kienlin, T. L. 2007. Von den Schmieden der Beile: Zu Verbreitung und Angleichung metallurgischen Wissens im Verlauf der Frühbronzezeit. *Praehistorische Zeitschrift* 82, 1: 1–22.
Kienlin, T. 2008. *Frühes Metall im Nordalpinen Raum. Eine Untersuchung zu technologischen und kognitiven Aspekten früher Metallurgie anhand der Gefüge frühbronzezeitlicher Beile. Universitätsforschungen zur Prähistorischen Archäologie*. Bonn: Rudolf Habelt.
Kienlin, T. L. 2010. *Traditions and Transformations: approaches to Eneolithic (Copper age) and Bronze Age metalworking and society in eastern Central Europe and the Carpathian basin*. Oxford: Archaeopress.
Kienlin, T. L., E. Bischoff, and H. Opielka. 2006. Copper and Bronze during the Eneolithic and Early Bronze Age: A Metallographic examination of axes from the Northalpine region. *Archaeometry* 48, 3: 453–468.
Kienlin, T. L., and T. Stöllner. 2009. "Singen Copper, Alpine Settlement and Early Bronze Age Mining: Is There a Need for Elites and Strongholds?," in T. L. Kienlin and B. W. Roberts (eds) *Metals and Societies. Studies in honour of Barbara S. Ottaway, Universitätsforschungen zur prähistorischen Archäologie*. 67–104. Bonn: Rudolf Habelt.
Kienlin, T. L., and A. Zimmermann (eds) in press. *Beyond Elites. Alternatives to Hierarchical Systems in Modelling Social Formations. International Conference at the Ruhr-Universität Bochum, Germany, October 22–24, 2009. Universitätsforschungen zur prähistorischen Archäologie*. Bonn: Rudolf Habelt.
Killick, D. 2004. Social Constructionist Approaches to the Study of Technology. *World Archaeology* 36, 4: 571–578.
Kristiansen, K. 1987. "From stone to bronze – the evolution of social complexity in Northern Europe, 2300–1200 BC," in E. M. Brumfiel and T. Earle (eds) *Specialization, exchange, and complex societies*. 30–51. Cambridge: Cambridge University Press.
Kristiansen, K., and B. Larsson. 2005. *The Rise of Bronze Age Society. Travels, transmissions and transformations*. Cambridge: Cambridge University Press.
Kuijpers, M. H. G. 2008. *Bronze Age Metalworking in the Netherlands (c. 2000–800BC). A research into the preservation of metallurgy related artefacts and the social position of the smith*. Leiden: Sidestone Press.
Kuijpers, M. H. G. in press. "Towards a deeper understanding of metalworking technology," in T. L. Kienlin and A. Zimmermann (eds) *Beyond Elites. Alternatives to Hierarchical Systems in Modelling Social Formations. International Conference at the Ruhr-Universität Bochum, Germany, October 22–24, 2009, Universitätsforschungen zur prähistorischen Archäologie*. Bonn: Rudolf Habelt.
Lehoërff, A. 2008. Les cuirasses de Marmesse (Haute-Marne), un artisanat d'exception. *Antiquités Nationales* 39: 95–106.
Mauss, M. 1973 [1935]. Techniques of the body. *Economy and Society* 2, 1: 70–88.
Merkl, M. B. 2010. Bell Beaker Metallurgy and the Emergence of Fahlore-copper Use in Central Europe. *Interdisciplinaria Archaeologica. Natural Sciences in Archaeology* I, 1–2: 19–27.
Meurkens, L. 2004. *A matter of Elites, Specialists and Ritual? Social and symbolic dimension of metalworking in the North-west European Bronze Age*. Unpublished Master Thesis, University of Leiden.
Needham, S. P. 2002. "Analytical implications for Beaker metallurgy in North-West Europe," in E. Pernicka and M. Bartelheim (eds) *The Beginnings of Metallurgy in the Old World* 99–113. Rahden: Marie Leidorf.
Needleman, C. 1979. *The Work of Craft: an inquiry into the nature of crafts and craftsmanship*. London: Routledge and Kegan Paul.
Nriagu, J. O. 2001. "Arsenic poisoning through the ages," in W. T. Frankenberger (ed.) *Environmental Chemistry of Arsenic*. 1–26. New York: CRC Press.
Olausson, D. 2008. Does Practice Make Perfect? Craft Expertise as a Factor in Aggrandizer Strategies. *Journal of Archaeological Method and Theory* 15, 1: 28–50.
Ottaway, B. S. 1994. *Prähistorische Archäometallurgie*. Espelkamp: Marie Leidorf.
Ottaway, B. S. 2001. Innovation, production and specialisation in early prehistoric copper metallurgy. *European Journal of Archaeology* 4, 1: 87–112.
Ottaway, B. S. 2002. "Towards interpretative archaeometallurgy," in M. Bartelheim, E. Pernicka, and R. Krause (eds) *Die Änfange der Metallurgie in der alten Welt. The beginnings of metallurgy in the old world*. 7–13. Rahden: Marie Leidorf.

Pearce, M. 2007. *Bright Blades and Red Metal. Essays on North Italian Prehistoric Metalwork. Accordia specialist studies on Italy 14*. London: Accordia Research Institute.

Pelegrin, J. 1990. Prehistoric Lithic Technology: Some aspects of research. *Archaeological Review from Cambridge* 9: 116–125.

Pfaffenberger, B. 1988. Fetishised Objects and Humanised Nature: Towards an Anthropology of Technology. *Man* 23, 2: 236–252.

Pfaffenberger, B. 1992. Social anthropology of Technology. *Annual Review of Anthropology* 21, 1: 491–516.

Piggott, S. 1965. *Ancient Europe from the beginnings of agriculture to classical antiquity: a survey*. Edinburgh: Edinburgh University Press.

Pye, D. 1968. *The Nature and Art of Workmanship*. Cambridge: Cambridge University Press.

Randsborg, K. 1986. A Bronze Age grave on Funen containing a Metal Worker's tools. *Acta Archaeologica* 55: 185–189.

Renfrew, C. 2004. "Towards a theory of material engagement," in E. DeMarrais, C. Gosden, and C. Renfrew (eds) *Rethinking materiality. The engagement of mind with the material world*. 23–31, Cambridge: McDonald Institute for Archaeological Research.

Reynolds, P. J. 1999. "The Nature of Experiment in Archaeology," in A.F. Harding (ed.) *Experiment and Design: Archaeological studies in honour of John Coles*. 156–162. Oxford: Oxbow.

Sennet, R. 2009. *The Craftsman*. London: Penguin books.

Skeates, R. 2010. *An Archaeology of the Senses: prehistoric Malta*. Oxford: Oxford University Press.

Sofaer, J. R. 2006. Pots, Houses and Metal: technological relations at the Bronze Age tell at Százhalombatta, Hungary. *Oxford Journal of Archaeology* 25, 2: 127–147.

Sofaer-Derevenski, J. R., and M. L. S. Sørensen. in press. "Technological change as social change: the introduction of metal in Europe," in M. Bartelheim and V. Heyd (eds) *Continuity-Discontinuity: Transition Periods in European Prehistory. Forschungen zur Archäometrie und Altertumswissenschaft*. Rahden: Marie Leidorf.

Sørensen, M. L. S. 1996. "Women as/and metalworkers," in A. Devonshire and B. Wood (eds) *Women in industry and technology from prehistory to the present day*. 45–52. London: Museum of London.

Sperber, L. 2000. Zum Grab eines spätbronzezeitlichen Metallhandwerkers von Lachen-Speyerdorf, Stadt Neustadt a. d. Weinstraße. *Archäologischen Korrespondenzblatt* 30: 383–402.

Stevens, F. 2008. Elemental interplay: the production, circulation and deposition of Bronze Age metalwork in Britain and Ireland. *World Archaeology* 40,2: 238–252.

Stout, D., and T. Chaminade. 2007. The evolutionary neuroscience of tool making. *Neuropsychologia* 45, 5: 1091–1100.

Thornton, C. P. 2009. "Archaeometallurgy: Evidence of a Paradigm Shift?" in B. Roberts and T. Kienlin (eds) *Metals and Societies. Studies in honour of Barbara S. Ottaway, Universitätsforschungen zur prähistorischen Archäologie 169*. 25–33. Bonn: Rudolf Habelt.

Tilley, C. 2004. Round Barrows and Dykes as Landscape Metaphors. *Cambridge Archaeological Journal* 14, 2: 185–203.

Tylecote, R. F. 1987. *The Early History of Metallurgy in Europe*. London: Longman.

Vandkilde, H. 2007. *Culture and Change in Central European Prehistory, 6th to 1st Millennium BC*. Aarhus: Aarhus University Press.

Vandkilde, H. 2010. "Metallurgy, Inequality and Globalization in the Bronze Age – discussant's commentary on the papers in the metallurgy session," in H. Meller and F. Bertemes (eds) *Der Griff nach den Sternen. Wie Europas Eliten zu Macht und Reichtum kamen. Internationales Symposium Halle/Saale 2005*. Halle (Saale): Landesamt für Denkmalpflege und Archäologie Sachsen-Anhalt. Landesmuseum für Vorgeschichte.

Williamson, D. 1990. *The role and status of the Bronze Age smith and the organisation of metallurgy*. Unpublished Master Thesis. University of Durham.

Winghart, S. 1998. "Produktion, Verarbeitung und Distribution – Zur Rolle spätbronzezeitlicher Eliten im Alpenvorland," in B. Hänsel (ed.) *Mensch und Umwelt in der Bronzezeit Europas. Man and environment in European Bronze Age*. 261–264. Kiel: Oetker-Voges.

Authors' short biographies and contact details

Sandy Budden-Hoskins is a Research Fellow at the University of Southampton. She is a ceramic specialist and combines theory and new methodologies to investigate notions of skill and making in the later prehistory of Central Europe. Her book 'Skill and Society in prehistory: a ceramic case study', based on her PhD and research as a Leverhulme Early Career Research Fellow, will be published by BAR in 2013. She is co-director of the ZSAP (Županja Museum and The University of Southampton Archaeological Project) in south-east Croatia. The first publications from this project are in preparation. Sandy is also a professional working potter and works on a consultancy basis for a number of academic projects, most recently for the AHRC funded project 'Can Egyptian Paste (Faience) techniques be used for 3d printed, solid, free-form fabrication of ceramics?' at The Centre for Fine Print Research, University of West England.

Dr Sandy Budden-Hoskins
Department of Archaeology
Faculty of Arts and Humanities
University of Southampton
Highfield
Southampton
SO17 1BF
United Kingdom
s.a.budden@soton.ac.uk

Jacob Copeman is a Lecturer in social anthropology at the University of Edinburgh, UK. He has conducted fieldwork in India on interrelations between blood donation, kinship, devotional religion and social reform movements. This has resulted in a number of articles and book chapters as well as two book-length works: "Veins of Devotion: Blood Donation and Religious Experience in North India" (Rutgers University Press, 2009), and an edited collection on blood donation and transfusion in different parts of the world: "Blood Donation, Bioeconomy, Culture" (Sage, 2009). His present work, an extension of his earlier interest in social reform, concerns rationalist anti-superstition activism in India. He recently completed an edited collection (with Dr Aya Ikegame), titled "The Guru in South Asia: New Interdisciplinary Perspectives" (2012 Routledge), which offers a fresh approach to the social lives and roles of gurus in South Asia.

Dr Jacob Copeman
Social Anthropology
School of Social and Political Science
University of Edinburgh
5.21 Chrystal Macmillan Building
15a George Square
Edinburgh
EH8 9LD
United Kingdom
jacob.copeman@ed.ac.uk

Helen Foxhall Forbes is a Lecturer in Medieval History at the University of Exeter. She studied Anglo-Saxon, Norse and Celtic at Trinity College, Cambridge, and Theology at the Albert-Ludwigs-Universität, Freiburg-im-Breisgau; she was then awarded a Leverhulme Early Career Fellowship which she held in the School of Historical Studies at the University of Leicester, where she was subsequently a Research Associate on the major Leverhulme-funded project, 'The Impact of Diasporas on the Making of Britain: Evidence, Memories,

Inventions' (http://www.le.ac.uk/diasporas). Her research focuses on the history of ideas and intellectual culture in the early middle ages, especially theology and belief, and how theological concepts were developed, communicated and adapted in their historical contexts.

Dr Helen Foxhall Forbes
History
University of Exeter
Amory Building
Rennes Drive
Exeter
EX4 4RJ
United Kingdom
H.G.Foxhall-Forbes@exeter.ac.uk

Harald Bentz Høgseth has been working as a carpenter from 1984 and was able to integrate his craftsman's skill and knowledge with his career as an archaeologist. In 2007, he completed a PhD thesis with the title "The Craftsman's Toolbox. An investigation of embodied knowledge as reflected in archaeological timbers dated to the 11th century AD". He has widely published on the perception of 'tacit knowledge' and 'knowledge in action'. He is currently employed as Associate Professor in Technical Building Protection and Restoration at the Department of Civil Engineering, Faculty of Technology, South-Trøndelag University College, Norway, developing a new interdisciplinary education between building protection, architecture and archaeology.

Dr Harald Bentz Høgseth
Domaine de Malcintat
11270 Orsans
France
harald.bentz.hogseth@hist.no

Mads Dengsø Jessen is a Project Researcher at the National Museum of Denmark. This position entails being head of the excavations at the Late Viking Age royal monument at Jelling, Jutland, and doing research into the introduction of Christianity in South Scandinavia. In his research he focusses on explaining the period of ideological transition through the study of material culture and in particular the description of the material grounding of religious behaviour. His research also includes theoretical issues such as the application of cognitive and evolutionary methods to archaeological assemblages and proto-historic periods. More recently, an uncanny interest in the spreading of watermilling technologies across Western Europe has surfaced. He is also co-editing the new journal *Danish Journal of Archaeology*.

Dr. Mads Dengsø Jessen
Nationalmuseet
Frederiksholms kanal 12
DK-1220 Copenhagen
Denmark
mads.dengso.jessen@natmus.dk

Lise Bender Jørgensen is Professor of Archaeology at the Norwegian University of Science and Technology in Trondheim, Norway. She is an internationally regarded expert in prehistoric textiles. A Danish national, she studied Archaeology at the University of Copenhagen. Before moving to Norway, she had a spell of teaching Archaeology at Göteborg University in Sweden, and in recent years has held an adjunct professorship in Textile Science at the University of Borås in Sweden, teaching philosophy of science and research methodology to students of handloom weaving, textile and fashion design and textile engineering. She has excavated in Scandinavia and Egypt and travelled widely in Europe recording archaeological textiles in almost 100 museums. She is the author of two monographs on Scandinavian and North European Textiles from the beginnings up to 1000 AD. A founding member of the North European Symposium for Archaeological Textiles (NESAT), she has organised several conferences on archaeological textiles and edited the proceedings. She has directed and participated in research projects on Roman textiles from Egypt, wool sails for Viking ships, and is currently engaged in a project on Creativity in Craft Production in Middle and Late Bronze Age Europe.

Professor Lise Bender Jørgensen
Department of Archaeology and Religious Studies,
Norwegian University of Science and Technology
N-7491 Trondheim
Norway
lise.bender@ntnu.no

Sheila Kohring is an Affiliated Lecturer in the Department of Archaeology and Anthropology at the University of Cambridge and a Research Associate of the McDonald Institute for Archaeological Research and Clare College. Her career has spanned both the academic and commercial world of archaeology and she has excavated in North America, Mesoamerica and Europe. Her current research interests explore Neolithic and Bronze Age pottery and representational technologies and communities along the Atlantic fringe of Europe, specifically in Iberia, the Channel Islands and southern Britain. She will be part of a new Wellcome Trust research project entitled 'The Anthropology of Health in Prehistoric Europe' from 2012.

Dr Sheila Kohring
Division of Archaeology
Department of Archaeology and Anthropology
Downing Street
Cambridge
CB2 3DZ
United Kingdom
sek34@cam.ac.uk

Maikel Kuijpers is a Ph.D. student at the University of Cambridge, under the supervision of Marie Louise Stig Sørensen. Originally from the Netherlands, he has completed a Research Master degree in Archaeology at Leiden University and a minor in Photography and Film at the Royal Art Academy in The Hague. His thesis "Bronze Age metalworking in the Netherlands" was awarded the bi-annual W.A. van Es prize by the Netherlands Institute for Heritage and published (Sidestone, 2008). Currently he is a Marie Curie fellow in the European project "Forging identities; the mobility of culture in the Bronze Age". Within this project he is studying the craft of metalworking in the Early Bronze Age. On a more theoretical level, he is inquiring into knowledge, skill, technology and craftsmanship.

Maikel Kuijpers
Division of Archaeology
Department of Archaeology and Anthropology
Downing Street
Cambridge
CB2 3DZ
United Kingdom
mhgk3@cam.ac.uk

Mary Laven is a Senior Lecturer in History at the University of Cambridge and a Fellow of Jesus College. Her publications include "Virgins of Venice: Enclosed Lives and Broken Vows in the Renaissance Convent" (2002), which was awarded the John Llewellyn Rhys Prize, and "Mission to China: Matteo Ricci and the Jesuit Encounter with the East" (2011). She is now embarking on a new project, 'Objects of Devotion: the Material Culture of Italian Renaissance piety, 1400–1600', supported by a Leverhulme Major Research Fellowship.

Dr Mary Laven
Jesus College
Cambridge
CB5 8BL
United Kingdom
mrl25@cam.ac.uk

Katharina Rebay-Salisbury received her Ph.D. in prehistoric archaeology from the University of Vienna, Austria, in 2005 and subsequently worked as a research associate in the Department of Archaeology, University of Cambridge, within the Leverhulme funded project 'Changing beliefs of the human body: A comparative social perspective.' Together with Marie Louise Stig Sørensen she investigated changing burial rites in the European Bronze Age, in particular focussing on the introduction of cremation. In 2009 she joined the School of Archaeology and Ancient History in Leicester as Leverhulme Research Associate within the project 'Tracing Networks: craft traditions in the ancient Mediterranean and beyond'. Her research at Leicester centres on studying human representations, identities and social relations in the Late Bronze and Iron Age of Central Europe.

Dr Katharina Rebay-Salisbury
School of Archaeology and Ancient History
University of Leicester
University Road
Leicester
LE1 7RH
United Kingdom
kcrs1@le.ac.uk

Joanna Sofaer is a Senior Lecturer in Archaeology at the University of Southampton. She has published widely on European Prehistory, archaeologies of craft and technology, and the archaeology of the body. She is the author of "The Body as Material Culture. A Theoretical Osteoarchaeology". She currently leads the HERA-funded project 'Creativity and Craft Production in Middle and Late Bronze Age Europe' (CinBA) and is involved in several other international collaborations including the major international excavation at Százhalombatta, Hungary, where she is responsible for analysing the extensive and varied pottery from the site.

Dr Joanna Sofaer
Department of Archaeology
Faculty of Arts and Humanities
University of Southampton
Highfield
Southampton
SO17 1BF
United Kingdom
jrsd@soton.ac.uk

Marie Louise Stig Sørensen is Reader in Prehistoric Archaeology at the University of Cambridge. She is an expert in the European Bronze Age and is currently a partner on several projects investigating creativity and material categories in the Bronze Age. She was a partner on the 'Changing Beliefs of the Human Body' project for which she (together with Dr Katharina Rebay-Salisbury) conducted research into the transition from inhumation to cremation burial practices during the Bronze Age. She has written extensively on Bronze Age Europe, on gender and materialisation, and on identity construction and heritage in the modern world. Her publications include "Gender Archaeology" (Polity Press 2000) and "Body Parts and Bodies Whole" with K. Rebay-Salisbury and J. Hughes (Oxbow 2009).

Dr Marie Louise Stig Sørensen
Division of Archaeology
Department of Archaeology and Anthropology
Downing Street
Cambridge
CB2 3DZ
United Kingdom
mlss@cam.ac.uk

Tim Flohr Sørensen is an assistant professor at Aarhus University. This is also where he received his PhD in 2010, and he subsequently worked at the University of Cambridge on the international Forging Identities project as a Marie Curie postdoctoral researcher. His research focuses on the relationship between material culture and movement, emotion and the senses. These topics have been investigated at prehistoric and modern cemeteries, while his research on Early Bronze Age metalwork has been linked to issues of copying and prototyping. He is currently associated with an inter-disciplinary project on 'Death, Materiality and the Origin of Time'.

Tim Flohr Sørensen
Section for Archaeology
Aarhus University
Moesgaard
DK-8270 Højbjerg
Denmark
farktfs@hum.au.dk